The Public Uses of Coercion and Force

The Public Uses of Coercion and Force

From Constitutionalism to War

Edited by

ESTER HERLIN-KARNELL AND ENZO ROSSI

OXFORD
UNIVERSITY PRESS

OXFORD
UNIVERSITY PRESS

Oxford University Press is a department of the University of Oxford. It furthers the University's objective of excellence in research, scholarship, and education by publishing worldwide. Oxford is a registered trade mark of Oxford University Press in the UK and certain other countries.

Published in the United States of America by Oxford University Press
198 Madison Avenue, New York, NY 10016, United States of America.

CIP data is on file at the Library of Congress
ISBN 978–0–19–751910–3

DOI: 10.1093/oso/9780197519103.001.0001

1 3 5 7 9 8 6 4 2

Printed by Integrated Books International, United States of America

Note to Readers

This publication is designed to provide accurate and authoritative information in regard to the subject matter covered. It is based upon sources believed to be accurate and reliable and is intended to be current as of the time it was written. It is sold with the understanding that the publisher is not engaged in rendering legal, accounting, or other professional services. If legal advice or other expert assistance is required, the services of a competent professional person should be sought. Also, to confirm that the information has not been affected or changed by recent developments, traditional legal research techniques should be used, including checking primary sources where appropriate.

(Based on the Declaration of Principles jointly adopted by a Committee of the American Bar Association and a Committee of Publishers and Associations.)

Table of Contents

List of Contributors

Yitzhak Benbaji, Professor of Philosophy and Law, Tel-Aviv University

Luigi Corrias, Assistant Professor of Legal theory, VU University Amsterdam

Katrin Flikschuh, Professor of Political Theory, London School of Economics

Rainer Forst, Professor of Political Theory and Philosophy, Johann Wolfgang Goethe University

Aravind Ganesh, Vice Chancellor's Research Fellow in Law at Oxford Brookes University

Alon Harel, Phillip and Estelle Mizock Chair in Administrative and Criminal Law, Hebrew University of Jerusalem

Ester Herlin-Karnell, Professor of EU Law and EU Criminal Law, University of Gothenburg

Thomas Mertens, Professor of Law and Philosophy, Radboud University Nijmegen

Peter Niesen, Professor of Political Theory, University of Hamburg

Johan Olsthoorn, Assistant Professor of Political Theory, University of Amsterdam

Alice Pinheiro Walla, Associate Professor in Philosophy of Law, MacMaster University

Massimo Renzo, Yeoh Professor of Politics, Philosophy and Law, Director of the Centre for Politics, Philosophy and Law, King's College London

Arthur Ripstein, Professor of Law and Philosophy, University of Toronto

Enzo Rossi, Associate Professor of Political Theory, University of Amsterdam

Anna Stilz, Laurance S. Rockefeller Professor of Politics and Human Values, Princeton University

Malcolm Thorburn, Professor of Law, University of Toronto

Bertjan Wolthuis, Assistant Professor of Legal Theory, VU University of Amsterdam

1
Introduction

Ester Herlin-Karnell and Enzo Rossi

The need to replace force with law is central to a Kantian theory of war. In such a theory the only ground for using force is the prevention of wrongful force. War is barbaric and wrongful. A key theme in Kantian political and legal philosophy is every individual's right to independence, which can only be secured through a legitimate state, which in turn is obliged to repudiate war and pursue peace. Yet the Kantian project of achieving perpetual peace among states seems (at best) an unfulfilled hope. The wider category of global justice does not fare much better, at least according to most theorists. The modern state is often portrayed as a major obstacle to both peace and global justice. Kantian theories are distinctive insofar as they try to carve out a relatively hopeful role for the state in international affairs, though one that is grounded in an analysis of the state's nature rather than being driven by completely abstract moral considerations. Modern states' authority claims and their exercise of power and sovereignty span a spectrum from the most stringently and explicitly codified—the constitutional level—to the most fluid and turbulent—acts of war. Inter alia, that suggests a specific connection between constitutionalism and just war theory, as both concern the justifiability of state action vis-à-vis individuals as well as states. This book aims to explore that connection through the lens of the relationship between law and (just) war theory from a Kantian perspective.

The various contributions in this book investigate both extremes of the spectrum: national and transnational constitutionalism and acts of war, and their relationship. The key questions considered—directly or indirectly—by all the contributors are the following: what, if any, are the normatively salient differences between states' internal coercion and the external use of force? Is it possible to isolate the constitutional level from other aspects of the state's coercive reach? How could that be done while at the same time guaranteeing a robust conception of human rights and adherence to the rule of law? Likewise, is war an extension of political practice or an alternative to it? New forms and technologies of warfare raise further fundamental questions about due process, individual responsibility, fairness, and broader questions pertaining to justice and the responsibility to protect. From a constitutional perspective, questions concern the justification for state action, the human rights framework, and the question of judicial review and proportionality reasoning in "emergency" contexts. The purpose of the book is to combine political theory on war, philosophy of law, criminal law theory, and constitutionalism

Ester Herlin-Karnell and Enzo Rossi, *Introduction* In: *The Public Uses of Coercion and Force*. Edited by: Ester Herlin-Karnell and Enzo Rossi, Oxford University Press. © Oxford University Press 2021. DOI: 10.1093/oso/9780197519103.003.0001

scholarship in order to provide a new platform for understanding the contemporary law of war through a Kantian prism.

1. War, Peace, and Politics

The main dividing line in current debates on the morality and legality of war is not the one Kant faced. In the eighteenth century, the main debate was between just war and regular war theories.[1] Just war theories were centered on an account of the acceptable reasons for war, and saw war as a mechanism for punishing wrongdoers, that is, those who go to war for the wrong reasons (Ripstein 2021; Frowe and Lazar 2018, 99–145).[2] Regular war theorists saw armed conflict as a dispute resolution mechanism for sovereigns. As Ripstein helps us see, Kant found a third way between those positions. His approach is centered on a novel understanding of the status of states, and puts the achievement of peace at the center of the picture: war is only permissible to enforce a prohibition against resolving disputes by force, and the conduct of war must be compatible with the possibility of a future peace. So Kant's view is that there are no just causes for war per se, since war is only admissible when it is the only way to secure peace, through hindrance of force in self-defense.

This view is of continuing relevance in the twenty-first century, even though the main dividing line is now between two wings of the just war tradition: traditionalists and revisionists (Ripstein manuscript; Benbaji and Statman 2019).[3] The former view is associated with Michael Walzer and others, the latter with Jeff McMahan and his followers (McMahan 2009; Frowe 2014). The basic assumption of traditionalist just war theory is that combatants should only be held responsible for whether they fight justly, that is, in accordance with the requirements of jus in bello, irrespectively of whether the war itself is just. According to Walzer, war has no equivalent in peacetime. It is an extreme situation, and needs to be understood in that context (Walzer 1977; Benbaji and Statman 2019; Noorda 2016; Bazargan and Rickless 2017). For Walzer the moral norms governing the initiation and conduct of war should be responsive to states' sovereignty, which inter alia gives special status to all state-fielded combatants (Walzer 1977). Revisionists like McMahan question the privileged moral standing of states, with wide-ranging consequences for the permissibility even of defensive war, a blurring of the line between combatants and noncombatants, and a number of other radical consequences, such as a privatization of the ethics of war, where the relevant decisions on whether to use force must be taken by each soldier (McMahan 2009).

Does Kant put forward a distinctive version of (traditionalist) just war theory or a new kind of theory altogether? It is hard to envision an answer to that question that

[1] Arthur Ripstein, *Kant and the Law of War*, forthcoming (Oxford University Press 2021). Referred to as Ripstein 2021.

[2] On the history of just war theory and regular war theory.

[3] Though there is, arguably, a third strand too, namely contractual just war theory, advocated by Yitzhak Benbaji and Daniel Statman. For them the society of states should organize itself in light of a basic aim that its members ought to pursue together, namely the preservation of peace. The legal rules regulating resort to and conduct of war express, in their view, an agreement between the parties on the international level to achieve this largely Kantian aim. See also Benbaji, "A Semi-Kantian Just War Theory," ch. 2 in this volume.

does not descend into a verbal dispute. Nonetheless one of the most striking and insightful aspects of Ripstein's interpretation of Kant on the law of war is indeed that the resulting view may still be seen as a third way between today's prevailing positions, just as it was a third way between just war and traditional war theories in the eighteenth century. That is because, unlike Kant, both traditionalist and revisionist just war theorists focus primarily on morally evaluating harms and wrongs whose nature is determined independently of a full theory of the proper aims of legitimate states. Whereas the Kantian view (as reconstructed by Ripstein) is centered not on pre-political moral commitments, but on preserving states' rightful condition, which in turn is a precondition of the right to independence (Ripstein 2009; Rostbøll 2017).[4] The upshot, as we will see below and as some of the contributions in this book as well as Ripstein's replies make clear, is a view that can countenance the distinctively political standing of states while avoiding the Scylla of giving in to a status quo bias in favor of sovereign prerogative, and the Charybdis of treating conflict among states as analogous to interpersonal disputes.

2. The Centrality of Law

Kant regards peace as the constituent through which everything about war—the grounds for going to war, the conduct of war, and what happens after a war—must be judged (Ripstein 2021: chap. 1). And it is the legal system that governs this. Kant's writings on peace are largely concerned with how the right of nations could be realized through a federation of nations and through a legal framework. For Kant, states have to enter a lawful condition through a universal association of states. As Ripstein explained already in *Force and Freedom*, each state's right against other states is purely defensive: to continue being the rightful condition it is, on its own territory (Ripstein 2009). Kant's solution to the just war theory problem of uniting just cause with the prospects of war has two parts: an account of the distinctively public nature of a state, and an account of peace as an essentially public condition under which disputes can be resolved on their merits (Ripstein 2021). Law is the key to Kant also in a global context, in terms of global institutions and cosmopolitan right, because on the Kantian view, legal institutions are not merely helpful tools in the service of achieving and maintaining peace; they are fundamentally constitutive of what it is to live in peace between nations (cf Kleingeld 2004; Maliks 2014). The same point about law generates Kant's claim in *Perpetual Peace* that no state can rightfully acquire another (Ripstein 2009). Defensive force for Kant is therefore about the political right to independence. In a state of nature each nation can only do what seems good and right to it. For Kant this is a problem as there may be good causes on both sides, which confirms the need for an international non-coercive legal order that secures the prospects of peace (Ripstein 2009).

As law is central to Kant and as war theory is seeking to achieve peace, Kant's war theory is also a constitutional law theory. For example, proportionality is important in

[4] On Kant's deviation from the centrality of pre-political moral commitments.

a Kantian war theory, yet as war is only permissible in self-defense there is no wrong that needs correcting by proportionality in this context. So for Kant the defender does not wrong the aggressor by using necessary but disproportionate force to prevent that wrong (Ripstein 2021: chap. 3). Proportionality is however still relevant to secure a broader condition of peace; as such, it must be committed to de-escalating conflict. Proportionality for Kant is then more of a global mechanism as it concerns questions about stability and the possibility to secure peace by insisting on proportionate action (Ripstein 2021, chap 3 and chap 5). Accordingly, while traditional just war theory is centered on the need for war to have be fought in a just way (jus in bello), it adds proportionality in the jus ad bellum test requiring weighing costs and benefits (Ripstein 2021, ibid). For Kant instead the proportionality assessment is a standard for securing peace. And as Ripstein reminds us, "War is the condition in which might makes right; peace is a condition in which, through the establishment of procedures, right guides might" (Ripstein 2016). So the right procedure is very important—a point taken up by several of the contributors in this book.[5]

As the proper authority for waging war is crucial in a Kantian war theory, there is also a clear constitutional dimension. Constitutional legal questions that are highlighted by a Kantian war theory are, inter alia: How and why can a decision regarding the authority to enter war be challenged? When is self-defense under Article 51 UN allowed? Is preventive force allowed? What is proportionate to secure peace? How does the law of war reconcile with human rights protection and its complicated relationship with international humanitarian law, and what is the impact of criminal law and international criminal law vis-à-vis the discussion of war? Other intriguing questions are multidisciplinary, for example, concerning the links between global justice and the right of war. These questions and other related matters are discussed throughout this book.

3. Structure of the Book

Arthur Ripstein's new book, *Kant and the Law of War*, published with the Tanner Lectures (Oxford University Press 2021), provides a focal point for the emphasis of the present volume. The contributions in this edited volume respond to the manuscript of *Kant and the Law of War* as of June 2018. The book is divided in three parts. The chapters in Part One engage most closely with *Kant and the Law of War*. The chapters in Part Two discuss Kantian theories of war and related questions more broadly. In the third part of the book Arthur Ripstein replies to all contributions.

3.A. Part One

In the first chapter, "A Semi-Kantian Just War Theory," **Yitzhak Benbaji** explores how the Kantian philosophy of the law of war and how it is based on two normative claims.

[5] See, e.g., the chapters by Alon Harel and Malcolm Thorburn.

First, states are independent of each other in virtue of their duty to provide a legal order in a territory which they rule. Second, any use of non-defensive force by a state in a territory which it does not actually rule is illegal. Benbaji shows that there is a deep tension between these claims, and he sets out to offer a contractarian theory of the crime of aggression, which he characterizes as semi-Kantian: states are fully legitimate only if their right to rule over their territory is recognized by all other states. Semi-Kantians argue that Ripstein's Kant misconstrues the standing of states vis-à-vis the territories over which they rule. Taking seriously the disanalogy between a state's (disputable) territorial right and the (indisputable) bodily rights of a natural person, semi-Kantians structure territorial rights as based on two factors. Benbaji argues that a state is entitled to rule over a territory if and only if (first) it secures the private freedom and the political participation of its inhabitants *and* (second) a treaty to which other states voluntarily entered obligates all actors not to use force in resolving future disputes over this territory.

Rainer Forst's chapter, "Might and Right: Ripstein, Kant and the Paradox of Peace," sets out to critically discuss the methodological approach that Ripstein chooses when reconstructing Kant's position and in particular how he defines and relates "constitutive" and "regulative" principles with respect to peace and public right. Forst focuses on what he calls a *paradox of peace*: peace is supposed to avoid, end or overcome war as the practice of might making right; but the principle of peace itself at crucial points seems to require us to accept that might makes right, at least as far as the past is concerned. But can this be reconciled with the principle of right that ought to guide our normative thinking about these matters? If we follow Kant, do we have to accept that the paradox of peace is unavoidable, as Ripstein implies? Specifically, Forst offers a different interpretation of the relevant constitutive principle in the realm of the practical, namely that of freedom, and he invites Ripstein to move beyond Kant in questioning rather than affirming the paradox of peace.

Next, in "Reading Kant's *Rechtslehre*: Some Observations on Ripstein's *Kant and the Law of War*," **Thomas Mertens** puts forward a semantic observation which he claims reflects not only Ripstein's Kantminterpretation, but also his own perspective as a long-term reader of Kant. Mertens observes that Kantian scholarship has become to a large extent an Anglo-Saxon affair, and Kant is read and interpreted against the background of political and legal problems of that world. History has shown that several readings of Kant are possible and Ripstein provides us with a new, powerful reading of Kant which is indebted to that Anglo-Saxon background. Mertens discusses several intriguing questions, inter alia, Ripstein's interpretation of Kant's view on the law of war is the distinction between the just war tradition and the regular war tradition and Kant's departure from both traditions. This enables Ripstein to carve out what he considers Kant's specific position, which consists in strictly limiting the grounds for a just war to defending the rightful condition of one's state (i.e., national defense as the only legitimate ground for going to war) and in emphasizing and broadly interpreting Kant's view on the law during war (the *ius in bello*). Mertens, however, is unsure that this solves the problem of whether or not Kant subscribes to a right to war.

Anna Stilz, in her chapter on "The Moral Basis of State Independence," takes as her starting point the central claim in Arthur Ripstein's defense of a Kantian approach to war, namely that each state has a right to be independent from the determining choice of other states. As she explains, Ripstein declares that the state is "sovereign—*sui iuris*" (16). The state's right to independence is the basis for its permission to use force in national defense, and also for *in bello* restrictions that limit the permissible means of waging war to those necessary to stop aggression. But what morally justifies the state's right to independence? And can this right be accounted for on Kantian grounds? Moreover, Stilz argues that according to her Kantian reductivist view, then, independence has two dimensions: a legitimate state must not only "rule on behalf" of its inhabitants (by protecting their *private independence*), it must also reflect the shared will of most cooperators on its territory to establish justice through a particular institutional configuration (thus protecting their *political independence*). Specifically, Stilz focuses on whether the Kantian view, as Ripstein reconstructs it, provides a philosophically satisfying basis for attributing a right to political independence to the state. In the final section, she outlines an alternative reading of Kant that may provide a more compelling moral foundation for this right.

Thereafter, **Peter Niesen**'s "Vulnerability, Space, Communication: Three Conditions of Adequacy for Cosmopolitan Right" argues that according to the Kantian completeness assumption, not only do we face a major unresolved systematic issue—all exercise of cosmopolitan right takes place in a state of nature between states and foreigners: a highly structured state of nature, to be sure, but a state of nature nonetheless—but we have too few options for fulfilling the conditions of explanatory adequacy. Niesen points out that in *Kant and the Law of War* no space is reserved for leaving normative issues undecided, to be resolved through politics and through additional, positive global institutions. This is why Niesen pleads for a fuller, less parsimonious account where Kant's provisional cosmopolitan right is to be completed via a process of institutionalization and lawmaking. He stresses that cosmopolitan lawmaking would have to chime with Kant's overall strategy of contractual, non-coercive international lawmaking, and cosmopolitan institutions would have to reflect international law's binding, yet nonviolent character.

Relatedly, in "Three Models of Territory: Arthur Ripstein on the Territorial Rights of States," **Alice Pinheiro Walla** argues that regardless of whether a legal order has been established over a territory, possession of land *itself* already imposes obligations on persons outside the territory to respect it. She points out that possession of land also imposes on the holders of territory duties that are global in scope. It is therefore not possible to reduce territorial rights to claims of juridical independence in virtue of a state's internal civil condition. The reason is a fundamental asymmetry between forms of private right that involve external objects, including land (*Sachenrecht*), and forms of private right involving only the wills of the parties. Pinheiro Walla argues that Kant recognized this asymmetry in his legal theory. As a result, rights to external objects require a form of consent that can only be achieved under a global rule of law. Therefore, the problem of territory (the fact that possession of land must impose an obligation on all others to respect that territory) is not yet solved in a world of "monadic" public legal orders coexisting in an international state of nature. A state that enforces a public legal order over a specific area solves the problem of consent only in regard to its own citizens. It does not yet solve the problem of possession of land and the obligations it gives rise to in regard to other states and foreign individuals.

Pinheiro Walla states that recognizing this problem does not commit us to adopting a conception of territory as the property of states, as Ripstein seems to conclude.

Subsequently, **Alon Harel** in his contribution, "A Kantian Defense of Remedial Wars," discusses how Ripstein differentiates among three types of wars: self-defense, remedial, and punitive. Harel then argues that Ripstein's reasons for rejecting a right to remedial wars fail. The underlying Kantian principles guiding Ripstein's own account dictate that remedial wars are permissible. There are very powerful consequentialist, intuitionist, and conventionalist arguments against recognizing a right to remedial wars. But the logic provided by Ripstein's account of Kant cannot justify the prohibition on such wars. Precisely as the right to conduct defensive wars follows inevitably from our understanding of states as independent of each other, the right to conduct remedial wars follows from this very same principle. Harel argues that punitive and remedial wars are fundamentally different and that while Ripstein's counterargument addresses successfully the case of punitive wars, it fails to address the case of remedial wars. Key to Harel's claim is that those two cases should be discussed separately and that the case for a right to remedial wars is much stronger than the case for a right to punitive wars.

Relatedly, **Massimo Renzo**, in his chapter on "National Defense and the Value of Independence" focuses on Ripstein's treatment of wars of national defense. Renzo points out that attempts by some of the most prominent philosophers working within this paradigm have either concluded that wars of national defense are impermissible,[4] or that they are permissible only in a restricted number of circumstances. If they are right, and if Ripstein's account does a better job in dealing with these cases, that would be a reason to favor the Kantian approach over the traditional just war model (assuming that we are reluctant to abandon the verdict of common-sense morality and international law on this issue). However, Renzo argues that neither of these claims is correct. He argues that Ripstein's view ultimately lacks the resources to justify a right to wage wars of national defense, and contemporary just war theorists are wrong in believing that their approach can at best ground a weak version of such right. According to Renzo this gives us some reason to prefer the just war approach over the Kantian one championed by Ripstein, as the former provides a suitably robust account of the right to wage wars of national defense, in the sense that the right is not conditional on further vital interests being threatened.

3.B. Part Two

In her contribution "Exactitude and Indemonstrability in Kant's *Doctrine of Right*," **Katrin Flikschuh** states that while the *Doctrine of Right* is often seen as Kant's contribution to legal theory, it is primarily a work in political philosophy. Flikschuh argues for a reading of the *Doctrine of Right* as a work that includes a concern with law but is not confined to that concern alone. She claims, however, that the state is a less comprehensive idea for Kant than it is for some of his immediate predecessors and successors on the one hand, and for most contemporary political philosophers on the other hand. Flikschuh emphasizes that the *Doctrine of Right* thus has a remit that is both wider than the philosophy of law and narrower than more comprehensive conceptions of political philosophy: it is concerned with the idea of the state and, correspondingly, with right as a distinctly associative political morality, but it is also the companion piece to the *Doctrine of Virtue*, which means that not everything of moral significance falls within the domain of law or that of political morality more generally. In her chapter she sets out to explain why this is the case.

In his "The Right to Wage Private Wars of Subsistence: Its Nature, Gounds, and Place in Revisionist Just War Theories," **Johan Olsthoorn** explores the controversial question of wars undertaken in defense of human rights to subsistence. Olsthoorn critically contrasts two ways to ground rights to wage so-called "subsistence wars": via "modern" duties of global justice and via "old" rights of necessity. He explains that Cécile Fabre and other revisionists have argued that subjection to extreme deprivation can in principle give the severely deprived a just cause for waging defensive subsistence wars against those flouting their duties of global justice. Therefore this traditional right of necessity makes for a rival ground for legitimate subsistence wars. Olsthoorn then proposes to conceptualize "rights to wage private wars in defence of basic rights" in terms of "remedial rights." His main contention is that which remedial rights persons acquire will depend on the nature and content of the first-order deontic notion whose violation or nonfulfillment generates the remedial right.

Next, in his chapter on "Between Wormholes and Blackholes: A Kantian (Ripsteinian) Account of Human Rights in War," **Aravind Ganesh** sets out to explore the question of how human rights can be safeguarded in war, against the background of a conception of law as an instrument for attaining moral ends. On this basis, Ganesh shows how the moral approach in line of a of cases finds human rights obligations to obtain despite the obligor state having little to no capacity to enforce its writ, due either to the territory being embroiled in an active civil war or under the control of another state. Subsequently, Ganesh introduces Kant's conception of rights and war, the hallmark of which lies in law being logically prior to morality. This results in a conception of "human rights" as possible only where there are political institutions capable of enforcing law, and a conception of war as a condition where there are no rights at all. Ganesh then argues that individuals may nevertheless make claims of human rights in time of war. Where an applicant can make out even a risible claim that a defendant state has held itself out as an authority over it, that state may not be heard to deny the concomitant obligations of human rights. Ganesh argues that the limit is reached, however, when the facts alleged by the applicant are *utterly* barbaric, disclosing not even a scintilla of the possibility of a rightful condition. In such situations, the applicant becomes a bare life: her inability to state a cause of action means that she can, in effect, never have any causes of action.

On the related theme of war crimes, **Malcolm Thorburn** in his chapter on "Kant and the Criminal Law of War," focuses on Ripstein's account of a doctrine that has caused a great deal of trouble to moral philosophers of law over the years: the equal criminal immunity of combatants. This is the doctrine according to which soldiers are held immune from criminal prosecution for their war-making activities irrespective of whether they fought for a just cause (defending their country) or as part of their country's unjustified aggression. This is probably the best example of a doctrine that appears to most moral philosophers to be either totally repugnant or else explicable only as a sort of depressing concession to reality. Thorburn sets out the contours of the Kantian approach to the morality of law, which begins with the relationship of public authority and only later proceeds to the evaluation of how that authority has been exercised. He then considers Ripstein's application of that approach to the legal equality of combatants in war. Although Ripstein's account suggests a comprehensive justification of the doctrine of equal criminal immunity, Thorburn shows that Ripstein's chapter does not spell this out fully, though he outlines a way in which Ripstein can do so. Thorburn concludes with reflections on the vision of international relations and of individual responsibility reflected in Ripstein's Kantian account and where he draws some lessons for his own field of study, criminal law theory.

Thereafter, **Ester Herlin-Karnell**, in her chapter on "EU Solidarity as Collective Self-Defense?: Constitutionalism and the Public Uses of Force" sets out to explore the notion of self-defense and how it is manifested in contemporary law of war. She links this debate to Ripstein's *Kant and the Law of War* by using the European Union (EU) as a test case. The chapter focuses on the imminence criteria for deciding on self-defense, and to the extent to which Kant completely ruled out any preventive use of force. Moreover, the EU's fight against terrorism and security regulation in general have many parallels with the "war" model and security enforcement practices and preventive justice model, so the chapter discusses the EU security context and its similarities to war. In doing so Herlin-Karnell compares just war theory to the extensive security regulation framework within the EU, and the chapter addresses some pertinent constitutional questions such as the EU solidarity and mutual assistance clauses in the context of collective self-defense and the rule of law.

Also related to the EU is the final chapter by **Bertjan Wolthuis** and **Luigi Corrias**, "Europe's Cosmopolitan Union: A Kantian Reading of the EU Internal Market Law and the Refugee Crisis." Wolthuis and Corrias argue that the EU is a union of cosmopolitan law, in Kant's sense of cosmopolitan law, with EU internal market law as presumably the most extensive system of positive cosmopolitan law in history. They argue that the EU should be viewed as a cosmopolitan union. They ask whether EU law, understood as positive cosmopolitan law, can be qualified as an extension of the legal condition, and whether it can be viewed as consistent with the other two parts of public law, especially with the freedom of EU member states which also depend on the possible connection to global, much less extensive, systems of positive cosmopolitan law such as migration law. Wolthuis and Corrias argue, inter alia, that the crucial question is whether EU migration law is consistent with universal cosmopolitan law, and that the answer is dependent on two major lines of thought, both constitutive of the Kantian framework. The first submits that every individual ought to reside within a state of law: every person is a member of a legal order. The second says that legal orders ought to respect each other's independence, which they explain more fully in their chapter.

3.C. Part Three

In the final part of the book, Arthur Ripstein in his "From Constitutionalism to War—and Back Again: A Reply" offers a response and close commentary of all the chapters. For Kant war is a mechanism that can be used only to securing peace and law and constitutionalism in this regard are essential in securing that order. Therefore, the public uses of coercion and force arrive with constitutionalism to war and back again.

4. Acknowledgments

The chapters in this volume were originally presented at two workshops organized by the editors at the University of Amsterdam's Department of Political Science. The first workshop, in February 2018, was co-funded by the VU University Research Chair scheme and Access Europe Amsterdam. The second workshop, in June 2019, was sponsored by the Amsterdam Centre for European Studies and the Dutch Research Council (grant nr 016.164.351). We are very grateful for all this support. We would also like to thank Roland Pierik at the Scholten Centre for Jurisprudence, University of Amsterdam, for kindly hosting the first workshop and to the University of Amsterdam Library for hosting the second one. We are also grateful to the secretaries at the VU University law school for their brilliant organizational help with the first workshop, and to Gordon Arlen for his extremely helpful assistance with various aspects and organizational matters related to the second workshop. Thanks also to Jamie Berezin at Oxford University Press for his guidance and for believing in this project, and to Arthur Ripstein for the great cooperation and for trusting us to edit a volume centered on his forthcoming book. We would also like to thank Charlotte Halloway and the OUP production team for their splendid support. Last but not least, we are grateful to Adrian Kreutz for his excellent research assistance.

References

Bazargan, S., and Rickless, S. (eds.) (2017). *The Ethics of War*. Oxford University Press.
Benbaji, Y., and Statman, D. (2019). *War by Agreement, a Contractarian Ethics of War*. Oxford University Press.
Frowe, H. (2014). *Defensive Killing*. Oxford University Press.
Frowe, H., and Lazar, S (eds.) (2018). *The Oxford Handbook of the Ethics of War*. Oxford University Press, 99–145.
Kleingeld, P (2004). "Approaching Perpetual Peace: Kant's Defence of a League of States and His Ideal of a World Federation," *European Journal of Philosophy* Vol. 12(3): 304–325.
McMahan, J. (2009). *Killing in War*. Oxford University Press.
Maliks, R. (2014). *Kant's Politics in Context*. Oxford University Press.
Noorda, H. (2016). *Thinking War in the 21st Century*. Amsterdam University Press, 35.
Ripstein, A. (2016). "Just War, Regular War, and Perpetual Peace," *Kant Studien* Vol. 107(1): 185.
Ripstein, A. (2009). *Force and Freedom*. Harvard University Press.
Ripstein, A. (2021). *Kant and the Law of War*. Oxford University Press.
Rostbøll, C. (2017). "Kant and the Critique of the Ethics-First Approach to Politics," *Critical Review of International Social and Political Philosophy* Vol. 22(1): 55–70.
Walzer, M. (1977). *Just and Unjust War*. Basic Books.

PART ONE

COMMENTARY ON KANT AND THE LAW OF WAR

2

A Semi-Kantian Just War Theory

Yitzhak Benbaji

Like most modern theories of international relations and international law, the Kantian theory of the law of war, which Arthur Ripstein has defended in his fascinating recent book, *Kant and the Law of War*, is structured on the basis of an analogy between natural and international persons (Ripstein 2021: chap. 1, section I). In the Kantian framework, the normative truth that no state is subject to another is understood by analogy to the normative truth underlying private law, that no person is subject or subordinated to another. Just like private law, whose basic role is (among other things) to preclude a person from violating the freedom of her neighbor, international law is set to preclude one state from violating the sovereignty of another. A person ought not to interfere with the body of another; analogously, a state ought not to violate the territorial integrity of another. The territorial rights of states that the law of nations secures and the bodily rights of persons that private law protects have similar structure, scope, and limits.

The organizing principle of the law of nations (to which states have been subject all along) was first articulated in the 1928 Kellogg-Briand Pact: nations must not use force to resolve "disputes or conflicts of whatever nature or of whatever origin they may be, which may arise among them." Kantian morality renders the first use of force of one legitimate state against another impermissible just like the UN Charter, which asserts that states ought to refrain from the threat or use of force "against the territorial integrity or political independence of any state."

Yet, unlike the normative truth that no person is subject to another, the normative truth that states are independent of each other is *derivative*, and so are the rights and duties that follow from this truth. Kant derives this norm from the states' basic duty to institutionalize the freedom of their citizens by establishing a rightful legal order. Indeed, as I shall elaborate later in this chapter, for Ripstein's Kant, the right of states to their sovereignty and territorial integrity follows from their duty to provide and maintain a "self-certifying" or a "closed" legal order in the territory over which they rule.

The argument of this essay starts by highlighting a distinction that Ripstein downplays, between the *legitimacy* of states and their *presumed* legitimacy. In effect, Ripstein's Kant deduces the fundamental rights of states from their legitimacy as well as from their presumed legitimacy. And, these two approaches are in deep tension with each other.

The fact that a state legitimately rules over a territory is *not* what *ultimately* grounds the prohibition to aggress against it. As I have just noted, the analogy to private law is

Yitzhak Benbaji, *A Semi-Kantian Just War Theory* In: *The Public Uses of Coercion and Force*. Edited by: Ester Herlin-Karnell and Enzo Rossi, Oxford University Press. © Oxford University Press 2021. DOI: 10.1093/oso/9780197519103.003.0002

misleading: the bodily rights of natural persons are indisputable; states, in contrast, are artificial entities and as such might rule over territories that other states understandably claim to be their own. Particularly, a state that is currently defending its territorial integrity might have conquered the land that it occupies by aggression, bribery, or illegal purchase. So, why does a formerly aggressive state possess a right against the first use of force by another state that was recently a victim of its aggression? Why is a failed state entitled to use defensive force to protect the territory that it merely formally holds? Ripstein's answer employs the notion of presumed legitimacy. For Kant, "any multitude of human beings must be assumed to be in a rightful condition" and therefore the use of force against it is "*a wrong against the people on whose territory it takes place*" (Ripstein 2014: 149). Conquering a territory is *presumably "contrary to the 'original right' of the people 'to unite itself into a commonwealth'"* (ibid.), since no state has the authority to judge another to be illegitimate. In the state of nature, a state has no authority to use force in a territory which it does not actually rule, even if its enemy took over the contested territory by aggression, or even if the enemy fails to provide a legal order in it.[1]

Unfortunately, this move from legitimacy to presumed legitimacy suggests two contradictory conceptions of territorial rights, of political societies, as well as two different moralities of state-building (that is, two opposing moralities of the exit from the state of barbarism).

In response to these difficulties, I will outline a semi-Kantian philosophy of the law of war, which is free from the tensions the Kantian law of war allows. Semi-Kantians argue that Ripstein's Kant misconstrues the standing of states vis-à-vis the territories over which they rule. Taking seriously the disanalogy between a state's (disputable) territorial right and the (indisputable) bodily rights of a natural person, semi-Kantians structure territorial rights is based on two factors. A state is entitled to rule over a territory if and only if (first) it secures the private freedom and the political participation of its inhabitants, *and* (second) a treaty to which other states voluntarily entered obligates all actors not to use force in resolving future disputes over this territory.[2]

I proceed as follows. In section 1, I describe central elements of Ripstein's Kantian theory of statehood and the law of nations. In section 2, I point to the significance and depth of the tensions that this theory allows for, by producing three hypotheticals with respect to which the Kantian framework suggests two contradictory verdicts. In section 3, I outline a political philosophy of international society, according to which states legitimately provide rightful conditions in their territories, only if they enter a treaty by which they recognize each other as the sole rulers in these territories.

[1] Ripstein (2021: chap. 1). I present Ripstein's reading of Kant, rather than Kant's own view, of which I cannot claim intimate acquaintance. For a different reading, see Stilz (2009: 185–213). According to her reading, while states do have the right to go to war, they must give it up, since wars "*jeopardizes individual freedom and independence* . . ." (ibid.: 207).

[2] According to Stilz's reading, unless domestic right is supplemented by peace among states, it "*is unavoidably undermined and must finally collapse*" (ibid.). I therefore conjecture that my semi-Kantian view has roots in Kant's writings.

1. Ripstein's Kant on the Moral Standing of States

1.A. The Independence of States from Each Other

According to Kant, "*freedom (viz., independence from being constrained by another's choice), insofar as it can coexist with the freedom of every other in accordance with a universal law, is the only original right belonging to every human being by virtue of his humanity*" (Ripstein, 2009: 13). Kant introduced another postulate of morality according to which "*[i]t is possible for me to have any external object of my choice as mine.*" Indeed, for him, "a maxim by which, if it were to become a law, an object of choice would in itself (objectively) have to belong to no one (*res nullius*) is contrary to right" (ibid.: 62). With no rights over external objects, individuals cannot legitimately pursue most ends that they can set for themselves.

Kant nevertheless insists that there are no property rights in the state of nature. This is because there is always a conceivable question regarding who owns a particular thing. Kant infers that the innate right to freedom can only be realized within a legal order that institutes and enforces private rights. Only a legal order can institutionalize, on behalf of all citizens governed by it, procedures of acquiring, abandoning, and exchanging external objects; and to specify under which conditions one person wrongfully uses another person's means, or wrongfully damages them (ibid.: 91). As far as the state acts on behalf of all citizens, the unilateral acts of acquisition and abandonment, as well as the bilateral act of transfer, are binding on other members of the political community. Individuals are therefore under a natural duty to establish a state that one of its fundamental duties is to provide a legal order structured in light of the republican "Idea of the Original Contract."

The other branch of a well-functioning legal system is public law. Public law regulates a state's provision and maintenance of the background conditions which enable citizens to realize their equal freedom. For example, the state should provide for "roads to freedom." That is, it "must make it possible to get from any piece of land to any other" (ibid.: 248–249). Additionally, it should provide public markets that make private interactions fully rightful. In pursuing "public" goals of this type, the state is entitled to use force against its citizens by taxing and punishing them. The state is entitled to determine when it is under attack and to recruit troops for national defense.

The independence of states follows from three normative facts. First, a world government is likely to be oppressive and tyrannical, hindering rather than enhancing the idea of the social contract.[3] Predictably, an opposition to the world government will engender incessant and intractable civil wars (Waltz, 1979). Equal private freedom and political participation can therefore be secured only in a world, which contains a plurality of states.[4]

[3] See Kant (1991).

[4] I am not sure whether in arguing for the plurality of states, Kant elaborates a non-ideal theory, or whether, for him, there is a principled reason why there shouldn't be a rightful condition with a single world government. The semi-Kantian view that I elaborate below is also indecisive in this respect. Generally, Kantians believe that the optimal size of nations is related to the idea of original contract; arguably, states are too big in the sense that the ability of its institutions to represent all citizens is compromised in virtue of the fact that it is under duty to represent too many individuals.

Second, states are legitimated by their internal functions rather than by any right that the law of nations confers on them. The source of state's legitimacy is the duty of its citizens to establish a state that provides a rightful condition. By fulfilling this duty, no individual wrongs another. In the same manner, "no nation wrongs another by providing a rightful condition to its inhabitants" (Ripstein 2021: chap. 2, section VII).

Third, adequate legal orders are closed or "self-certifying." Rights are "indisputable" only if they are secured by a legal order that includes a mechanism for resolving any dispute with respect to the scope of the private rights that individuals possess. Accordingly, a legal order should empower judges to determine which things you already have, whether another person used them without your permission, and whether you used them in a way that is inconsistent with the freedom of another person. As importantly, a legal order empowers legal officials to determine which public ends the state should pursue, given the fact that there are many things that the state can permissibly do in protecting and enhancing the freedom of its citizens. Finally, in pursuing public ends, a state must have the power to tax, punish, recruit troops for national defense, etc. Therefore, a legal order should create institutions that empower officials to address, on behalf of all citizens, the question of proportionate punishment as well as distribution problems in the contexts of taxation and recruitment.

Ripstein (2021: chap. 3, section V) argues that "*self-certification leads directly to political independence and territorial integrity*":

> The only way a plurality of closure-imposing institutions can coexist is if none is subject to the approval of the others. A legal system's right to non-interference is the right to self-certification: ... each [legal order] has a mandate to provide a rightful condition for the human beings subject to it, but not for those outside it—. . . one legal order wrongs another by subordinating the other legal order to its own." (ibid.)

It follows that the recognition of another state ought not to be an affirmative and voluntary act. "*Instead, recognition must be a part of the peremptory law of nations rather than of the voluntary law of nations*" (ibid.); it is not up to another state to negate the right of the citizens to be governed by their own state.

Two further notes are in place here. First, and most importantly, the Kantian view does not base states' right to independence on their meeting more than the minimal demands of legitimacy. This minimum includes practically everything short of what Kant calls "barbarism" (which he defines as force without freedom or law). Indeed, a state might be ruled by a tyrant, who fails to act on behalf of its citizens. Such a state is legitimate, as far as it provides the minimal basis of a rightful condition for its inhabitants, protecting them against each other's violence and unilateral enforcement of private claims. It follows that states might have a right to independence against other states, even if they are gravely defective and thus subject to an internal obligation to reform themselves.[5]

[5] This point is debated between Kant's scholars. Ripstein and Stilz (2009: 203) argue that since for Kant "*political authority is a necessary prerequisite to justice's establishment ...we must defer to existing political authorities, in the hope that justice will eventually be brought about through them.*" Thus, Kant infers that "*the presently existing legislative authority ought to be obeyed, whatever its origin*" (6: 319). Cf. Hill (2002), Holtman (2002).

Second, "it is futile to inquire into the historical documentation of the mechanism of government." Even if "power came first and law arrived only afterwards," this has no moral bearing since the state's right to rule over a territory is not the result of it having come about through appropriate procedures.[6] A state legitimately rules over a territory even if it has just conquered the land over which it rules by aggression. In other words, the law of nations (just like some domestic legal systems) allows adverse possession. Indeed, past wrongs might generate good title.

To summarize, states' independence of each other, namely, their duty to respect the sovereignty of other states, follow from three related facts. At least in our non-ideal world, a plurality of states is a moral necessity; states are legitimated by their internal functions, rather than by the law of nations, or by the recognition of other states; states can fulfill their fundamental duty to lawfully rule over a territory only if their legal system is self-certifying.

The core argument of this part is based on a further move that Ripstein attributes to Kant, namely that the right of states against aggression is independent of their legitimacy-based right to independence and sovereignty. Crucially for my purpose here, the fact that a state legitimately rules over a territory is not what ultimately grounds its right to use force in defense of its borders. Even "failed states"—namely, states that fail to provide a legal order in certain territories, and thus leave their inhabitants in the condition of barbarism—possess a right against aggression as well as a right to protect their borders by using force. Indeed, the Kantian argument for the inherent right to defensive force is based on a further assumption: a state's judgment that its neighbor is a failed state cannot justify a coerced temporary occupation or a coerced permanent annexation. The right to national defense is not only a right to defend a rightful condition; it might be nothing more than the right to prevent an unauthorized use of force. The right of self-defense is the right to defend a rightful *or* a presumably rightful condition.

Let me put the argument in a slightly different way. The Kantian law of nations renders the defender's claim over the territory that it holds "presumptively" good. This is because its title is "presumed good until someone has successfully contested it before competent institutions" (Ripstein 2021: chap. 3, section IV). The presumption is indefeasible due to the state of nature between nations: states might be legitimate or illegitimate on the one hand and yet, there is no legal procedure by which it is legitimated or de-legitimated, on the other hand. It follows that a failed state is presumptively legitimate since no one is authorized to deny its right to rule in the territory that they actually hold. No one is in a position to judge their legal order as illegitimate. Indeed, "*a system of independent nations presupposes that no nation has standing to contest another's title*" (Ripstein 2021: chap. 3, section VI).

Kant's critique of European colonialism in Africa, Asia, and the Americas expresses the same conviction: even if the colonial powers were right about the "savages" inhabiting those lands, and about those lands being lawless, states are not entitled to take them by force. They had the "right to visit," namely, the right to arrive on another's shores and propose terms of trade with them. But they are not allowed to

[6] See Ripstein (2021: chap. 2, section VI). Quoting Kant (6: 340) and (6: 318) respectively.

annex territories over which they do not rule since they have no right to settle or appropriate them by force, based on their judgment of their moral standing.[7]

This approach to states' use of force implies one fundamental restriction on the aims of permissible defensive wars. The defender should aim at stopping the instrumentality (that is, the armed forces of the aggressor) through which the incursion takes place. The domestic analogy to cases of individual self-defense is illuminating: if I am attacking you, you can stop my attack. You are not allowed to prevent future attacks by using more force than is required to stop the immediate attack. States' defensive force is not used with the aim of achieving longer-term security, but rather focused exclusively on the immediate threat. Its rationale is no more than the upholding of the respective independence of the parties (Ripstein 2021: chap. 3, sections VI–VII).

1.B. Ripstein's Kantian Framework: A Summary in Five Propositions

In the next section, I will offer a series of cases, which expose a fundamental tension between the two essential elements in the Kantian theory of the moral standing of states, namely, its conception of legitimacy and its argument for a blanket prohibition on the use of non-defensive force. To better understand this tension, I should end this expository section by summarizing Ripstein's Kantian framework.

I have already presented the first propositions, which can be summarized as follows.

1. Legitimacy: States are legitimate in virtue of the legal order that they provide. States' power to provide and to enforce a legal order is grounded in the duty of individuals to establish a rightful condition by which they give law to themselves, and thereby secure their private rights. By lawfully ruling over their subjects, states wrong no other state and no other individuals.
2. Territoriality: Human beings must organize themselves into political units that are territorially (that is, spatially) individuated. The territorial right of a state—namely, its right to rule over a territory by enforcing a legal order there—is grounded in the duty of its inhabitants to establish a rightful condition. A territorial state wrongs no other state by ruling in the territory that they actually hold, and therefore, they are entitled to be recognized by other states, whatever the borders of this territory are.

One tension that I will discuss in the next section manifested itself due to another feature of the moral standing of states, which I haven't discussed thus far. States are entitled to pursue public ends by shaping contractual relations with other states. Moreover, they might impermissibly pursue "private" ends—like political clout in the

[7] See Ripstein (2014). Ripstein's reading of Kant is controversial. Stilz (2014: 209) argues that Kant does not rule out territorial annexations. "*When there is no collective agent capable of constituting a state to secure basic rights, it may not be wrong for another state to annex the territory, if they can commit to ruling legitimately.*" She admits that unilateralism does impose a problem, but if the annexation is "internationally authorized" then this problem might be resolvable.

international community—by entering treaties that are nevertheless binding. I call this proposition,

3. Sovereignty:
   a. Treaties between states are binding. In particular, states are entitled to agree on redrawing their borders. Such agreements are appropriate especially if a division/secession/purchase of land etc. will enhance the ideal of political participation. "*In circumstances in which the residues of war or colonialism create units in which divisions make it difficult for institutions to properly represent everyone, it is within the power of a legal order to modify its borders*" (Ripstein 2021: chap. 2, section VI).
   b. States' mistakes and misbehavior might nevertheless be normatively effective: "there was no rightful basis for France to sell Louisiana, or Russia to sell Alaska, or for the United States to purchase either territory" (Ripstein, ibid.). Still, like adverse possession, these treaties might be binding.[8]

A fourth proposition, which I call Regulated Anarchy is important, in light of a truth that Sovereignty should make clear: reasonable disagreements between fully legitimate states are possible. For example, they might reach a treaty and later disagree on its interpretation.

4. Regulated Anarchy: the state of nature between nations is defective because it lacks a common standpoint through which competing claims can be assessed and coercion can be authorized. The society of states ought to found international institutions that act on behalf of all states, whose constitutive task is to prevent unilateral state action (Ripstein 2021: chap. 2, section I).

I call the first four propositions Ripstein's "Legitimacy-based" propositions, since they all follow from the Kantian conception of the duties of the state and its freedom to violate them. The fifth proposition goes beyond legitimacy. Even in the absence of the international institutions that Regulated Anarchy requires, States' non-defensive force is impermissible; indeed, force exercised against a failed state or against individuals who live in the state of nature is illegal.

5. Force: One state uses force against another, if and only if it violates its territorial integrity. A state violates the territorial integrity of another state by an armed invasion of its territory, or by an imminent threat to do so. A state ought to presume that a territory over which it does not currently rule is governed by another, independent, rightful condition.

Hence, a state's use of force is impermissible in any territory over which it does not currently rule, unless it is necessary for self-defense. European colonialism in Africa, Asia, and the Americas is therefore impermissible. Even if the "savages" inhabiting

[8] There are, though, a priori limitations: "*authorizing slavery or genocide would be the international equivalent of an individual human being selling him- or herself into slavery*" (Ripstein, ibid.: 57). Moreover, arguably, third parties and international organizations (EU and Nato) might try to block normatively defective deals like the US purchase of Greenland from Denmark. Nevertheless, according to Sovereignty, once the agreements are reached, they are binding.

these lands are lawless, invaders have no right to take by force the territory in which these "savages" live.

1.C. Is the Kantian Theory of Territorial Rights Coherent?

This section shows, through a series of cases, that Force (on the one hand), and the Legitimacy-based propositions (on the other) convey opposing conceptions of territorial rights, of political societies, and of the ethics of state-building.

The First Hypothetical: Two Conceptions of Territorial Rights

The first hypothetical—I call it the "Warlord hypothetical"—exposes the basic Legitimacy/Force tension. The very idea of borders of a territorial state tolerates two different interpretations. Force prohibits any unilateral expansion or annexation; from Force's perspective, like bodily rights, territorial rights are fixed by the normless status quo. In contrast, Legitimacy suggests that the scope of states' territorial rights is sensitive to their duty to provide a rightful condition as well as to their duty to represent everyone who is subject to their rule.

The Warlord hypothetical illustrates the tension between Force and Legitimacy since in this case a unilateral annexation is necessary for maintaining the rightful condition of the expanding state as well as welcomed by the annexed community and its local leaders. Consider a failed state ("Failed") that has no effective control over part of the territory that it is under a duty to rule by law. De facto, a hostile warlord ("Warlord"), who illicitly denies the authority of Failed to rule over the relevant territory, controls it by force. Warlord's territory is lawless; there is no effective legal order by which the inhabitants of this territory are governed. Worse, Warlord leads a group of militants who permanently pose threats to one of Failed's neighbors, Well-Ordered. The Warlord's gang targets some of Well-Ordered's citizens, dig tunnels into its territory, etc., and thus undermines the rightful condition that Well-Ordered provides to these citizens. Failed has no resources to enforce the law in Warlord's territory. Moreover, military occupation would not solve the problem. The only way Well-Ordered can maintain the rightful condition that it used to provide to its own citizens is annexing Warlord's territory.

The annexation is bloodless. After the bloodless invasion, Well-Ordered provides a legal order in a territory, which previously was lawless, and whose inhabitants had lived under a condition of barbarism. Moreover, the inhabitants of Warlord's territory voluntarily join Well-Ordered's civil society. The intentions of Well-Ordered are rightful; it would have been ready to give Warlord's territory back to Failed; it annexes it only because the Warlord gang will certainly reoccupy Warlord's territory immediately after the disengagement.

The Kantian framework strongly suggests two different resolutions of the conflict between Failed and Well-Ordered. By annexing Warlord's territory, Well-Ordered acts as a public entity; unlike typical aggressors, its leaders do not pursue prosperity or political clout; they aim to fulfill their duty to maintain the necessary background conditions for the private freedom of all their citizens. Moreover, in a sense, the annexation is consensual; the inhabitants of the disputed territory and their local leaders are happy to join Well-Ordered.

Viewed from Force's standpoint, things look very different. In a clear sense, Well-Ordered's bloodless annexation was unilateral; as such, it "substitutes force for right." Failed protests against the annexation. It refuses to recognize the new legal arrangement that Well-Ordered imposed in the annexed territory. The public officials whose role was to enforce Failed's law in Warlord's territory had the right (viz., the normative power) to rule over this territory; the invaders used force to take this right from them. Thus, if Force is true, Well-Ordered has no authority to use force based on its judgment that Warlord's territory is lawless (in effect, even the inhabitants of Warlord's territory have no right to rebel against Failed; instead, they have the right to pacifically struggle for a change *in* Failed). Furthermore, no one but Failed is entitled to use force in solving the problem posed by Warlord's gang. Force's opposition to colonialism entails an indefeasible presumption against other states taking it upon themselves to provide legal order in a territory they do not hold. The presumption is not defeated by the fact that those territories are lawless and that their inhabitants are in a condition of barbarism.

In sum, the Warlord hypothetical shows that Ripstein's Kantian framework offers two different conceptions for states' territorial rights. Force suggests that states have no right to unilaterally change the normless status quo. In contrast, Legitimacy strongly suggests that a state's right to unilaterally annex another territory that is necessary for its capacity to provide a rightful condition to its citizens depends on the standing of the territory (it should be lawless or empty) and on the consent of the inhabitants of this territory, a consent that should be mediated by their local leaders.

The Second Hypothetical: An Imagined History of China, Britain, and Hong Kong

The second hypothetical will show that Force and the Legitimacy-based propositions offer opposing conceptions of political society, namely, the groups of individuals that the law of nations treat as societies entitled to political sovereignty. Force imposes a general duty: a nation ought to treat any multitude of individuals, which it does not currently rule, as a political society that possesses a right to self-rule and independence. In contrast, the Legitimacy-based propositions offer a much more nuanced and complex conception of political society.

The hypothetical is structured in light of the history of Hong Kong. Before elaborating an imagined history, I sketch, in the following three paragraphs, its actual history.[9] The first Anglo-Chinese war—an economic war that Great Britain fought against China—ended on August 29, 1842, with the Treaty of Nanking, by which Hong Kong Island became formally ceded to the United Kingdom. More than 50 years later, in 1898, China and Britain were engaged in another economic war—the Second Anglo-Chinese War—which ended with a further imperial acquisition. Britain leased the Kowloon peninsula (a total area of about 370 square miles) for 99 years.

Hong Kong, China, and Britain dramatically changed in these years. In 1949, the Chinese Civil War resulted in the Communist Party taking control of mainland

[9] See Tsang (2004: 9–16, 29–45, 218–225, 269–273).

China. During the war, Hong Kong became an asylum for Chinese refugees—many of whom were skilled workers—who opposed communism. In its turn, Britain abandoned its colonial and imperial ambitions, and gradually committed itself (at least formally) to the principle of national self-determination. In light of its allegiance to liberalism and democracy, the British cabinet proclaimed in response to the establishment of the People's Republic of China (PRC) a new policy regarding Hong Kong: Britain "*should not be prepared to discuss the future of Hong Kong with the new Government unless it were friendly, democratic, stable and in control of a united China*" (Tsang, 2004: 156).

This policy did not endure mainly because China stabilized and became a significant world power. Rather than insisting on democracy in Hong Kong, the United Kingdom and the PRC drafted the 1984 Sino-British Joint Declaration, which reported that both sides agreed that sovereignty over the whole of Hong Kong would be transferred from Britain to the PRC on July 1, 1997. On this day, Hong Kong, the last great imperial possession of the British Empire, became the first Special Administrative Region of the PRC. In modern Chinese history, "*this marked the completion of a full circle, from Hong Kong being ceded by the declining Chinese Empire in 1842 to its peaceful and successful retrocession to resurgent and powerful China 155 years later*" (ibid.: 268).[10]

Let us turn to what I call "the HK hypothetical"; the HK hypothetical is supposed to uncover a tension between two ways of individuating political societies that, under the Kantian law of nations, are entitled to self-rule. Imagine a somewhat different history: after China becomes a communist country, citizens of Hong Kong form a unified political society (hereinafter, "HK political society" or for short "HK"). HK was adequately represented by elected leaders (the HK leaders). They demand that the British government supports a new, independent state in Hong Kong and protect the newly established state from Chinese aggression. HK leaders base their requirement for independence on a simple (and obviously Kantian) argument. A local independent government would be better positioned to act on behalf of individuals living in Hong Kong, since China is too big, and the Chinese central government is too remote. An independent government is more likely to secure political participation for the simple reason that self-rule is easier to manage in a smaller unit. Moreover, the great majority of Hong Kong's inhabitants oppose communism and the Communist Party of China. Without contesting the very legitimacy of the regime that the PRC institutes, HK leaders argue that, as a matter of fact, a democratic regime better fits the political traditions, beliefs, and preferences of the current residents of Hong Kong. British officials tend to agree. They secretly hope that Hong Kong will stay democratic and, accordingly, did not prevent HK leaders from effectively governing HK and from organizing a military force that will protect it. By the end of June 1997, just before Britain left Hong Kong, HK openly positioned military forces on the borders of Hong Kong asserting its right over this territory.

Responding to the argument advanced by HK leaders, China argues that the PRC is a legitimate state that reached a binding treaty with another sovereign state. It argues,

[10] The current political situation in Hong Kong shows how problematic it was for the United Kingdom to allow a non-democratic regime to rule over a community whose democratic tradition was quite solid.

in other words, that the Sino-British Joint Declaration is based on a binding treaty that China and Britain were entitled to enter. It further argues that the agreement itself is part of the process of a just decolonization of Hong Kong. Based on the 1898 lease agreement and the history of Hong Kong, China claims that HK's attempt to establish a new state is an illegal revolution. The use of force of the individuals positioned on the borders of Hong Kong is a violation of the Chinese law to which these individuals are subject by the end of June 1997.

The Kantian framework strongly suggests two contradictory resolutions of the PRC/HK dispute. The Legitimacy-based propositions leave it open whether, in our hypothetical, China violated the territorial right of the state that HK successfully established, or whether some residents of Hong Kong illegally (or in fact, criminally) revolted against China. Was HK a political society, or was it a group of individuals who were subjects of the Chinese rule post July 1, 1997? Could the state for which HK leaders struggled be established despite the 1984 British rectification of the Chinese territorial right over Hong Kong? In a well-ordered society of states, legal officials would have the authority to decide which party to this dispute prevails. They would address the dispute by weighing backward-looking considerations as to the validity of the treaty between China and Britain, the standing of Britain to reach it, the legitimacy of Chinese rule over a territory that it does not hold, etc.

In contrast, Force offers a determined verdict regarding China's invasion. China had not ruled Hong Kong before July 1, 1997. On July 1, China was using force in enforcing a legal system over individuals who were not subject to it beforehand. China *coerced* non-citizens to become subjects of a legal system that it established in Hong Kong. As far as Force is concerned, a state's use of force against people who are not currently subjects of its legal system, in a territory that this state does not actually rule, is illegal. This is because, presumably, HK *is* a political society entitled to sovereignty. Hence, Force strongly suggests that China should change the status quo without using force.

The Third Hypothetical: An Ethnic Conflict over Holy

A third hypothetical exposes a tension in the approach of the ethics of state-building, and the role of force and coercion in it. While Legitimacy suggests that, in condition of barbarism, there is no procedure that governs this process—in particular, there is no legal or illegal way for individuals to exit the condition of barbarism—Force suggests that individuals in the state of nature (and not only states) ought to avoid unilateralism and force.

Suppose that a former colonialist power leaves a territory (hereinafter "Holy") without establishing an alternative legal order. Members of two ethnic nations, the A-Nation and the B-Nation, remain with no legal order to secure their standing in Holy. Its current residents advance four different solutions for Holy. A- and B-Minimalists believe that a division of the land to two states, or a multinational state, are the desirable solutions for Holy. But A-Maximalists and B-Maximalists argue against the minimalist one-state solution: due to cultural differences and ethnic tensions between the As and the Bs, it would be impossible for institutions of such a state to rule over Holy on behalf of both nations. Maximalists further argue against a two-state solution: Holy is too small to be divided. It should come as no surprise that the A-Maximalists argue that the Bs should leave to other countries and that the As should compensate the Bs for the inconveniences involved in transferring them, whereas the B-Maximalists insist that the As should be transferred out of Holy. Both maximalist solutions entail "voluntary" displacement and transfer.

Since there is no legal order that empowers officials to authoritatively resolve this dispute, an important feature of the Legitimacy-based approach to statehood manifests itself: the process of a state's coming into being is extralegal. Hence, as far as the As do not violate preexisting rights that the Bs possess, it would not be illegal for the As to unilaterally establish an A-State that will rule over all of Holy. The A-State that the A-Maximalists will struggle for will do whatever it legally can to push the Bs out of Holy, by permissibly limiting their property rights and citizenship. (The extent to which a legal order is free to distribute property rights is debated between Kant's scholars.[11]) Of course, while establishing such a state is not illegal, it is not legal either. Hence, the Bs do no wrong by unilaterally acting against those who try to transfer them out of Holy by establishing a maximalist A-State.

Yet, Force suggests a different approach to the Holy conflict.[12] This proposition expresses a general condemnation of private, unauthorized, use of force. Naturally extended, it implies that like illegal use of force, extralegal use of force is barbaric—even if it involves no violence and no bloodshed. Any action that substitutes right with force is impermissible. Ideally, then, all disputes should be resolved by legal officials who represent the parties to the dispute and look into the merits of the case. Hence, in a case in which implementing the best arrangement involves "extralegal force," acting unilaterally in promoting this arrangement is impermissible.

Accordingly, the first duty of the As and the Bs is to establish a legal order that empowers legal officials to resolve the dispute on behalf of members of both ethnic nations. According to this view, the As and the Bs should enter a civic condition whose borders are dictated by the fact that Holy is lawless. Indeed, "*a person ... is permitted [arguably, he ought] to coerce his neighbors to enter a state along with him*" (Stilz, 2014: 207). This is because "*[p]eople who regularly interact [and thus] come into conflict about rights" should establish "a system of public law in order to resolve these conflicts in a way that safeguards their freedom and independence*" (Stilz, 2014: 207). Of course, nothing makes Holy the desirable territory of a single state in which both As and Bs enjoy equal civil rights. Yet, since the use of force is wrongful unless it is subject to a legal oversight, and since the As and Bs do not agree on how to fulfill their duty to establish a civic condition, they should establish a state whose legal officials will determine the right solution for Holy on behalf of both communities.[13] This duty holds even if the presence of maximalists makes it de facto impossible for such a state to endure.

[11] Thus, Stilz (2011) believes that Kant introduces a diluted form of property rights in the state of nature, and that therefore, denying property rights form the A-residents of Holy would be legally impossible.

[12] For an elaboration of this aspect of Kantian ethics of state-building, see Ypi (2013).

[13] Ripstein might argue that the Kantian framework is elaborated on the assumption that these hypotheticals are exceptional. The Kantian framework proceeds on the basis of the presumption that states are legitimate is true.

2. A Contractarian, Semi-Kantian Theory of Territorial Rights and the Crime of Aggression

The contractarian theory of territorial rights and the crime of aggression that I will elaborate in this section is free from the tensions that the three hypotheticals expose. I characterize the view that I elaborate here as "semi-Kantian," since its ingredients are all taken from the Kantian theory on the moral standing of states. In particular, the semi-Kantian view is based on Ripstein's Kantian approach to the duty of states to solve the problem of right for their citizens, and the idea that in order to do so, they must be independent of each other.

The most significant difference between the Kantian and the semi-Kantian views is that for Ripstein's Kant, states gain their moral independence from each other independently of each other, in virtue of their duty to provide a closed legal order in the territories that they hold. In sharp contrast, in the semi-Kantian view, pre-1928 Kellogg-Briand Pact, states only partly fulfilled their fundamental duty to provide a rightful condition to their citizens. An explicit treaty, in which states undertake the duty not to use force in solving their disputes, generates (rather than expresses a recognition in) their right to rule over a territory by a self-certifying legal order. In other words, states' power to rule by law is defective before a treaty, which secures states' right to be the sole rulers in their territories, has been voluntarily accepted by all states. Pre-1928, states should have reached such a treaty, but they haven't been bounded by it before it was actually reached.

2.A. Territorial Rights and the Independence of States

I start by drawing a distinction between Ripstein's Kantian critique of positive international law pre-1928 and the semi-Kantian critique of this legal system. As it is standardly understood, the positive international law pre-1928 conferred on states a right to go to non-defensive wars; this right was conceived as a "right inherent in sovereignty itself."[14] The legal freedom to wage war manifested itself through the constitutions of quite a few countries that openly distinguished different procedures to the initiation of offensive and defensive wars. Now, according to the Kantian critique, a legal system that fails to prohibit non-defensive wars is self-contradictory. The law of nations is based on respect for the sovereignty of states whereas conferring on each state a liberty right to violate the sovereignty of other states expresses disrespect for states' sovereignty. States and statesmen in the nineteenth (and early twentieth) century inadequately disavowed this "logical impossibility." They fail to consider it as "*inconceivable that, in the name of sovereignty, each State was empowered to challenge the sovereignty of other States*" (Dinstein, 2014: 71).

The semi-Kantian critique of the pre-1928 legal system is very different. It concedes that the absence of a legal prohibition on non-defensive wars is an anomaly. It takes Grotius's understanding of public war as a legal procedure by which states resolve

[14] I borrow this analysis from Dinstein (2014: 71).

their disputes as deeply confused.[15] Kant correctly argues that in the state of nature, there is no legal permission to go to a non-defensive war even in cases in which such a war is necessary for maintaining a rightful condition of an existing state. After all, wars are barbaric; they substitute might for right. Semi-Kantians insist that, nevertheless, non-defensive wars are *not il*legal. This fact is a feature of the diluted "legal order" that governs the state of nature between nations. As noted, the analogy between territorial rights and bodily rights is limited: unlike human bodies, the territories of states have no natural borders. Therefore, like property rights, the territorial rights of states might be disputed. It follows from this feature of the normative relation between states and the territories that they rule that, while the law of nations should not permit non-defensive wars, it has no moral justification to prohibit such wars either. Wars should be outlawed, but they can be justifiably outlawed only after states recognize the right of other states to rule over the territories that they hold. Up until then, the contradictions in the legal system that governs the society of states cannot be avoided.

To see why more clearly, let us return to the Warlord hypothetical. Some of the territories held by Failed are the Well-Ordered's necessary means for maintaining the rightful condition that it provides to its citizens. A bloodless invasion (which nevertheless involves coercive unilateral state actions) whose aim is taking over part of Failed's territory is the only way that Well-Ordered can successfully fulfill its basic duties toward these citizens.[16] In the decentralized society of states, there is no legal official entitled to authoritatively assess the merits of Well-Ordered's claim over the territory that Failed formally holds. Semi-Kantians infer that absent a legal order that regulates the distribution of territories among states, a law of nations has no authority to prohibit Well-Ordered's unilateral expansion. Notwithstanding, since Failed did not wrong Well-Ordered by formally ruling over the territory that Well-Ordered allegedly needs, it is not legally impermissible for it to wage war in defense of this territory. The state of nature between nations is defective precisely because of this anomaly: both wars are not illegal nor are they legal.

For semi-Kantians, this defect has another crucial consequence. It entails that neither state can fully secure the private freedom of their citizens. If coerced annexation of a state's territory is not illegal, the private rights that states institute are defective; their legal system fails to make legally defined acquisition, abandonment, and exchange of external objects binding. The rights states confer on their citizens might still be disputed by other states and their citizens; a state might invade the territory over which another rules and impose a different legal system, without violating any law to which nations are subject. True, a state S might be quite strong, such that no other state has a chance to undermine the property rights that S secures. Moreover, states might be "well-disposed and right-loving" and thus might enter no dispute. But this is irrelevant since, "*the problem is conceptual: the idea of a rightful condition contrasts*

[15] Needless to say, according to Kantians and semi-Kantians, Grotius's construal of *private* war as a procedure by which private people (who are governed by different legal) can resolve their disputes is also deeply confused.

[16] Thus, while Stilz's Kant is right that non-defensive wars typically "*jeopardizes individual freedom and independence* . . . " (ibid. fn. 3), such wars might be necessary for securing freedom and independence. This implies that with no actual mutual recognition between states, they fail to fully secure the private rights of their citizens.

with 'savage violence' because in the former, disputes are resolved by law, and in the latter, by force" (Ripstein, 2009: 179). The private rights that states ought to institute and secure remain contingent on how forceful they are.

How, then, should states overcome this defect? Following Kant, semi-Kantians realize that due to its size, a global state won't be able to act on behalf of all its citizens, and will (most probably) fail to implement the idea of the social contract. At least in our world, plurality of states is a moral necessity. Semi-Kantians therefore suggest that, as part of their duty to secure the private freedom of their citizens, states ought to enter a treaty in which they recognize the territorial rights of all other states and thereby subject themselves to a legal norm that prohibits the use of force against other states. In other words, states ought to form a society of mutually respectful states, by committing themselves to something along the lines of the 1928 Kellogg-Briand Pact. They reject Kant's idea that the moral right to territorial integrity is not an acquired right. For them, peace, is the result of contract (*pacto*) rather than status (*lege*).

Now, the historical Pact was incomplete and imperfect. Ideally, the society of states that this treaty forms ought to be maximally inclusive: the negotiations should result in an outcome where all states recognize the territorial integrity of all other states. Hence, ideally, all parties to the treaty should be able to fulfill their basic duty to provide rightful conditions to their citizens in the territory that the treaty allocates to them. That is, before agreeing on the terms of the treaty, states ought to ensure that the treaty secures the background conditions, which would enable them to provide and maintain a rightful condition. The treaty should be maximally inclusive in another sense: it should leave no individual person stateless.

The "peace treaty" that semi-Kantians envision is constitutive: states are fully legitimate only if their right to rule over a territory becomes strictly indisputable thanks to the treaty. Thus, the actual treaty partly generates, and is therefore partly constitutive of the status of states as the providers of a rightful condition in their territories. This feature of the treaty renders states' commitment to peace irreversible. States are not free to abandon it. True, if some constraints are satisfied, international law should allow (and in some contexts, it should even require) secessions, unifications, divisions, and annexations. Yet, the treaty does not license a state to require a change of the status quo ante as a condition of its very membership in the society of states. After the treaty has been reached, the commitment of the parties to peaceful bargaining and peaceful negotiations is not open to negotiations. Another individuals/states analogy might illuminate this feature of the treaty that semi-Kantians envision: states allow individuals to leave to other states, but it is conceptually impossible to entitle individuals to return to the state of nature. States have the power to divide themselves, sell some of their territories, etc., but they have no right to leave the society of states.

Of course, as a matter of fact (not norm), individuals and states might fall back to barbarism. Probably, people living under the Nazi regime in Germany were not living in a rightful condition. Due to its denial of the very idea of territorial integrity, Nazi Germany itself fails to uphold its membership in the society of states. The semi-Kantian response to the phenomenon of rogue states is simple. As part of the obligation to reach an inclusive peace treaty, the international community is required to ensure that Germany will be a member of the society of states once again, and that individuals living in the territory in which the Nazis lawlessly ruled should not remain stateless. The ideal treaty dictates a further instruction: the territory and the population of the new states that replaced Nazi Germany should be established by peaceful negotiations, with the help of international institutions.

Thus, while for Ripstein's Kant, the 1928 treaty positivized a law to which states were already subject, the semi-Kantian view treats the 1928 treaty as the beginning of a process in which a society of states—a society governed by a law that secures states' independence of each other—was formed. States fully safeguard the closure of the legal order that they provide only at the end of this process, where all individuals are protected by states and all states are members of an international society committed to peace.

The semi-Kantian view does not entail determined resolutions to the three hypotheticals offered in section 2. Its advantage is that unlike the Kantian framework that suggests contradictory verdicts, the semi-Kantian view stresses that the very ideas of territorial integrity, political societies (entitled to sovereignty and independence), and use of force are fully defined only given an explicit treaty under which states recognize each other as sovereigns in the land over which they rule by law. It is the treaty that renders states' territorial rights indisputable, define the individuals that inhabit their territories as political societies, and render unilateral change of the status quo ante, an instance of use of force. The duty of states to provide a self-certifying legal order entails a duty to enter a maximally inclusive peace treaty, which, in its turn, obligates them to ensure that the Failed/Well-Ordered dispute, HK/China dispute, and the Holy dispute are resolved one way or another. The treaty itself contains no normative material, which dictates a specific resolution, but it does offer two general constraints on any resolution that states might reach. First, under no circumstances, are people to be left in a barbaric condition, and second, the newly established states should respect the independence of all other states.

2.B. The Duty to Undertake a Duty

An important objection to this analysis merits attention. In a nutshell, the objector argues that the semi-Kantian view collapses into the Kantian view. Even according to the semi-Kantian view, states should respect the sovereignty and the territorial integrity of other states, whether or not they voluntarily subjected themselves to a treaty that obligates them to do so. After all, the objector observes, semi-Kantians concede that states ought to subject themselves to a treaty that prohibits the use of force against other states. But if states are duty-bound to enter such a treaty, then (so runs the objection) the treaty does not generate states' obligation to respect each other's right to rule in the territory that they hold. Indeed, "*[the] particulars [of the treaty] may be negotiated, but its basic form is mandatory, because participation in it is interdependent but non-consensual*" (Ripstein 2021: chap. 2, section VI). If that is how the semi-Kantian account is supposed to generate territorial rights, it would seem to already presuppose them; the treaty is pursuant to an antecedent duty.

I would like to make two points in response to this objection. First, I will show that in many cases, the parties to an agreement are not bound by its terms before they enter it, even if they ought to try to enter the contract and undertake the duties to which it subjects them. Second, I will argue that, in some circumstances, a contract does not bind one party, if it can be known that the other party will fail to fulfill its duty to enter it. The peace treaty that (according to the semi-Kantian view) establishes the society of states, instantiates both of these features.

To see that some contracts are not binding even if the parties ought to try to enter them, let us imagine a case where you and I should collaborate in rescuing a shared friend (hereinafter "Friend") from drowning to death. I cannot rescue her by myself, nor can you, hence (by the ought-implies-can axiom) neither I nor you are under a duty to rescue Friend by ourselves. We can however save her together, but only if we divide the labor wisely: I should swim toward her and hold her head above the water level, while you bring the boat by which we will bring her to the shore. We will fail to rescue her if we divide the labor unwisely: if you swim toward Friend while I bring the boat. I am a better swimmer than you whereas you know better than I how to handle the boat.

Let us further suppose that, at the relevant time t_0, there is no fact of the matter as to whether our interaction will end with the right division of labor. Despite our sincere efforts we might fail to find the right arrangement. In other words, at t_0, it is indeterminate whether we will divide the labor appropriately or fail to do so, through no fault of our own. It follows that at t_0, it is indeterminate (in the fact relative sense) whether we ought to rescue Friend. At t_0, we are under a duty to do whatever we can to reach a division of labor that will enable us to do so. Now, if at t_1, it turns out that we cannot divide the labor wisely—either because we (faultlessly) do not trust each other or because we are (faultlessly) unaware of each other's skills and weaknesses—we are under no duty to rescue Friend at t_1. If, however, it turns out that we can reach the appropriate division of labor, we become subject to a duty to which we were not subject to at t_0. It follows that it is the agreement that we ought to try to reach at t_0, or—alternatively—the fact that it turns out that we can reach this agreement, that generates our duty to rescue Friend. Thus, the objector's assumption—that if at t_0, we ought to try to reach an agreement that obligates us to *v*, we were already under a duty to *v* at t_0—is false.

Turn to the pact that generates the territorial rights of states, according to the semi-Kantian view. Since territorial rights are disputable, they can be secured only through a pact by which states recognize their independence of each other. As public entities whose prime duty is to institute private freedom, states ought to try to enter such a pact. Still, they are not subject to it—and therefore, they are not independent of each other—before they actually agree on the terms of the pact. Due to problems of trust, good-willing states might be uncertain as to their ability to reach such a contract with other states. Indeed, due to coordination problems and collective actions problems, it is objectively indeterminate whether or not states can reach the treaty that confers on them the right to be the sole rulers in the territories that they occupy. Hence, the treaty is not binding unless all parties can know that all other parties are ready to fulfill their duty to submit to it, before, that is, they actually entered the contract.

3. Conclusion

The debate between the Semi-Kantian and the Kantian views is focused on two of the five propositions that constitute the Kantian framework. They agree on Legitimacy, Sovereignty, and Regulated Anarchy. The semi-Kantian view replaces Territoriality with Territoriality* and Force with Force*.

> Territoriality*: Human beings must organize themselves into political units that are territorially (that is, spatially) individuated. Hence, a state's right to rule in a territory is grounded in the duty of its inhabitants to establish a legal order there, and in the recognition of all states that no other state is entitled to enforce a different legal order in the same territory.

> Force*: A state uses force against another, if and only if it violates its territorial integrity. A state violates the territorial integrity of another state by an armed invasion of its territory or by an imminent threat to do so. State's use of force is impermissible if (*i.*) it is unnecessary in order to secure the rightful condition of its citizens, or if (*ii.*) it is bound by a voluntary treaty to bargain rather than use force in pursuing its territorial claims.

According to the view elaborated here, in circumstances in which the distribution of territories among states is disputable, the question, whether the territorial integrity of a state has been violated, as well as the question whether a state's use of force constitutes a crime of aggression, make no definitive sense. This view is superior to the Kantian view as it offers a more coherent understanding of the right of states to rule over a territory whose borders are arbitrarily drawn, and as such might be in dispute between different political societies.

References

Dinstein, Y. (2014). *War, Aggression and Self-Defence*. 3rd ed. Cambridge: Cambridge University Press.

Hill, T. (2002). "Questions about Kant's Opposition to Revolution." *Journal of Value Inquiry*, 36(1), pp. 283–298.

Holtmann, S. (2002). Revolution, Contradiction, and Kantian Citizenship. In Timmons, M., ed. *Kant's Metaphysics of Morals: Interpretative Essays*. Oxford: Oxford University Press, pp. 210–231.

Kant, I. (1991). Idea for a Universal History with a Cosmopolitan Purpose and Perpetual Peace: A Philosophical Sketch. In Reis, H. ed., Nisbet, H. trans. *Political Writings*. 2nd ed. Cambridge: Cambridge University Press, pp. 41–53, 93–130.

Ripstein, A. (2009). *Force and Freedom*. Cambridge, MA: Harvard University Press.

Ripstein, A. (2014). Kant's Juridical Theory of Colonialism. In Flikschuh, K., and Ypi, L., eds. *Kant and Colonialism: Historical and Critical Perspectives*. Oxford: Oxford University Press, pp. 147–168.

Ripstein, A. (forthcoming). *Kant and the Law of War*.

Stilz, A. (2009). "Why Do States have Territorial Rights?" *International Theory*, 1(1), pp. 185–213.

Stilz, A. (2011). "Nations, States, and Territory." *Ethics*, 121(1), pp. 572–601.

Stilz, A. (2014). Provisional Rights and Non-State Peoples. In Flikschuh, K., and Ypi, L., eds. *Kant and Colonialism: Historical and Critical Perspectives*. Oxford: Oxford University Press, pp. 199–219.

Tsang, S. (2004). *Modern History of Hong Kong*. London and New York: I.B. Tauris.

Ypi, L. (2013). "What's Wrong with Colonialism." Philosophy & Public Affairs, 41(1), pp. 158–191.

Waltz, K. (1979). *Theory of International Politics*. Toronto: McGraw-Hill Publishing.

3

Might and Right

Ripstein, Kant and the Paradox of Peace

Rainer Forst

> *The only way to prohibit the principle that might makes right with respect to territory is to concede that might made right in the past, and thus to outlaw all future acquisitions through might.* (Ripstein 2021: chap. 2, section VI)[1]

Arthur Ripstein's interpretations of Kant's legal and political philosophy are unrivaled in their systematic power and elegance, given his ability to develop a complex system of thought on the basis of a transparent set of basic principles. Simply put, after reading Ripstein one knows much better what Kant thought and why he did so—and what to think about the things Kant thought about. While Ripstein showed in his *Force and Freedom* how Kant's conception of political order for one nation can be developed on the basis of the "innate right" of persons to independence from the choices of others,[2] in *Kant and the Law of War* he shows how that idea of persons being "*sui iuris*" relates to a system of right and peace between states that are also "*sui iuris*," yet are bound to form a federal league of nations to promote an international legal order to secure peace.

In developing this argument, Ripstein suggests an intriguing way to resolve the dilemma that has haunted the debates about Kant's proposal for such a system of international right. The main question is why Kant thought that, unlike the case of persons leaving the "rightless" state of nature to enter a system of public law in accordance with a "postulate of public right" (MM VI 307),[3] the international state of nature between states or nations ought not to lead to a full-blown "world state" but only to a federation of states that falls far short of a state-like entity. There is an analogy between the two states of nature, but in Kant's view only a limited one, since states *already* are in a condition of public right, and requiring them to leave it to form a superstate would put that achievement at risk, both in the process of reaching that end as well as given the most likely outcomes of such a process. The first creates problems of war, the second

[1] In the following, I refer to the June 2018 draft of *Kant and the Law of War*. The references have been updated to reflect the 2021 version of the manuscript as far as possible.

[2] For a discussion of Ripstein (2009), see Forst (2020).

[3] Kant's works are cited according to the Academy edition pagination and the Cambridge translation, abbreviated as: CJ—*Critique of the Power of Judgement*, CPR—*Critique of Pure Reason*, CPrR—*Critique of Practical Reason*, MM—*The Metaphysics of Morals*, PP—*Toward Perpetual Peace*.

Rainer Forst, *Might and Right* In: *The Public Uses of Coercion and Force.* Edited by: Ester Herlin-Karnell and Enzo Rossi, Oxford University Press. © Oxford University Press 2021. DOI: 10.1093/oso/9780197519103.003.0003

harbors the danger of a "soulless despotism" (PP VIII 367) of a global state. The secret to understanding this argument, Ripstein forcefully argues, is to understand the role that the principle of peace plays in Kant's thinking in connection with the constitutive principle of public right. So this is what I will focus on in my brief remarks, namely the relation between peace and public right.

I will first critically discuss the methodological approach that Ripstein chooses when reconstructing Kant's position and in particular how he defines and relates "constitutive" and "regulative" principles with respect to peace and public right. In a second step, I will discuss some of the problems to which this reconstruction leads, especially what I call the *paradox of peace*: peace is supposed to avoid, end, or overcome war as the practice of might making right; but the principle of peace itself at crucial points seems to require us to accept that might makes right, at least as far as the past is concerned. But can this be reconciled with the principle of right that ought to guide our normative thinking about these matters? If we follow Kant, do we have to accept that the paradox of peace is unavoidable, as Ripstein implies? I would like to express some doubts concerning that view. By offering a different interpretation of the relevant constitutive principle in the realm of the practical, namely that of freedom, I will extend a Kantian invitation to Ripstein to move beyond Kant in questioning rather than affirming the paradox of peace.

1. Constitutive and Regulative Principles

Ripstein makes extensive use of the distinction between constitutive and regulative principles in Kant, arguing, for example, that "peace is the constitutive principle of the morality and law of war" (Ripstein 2021: chap. 1, section I) and that this is so because, on Kant's "distinctively public conception" (ibid.) of peace, it represents an imperative to establish the condition for public right, both within and between states. Ripstein argues that Kant makes a contrast between the "postulate of public right" as a *constitutive* principle of public legal institutions and the "idea of the original contract" as a merely *regulative* principle (ibid.). I expressly say "merely" here because for Ripstein this distinction amounts to establishing a "priority" (Ripstein 2021: chap. 8, section I) of the constitutive over the regulative principle, which has important implications for Kant's conception of war and peace in Ripstein's view.

This is an intriguing argument, yet it might puzzle some readers of Kant, since he does not—at least not explicitly—make the distinction in the exact way argued for by Ripstein. In fact, one might wonder what use the distinction between constitutive and regulative principles has at all in the practical realm, since Kant develops it in his theoretical philosophy and makes rather sparse use of it in his practical philosophy. And where he does use it, the regulative character of transcendental ideas of reason is most prominent, including his view of perpetual peace as a moral duty and an idea that guides political action, even though such a condition of peace may never be fully realized (MM VI 350; PP VIII 386). For many (see, for instance, Kleingeld, 2008), this makes it into an idea that reasonably regulates our action in a similar way to others, such as historical progress toward a republican state—that is, a regulative idea. So what is peace: a constitutive principle, as Ripstein argues, or a regulative idea or both?

Let us briefly retrace some of Kant's steps in that regard. Kant first introduced the distinction between constitutive and regulative principles as principles of the understanding in the Transcendental Analytic of the *Critique of Pure Reason*. Constitutive principles are used to construct objects of appearances of which we can have secure knowledge, such as in mathematical reasoning, whereas regulative principles are heuristic rules for seeking a well-grounded unity of experience, for example, in reasoning by analogy (CPR A 179–181, B 221–223). Regulative principles connect appearances but do not constitute them.

Kant returns to the distinction in a different context in the Transcendental Dialectic, especially in the Appendix where he discusses the regulative *use* of ideas of pure reason. Through the regulative use of these ideas, reason works on the understanding in such a way that it seeks unity in the concepts and cognitions of the understanding without creating new concepts or objects (CPR A 643, B 671). The use of regulative ideas is hypothetical in nature in that it seeks systematic unity of knowledge by providing a "focus imaginarius" (CPR A 644, B 672) that is not itself an object of experience but helps to establish a meaningful unity of experience (CPR A 671, B 699).[4] Transcendental ideas can thus never be used constitutively in the theoretical realm (CPR A 644, 672), since that would amount to an illegitimate speculative use of such ideas.

In the *Critique of Practical Reason*, Kant applies the distinction to the realm of moral action. He does so in the context of a discussion of the "objective reality" (CPrR V 134) of the highest good. For the categorical imperative to determine the will, the ideas of freedom, immortality, and God must enjoy such objective reality, although they cannot be objects of theoretical knowledge. Kant argues that practical reason has to assume that these ideas are real and therefore that "by this the theoretical cognition of pure reason certainly receives an increment [*Zuwachs*]" (ibid.), but not in a speculative sense that constructs transcendent objects. Rather, they are *practically* postulated as real, as necessary conditions of the possibility of acting according to the moral law. Thus, in respect of theoretical knowledge they remain regulative, but in the realm of the practical "they become *immanent* and *constitutive* inasmuch as they are grounds of the possibility of *making real the necessary object* of pure practical reason (the highest good)" (CPrR V 135).[5] Used constitutively, they do not produce new objects of knowledge but presuppose *and* produce a *moral reality*, the reality in which free and moral action has its place and effect. In short, reason presupposes that these ideas have a moral reality in order to be able to think of a reality in which moral action is possible.

Kant expands on this in §88 of the *Critique of Judgment*. Pure practical reason as the capacity to have causal effects by being guided by ideas of reason contains not just the regulative principle of the moral law, but also a "subjectively constitutive" (CJ

[4] For an illuminating discussion of this distinction within the Transcendental Dialectic, see Willaschek (2018: 110–118).

[5] Orig.: "immanent und konstitutiv, indem sie Gründe der Möglichkeit sind, das nothwendige Object der reinen praktischen Vernunft (das höchste Gut) wirklich zu machen."

V 453) principle that makes free and moral action a real possibility: "The idea of a final end in the use of freedom in accordance with moral laws thus has subjectively *practical* reality" (ibid.). The idea of freedom constitutes its possibility as a real one for moral action; thus Kant concludes: "But when it comes to the practical sphere, such a *regulative* principle ... —namely, to act in conformity with something, as an end, which given the constitution of our cognitive faculties can only be conceived by us as possible in a certain way—is at the same time *constitutive*, i.e., practically determining" (CJ V 457).[6] In the realm of freedom, the idea of freedom *constitutes* a world of possible moral action. As free agents, we cannot but think of such a world as a reality, that is, as one where the final end of morality is worth aiming at.

The only place in the *Metaphysics of Morals* where Kant takes this up, as far as I can see, is in the Introduction, in the context of the idea of freedom as a pure concept of reason. He stresses again that with respect to theoretical philosophy it is a regulative and also negative principle, as it does not give rise to new objects for speculative knowledge. Yet at the same time he emphasizes a positive, constitutive, practical use of the idea of freedom, which "proves" (MM VI 221) its reality by practical principles generating a causality of pure reason through a moral will. That kind of freedom constitutes morality as a system of categorical imperatives. And with respect to the external regulation of freedom, it leads to the principle of right (MM VI 230) and the notion of an innate right of freedom under a system of law that treats each person as *sui iuris* (MM VI 237f.). We could also say that the idea of freedom *constitutes* the principle of right, practically speaking.

Viewed in this light, it seems doubtful whether Kant's legal philosophy contains a distinction and hierarchy between the postulate of public right and the original contract as a constitutive and a regulative principle, respectively, in the way Ripstein thinks it does. Rather, both principles rest on the transcendental concept of freedom and form a unity in constituting a system of right as a practical task. The principle of right can only be realized in a state that is legitimate in light of the idea of an original contract, according to which "everyone (*omnes et singuli*) within a *people* gives up his external freedom in order to take it up again immediately as a member of a commonwealth, that is, of a people considered as a state (*universi*)" (MM VI 315).[7] And as Kant explains in the first definitive article of *Toward Perpetual Peace*, the *sole* constitution "that issues from the idea of the original contract, on which all rightful legislation of a people must be based" (PP VIII 350), is the republican one.[8] Thus, *pace* Ripstein, there does not seem to be any reason to assume that Kant accords a constitution of right priority over a republican form of constitution, as implied by the constitutive-regulative distinction. In the footnote to the quoted passage, Kant emphasizes that external legal

[6] Orig.: "Wenn es aber auf das Praktische ankommt, so ist ein solches regulatives Prinzip [...]: dem, was nach Beschaffenheit unserer Erkenntnisvermögen von uns auf gewisse Weise allein als möglich gedacht werden kann, als Zwecke gemäß zu handeln, zugleich constitutiv, d.i. praktisch bestimmend."

[7] Orig.: "nach welchem alle (*omnes et singuli*) im Volk ihre äußere Freiheit aufgeben, um sie als Glieder eines gemeinen Wesens, d.i. des Volks als Staat betrachtet (*universi*), sofort wieder aufzunehmen."

[8] Orig.: "welche aus der Idee des ursprünglichen Vertrags hervorgeht, auf der alle rechtliche Gesetzgebung eines Volks gegründet sein muß."

freedom just means "to obey no other external laws than those to which I could give my consent" (ibid., fn., tr. altered).[9]

So how does Ripstein argue for his interpretation of the priority of the postulate of public right as a constitutive principle over the idea of the original contract as a regulative one? In his text, he does not refer to any of the passages I cited above but instead to a general discussion of ideas in the Transcendental Dialectic in which Kant interprets and, to some extent, defends Plato's notion of ideas. An idea like that of virtue, Kant argues, is never encountered in perfect form in reality; but that is no reason to reject it, because ideas cannot be taken from experience (CPR A 315, B 371f.). With respect to a legal constitution of freedom, Kant argues for his principle of right as a necessary idea that should guide all legislation (CPR A 316, B 373). Ripstein, however, interprets Kant as asserting that "the regulative principle for a thing is to be found in its nature" (Ripstein 2021: chap. 1, section V) and goes on to say that even in war that approach can "find right" (ibid). In support of the latter interpretation, he cites a passage from the *Metaphysics of Morals* in which Kant says that when war breaks out without a formal declaration ending peace, one must assume an implicit contract to the effect that both parties want to seek their right by means of war, if one wants to find right in a state of war ("to find a right in a condition of war," MM VI 346).[10] I am not sure whether this argument is sufficient to substantiate the topical, Kantian-Platonic approach to finding a regulative idea in a practice of war; and apart from that, it leaves open how we can identify regulative or, for that matter, constitutive principles in the practical realm.

So my conclusion is that questions remain concerning the interpretive position Ripstein adopts with respect to Kant's use of the constitutive-regulative distinction. I will now examine how Ripstein uses that distinction within his overall argument.

2. Ripstein's Reconstruction of the Law of War in Kant

Ripstein is, of course, on solid Kantian ground when he locates the question of war and peace in the realm of practical freedom and right and rests his case on the principles of reason that apply there. And as already indicated, the system of principles he elaborates is complex. As Ripstein persuasively argues, in contrast to the just war tradition (which focuses on a moral notion of justice) and the regular war approaches (which focus on the rights of sovereigns), Kant defends a "distinctively public conception" (Ripstein 2021: chap. 1, section I) of the state and of questions of war and peace. Even though nations have a duty to establish a legal system of perpetual peace, they are horizontally "*sui iuris*" (ibid.) and remain so "even when international law becomes more nearly public" (ibid.). This is the most important implication of the difference that Ripstein constructs between the constitutive principle of public right and the regulative principle of the original contract. According to the former, nation-states remain independent realizations of public right, and international law aiming

[9] Orig.: "keinen äußeren Gesetzen zu gehorchen, als zu denen ich meine Beistimmung habe geben können."

[10] Orig.: "ein Recht im Kriegszustande finden."

at peace has to heed and protect this independence; thus, the second contract (which holds between states) remains limited.

According to Kant, Ripstein argues, "war is the condition in which might makes right" while "peace is a condition in which, through the establishment of procedures, right guides might" (Ripstein 2021: chap. 1, section I). Following the "idea of the original contract" (MM VI 315), the "sole purpose" (Ripstein 2021: chap. 1, section I) of the state is "the provision of a rightful condition for its inhabitants" (ibid.). Hence "a state's right to territorial integrity and political independence follows from its internal provision of a rightful condition, rather than any process or procedure of acquisition" (ibid.). That means that no other state has any authority over it (Ripstein 2021: chap. 1, section III). It follows that the only legitimate reason to go to war for a state is to defend its "existing rightful condition" (ibid.), not to defend its property or act on behalf of universal principles of justice, as the other traditions would have it. Both the reasons for and the conduct of such a defensive war (which I will not go into in detail here) "must be consistent with future peaceful resolution" (Ripstein 2021: chap. 1, section IV), that is, they must be such as to lead to a legal condition which prevents future war between nations.

Maintaining and working toward the possibility of peace is the major imperative of state action in the international realm, and that has important consequences. Ripstein argues that "if the possibility of a future peace is the constitutive principle of war, then its regulative principle must reflect that" (ibid.). Given that Ripstein has argued that peace is an imperative based on the principle of right to ensure that "right guides might" (Ripstein 2021: chap. 1, section II), we ought to read this as saying that the truly constitutive principle of peace (and thus of ending war) is the postulate of public right, and thus that peace must never be used to turn might into right, since this would be contrary to its constitutive principle. However, Ripstein draws precisely the opposite conclusion when he argues that the—as I assume, regulative—principle of seeking peace implies that "peace can only be achieved if the past is taken by all to be settled" (Ripstein 2021: chap. 1, section IV). In support of this conclusion, he appeals both to the first Preliminary Article of *Toward Perpetual Peace*, which states that a peace treaty must annihilate all causes for a future war (PP VIII 343f.), especially territorial disputes, and to the important argument in the *Metaphysics of Morals* that, because war is the practice of seeking to make right through might, the victorious state cannot claim compensation from an unjust opponent who started a war, since that claim would presuppose what does not exist, namely an order of right that would determine who acted justly or unjustly in starting a war (MM VI 348). Thus, the victorious state has no claim to compensation, and asserting such a claim would turn the war, even if it was engaged in for purely defensive reasons, into a *punitive* war, something which Kant rules out (MM VI 347).

This is why Ripstein argues for what I call the *paradox of peace*: "Peace is only possible if the results of war are accepted. Future disputes can only be resolved on their merits if past disputes are regarded as settled. Peace is the constitutive principle through which war can have a regulative principle." (Ripstein 2021: chap. 1, section IV). And that latter principle states that the grounds and conduct of war have to serve the end of the possibility of peace as demanded by right. But note the paradox here. If war is the practice of trying to make right through might, and peace is the reverse, then accepting the results of war—of right by might—is *necessary* for peace as guiding and constraining might through right. Yet what exactly is the "necessity" here—is it of a prudential or, as Ripstein argues, of a principled nature?

If we follow Kant, there is a prudential reason and a principled reason. The prudential one is in line with the first Preliminary Article, according to which one should avoid burdening peace with reservations and unsolved conflicts that could lead to further strife and new wars. Kant describes this as an enlightened sense of "political prudence (*Staatsklugheit*)" (PP VIII 344). Principled reasons, by contrast, reflect the fact that war amounts to a general state of injustice without a relation of right that could determine which party acts justly and which does not (MM VI 347–349).

Ripstein emphasizes the principled argument. Not only does peace serve the end of establishing an international order of right, it also has to respect the limits of that order and the continuing right of states "to political independence and territorial integrity" (Ripstein 2021: chap. 1, section IV). By an analogy to private law relations, juridical relations between states will be limited because they do not "give rise to any new *private* rights as between nations" (ibid.). Ripstein reads Kant's emphasis on the "provisional" nature of private right claims prior to a state of public right as implying that "the moral norms governing private rights are only enforceable through public institutions, capable of making, applying, and enforcing law on behalf of everyone" (ibid.). Thus rights to property only exist in a public condition with a system of "omnilateral authorization" by an "omnilateral will" (ibid.). Ultimately, only a fully republican legal and political order can overcome the provisional character of private possessions.

But this is where the analogy between private law and juridical relations between states ends for Ripstein.[11] While the acquisition of property in the state of nature can only be made right by being transformed into law within a condition of public law and general justification, things are different in the international realm because "the birth of a state does not stop another state from doing anything it was previously entitled to do"—that is, as long as the latter state did not have jurisdiction over the territory of the newly created state (Ripstein 2021: chap. 2, section IV). This is analogous to the difference between "wronging a person and merely changing the context in which that person acts" (ibid.). This is indeed how Kant views the matter—but note that there is an important proviso, namely that state formation is legitimate only if it does not violate the rights of another state. States are legal equals, and no system of public law must subordinate another state (ibid.). States are obliged to end the rightless international state of nature and form a "league of nations in accordance with the idea of an original social contract" (MM VI 344);[12] but this must take the form of a federation that does not establish a new sovereign and leaves national independence under law intact.

Ripstein infers from this that a state does not have to justify its claim to a certain territory within a new international public order, and thus that the analogy between acquiring property and acquiring territory breaks down (Ripstein 2021: chap. 2, section IV): "There is no backward-looking procedure that counts as the rightful way of acquiring territory, only a forward-looking prohibition on wrongful acquisitions." And Ripstein includes acquisitions of territory made by war, since they were made in a

[11] For different interpretations of the analogy, see Ypi (2014) and Huber (2017).
[12] Orig.: "Völkerbund nach der Idee eines ursprünglichen gesellschaftlichen Vertrages."

rightless condition, and rectifying them would, or at least could, lead to war. Thus the "only way to prohibit the principle that might makes right with respect to territory is to concede that might made right in the past, and thus to outlaw all future acquisitions through might" (ibid.).

Ripstein is, of course, aware of the paradox of accepting past instances of might making right in order to prevent precisely the same thing in the future. He revisits the issue in the context of his discussion of colonialism and mentions the obvious objection that the argument for accepting the results of past aggressions might be a major incentive for future war rather than for peace (Ripstein 2021: chap. 8, section II). He concedes the plausibility of that objection, but objects that "the point of the philosophical theory of wrongdoing is not to prevent wrongdoing" (ibid.). This is true in general, yet in the context of a discussion of the forward-looking principles of establishing perpetual peace, where the argument (as noted above) is the at least partially prudential assertion that punishing the unjust aggressor state would lead to insult and resentment and thus could cause further war, the truth of the matter lies in the social circumstances. Failing to impose sanctions on or to demand reparations from a state that had attacked another unjustly could equally lead to a more unstable situation of right.

But these considerations are not sufficient, as Ripstein argues; ultimately, we must follow reasons of principle. I agree, and I think that there are two reasons available, both of which can be used to support a different perspective on the paradox. First, imagine the case of a defensive war (which is rightful according to Ripstein's argument for the *sui iuris* natural status of states—(Ripstein 2021: chap. 2, section V and Kant on MM VI 346) that was eventually successful against a brutal opponent which waged an aggressive war that cost a lot of lives and destroyed large parts of a country's infrastructure and resources. Does demanding compensation in the context of a peace treaty on terms that ensure that the former aggressor state can remain stable and is not exploited or colonized really turn the defensive war into a punitive war, as Ripstein argues (Ripstein 2021: chap. 8, section III) and Kant also seems to say (MM VI 348)? Can it be an offense, morally speaking, and a contradiction, conceptually speaking, to describe what happened in normative terms—that is, to specify which side acted justly or unjustly in what ways when entering and conducting the war in question? There must be a difference between doing this unilaterally, such as when the victorious side declares itself to be the supreme arbiter of justice, which is the case Ripstein and Kant rightly criticize, and doing it *within the multilateral process of setting up a peace treaty as part of the formation of a durable condition of public right and peace.* Furthermore, if public right is the constitutive principle of such a new international order, it should affirm and respect the reasons for establishing that order in the first place. Doing so seems essential for an approach of *jus post bellum* that "focuses exclusively on institutional structures to prevent future war" (Ripstein 2021: chap. 8, section III). To be sure, I am not objecting to Ripstein's reading of Kant here, only pointing out that the argument fails to take account of a difference that should be made within the Kantian framework, namely the difference between an illegitimate imposition of justice by the victorious and one that responds appropriately to past injustice when establishing future peace terms. To declare both acts to be *similarly* unjust runs the risk of ignoring the notion of an "*unjust enemy*" (MM VI 349), which Kant defines as one whose maxim, "if it were made a general rule" (ibid.), would render a condition of peace among nations impossible. Ignoring that notion when setting up a postwar framework of right and peace also seems to be in danger of making peace impossible and, ultimately, of giving might priority over right and enshrining it.

The second principled reason for my skepticism about Ripstein's argument stems from my reading of the final paragraph (MM § 61) of the section on the law of peoples in the *Rechtslehre*. There Kant draws a stronger analogy between establishing a system of public right within and between states than Ripstein allows for—at least in the postwar context, which is relevant here. Kant argues: "Since a state of nature among nations, like a state of nature among individual human beings, is a condition that one ought to leave in order to enter a lawful condition, before this happens any rights of nations, and anything external that is mine or yours which states can acquire or retain by war, are merely *provisional*. Only in a universal *association of states* (analogous to that by which a people becomes a state) can rights come to hold *conclusively* and a true *condition of peace* come about" (MM VI 350).[13] This I take to be the proper conclusion from the relation between public right, the original contract, and peace, a conclusion that does not paradoxically accept past results of might turning into right in order to ensure that right overcomes might. Rather, what might be achieved by war should not be declared to be right, or, if so, then strictly conditional on its general justification within a true public condition of international law. That is the main reason why I remain somewhat skeptical concerning Ripstein's general argument about the "priority of the Postulate of Public Right as a constitutive principle over the Idea of the Original Contract as a regulative one" (Ripstein 2021: chap. 9, section I). A fully public condition of international peace must aim to create a public order that does not condone rightless and unjust acts of war as constitutive of an order of peace. That Kant, according to the quoted passage, declares perpetual peace to be an "unachievable idea (*unausführbare Idee*)" (MM VI 350) due to the difficulties of establishing such a "state made up of nations" globally does not alter his conviction that it serves as a regulative idea for a "continual *approximation*" to a condition of perpetual peace through the formation of alliances of states (*Verbindungen der Staaten*).

I concur with Ripstein that the aim of peace may often require the "surrogate" of accepting the results of past injustice as normative facts in order to establish a lasting public order of peace, because undoing past developments might be too difficult and dangerous and lead to aggression and war. But that is an argument which combines principle and prudence, emphasizing the latter, and I do not regard it as a necessary implication of the principled argument for public right that would be fully realized only in an *association of states* (*allgemeiner Staatenverein*) of the kind described as ideal in §61.

[13] Orig.: "Da der Naturzustand der Völker eben so wohl als einzelner Menschen ein Zustand ist, aus dem man herausgehen soll, um in einen gesetzlichen zu treten: so ist vor diesem Ereignis alles Recht der Völker und alles durch den Krieg erwerbliche oder erhaltbare äußere Mein und Dein der Staaten bloß *provisorisch* und kann nur in einem allgemeinen *Staatenverein* (analogisch mit dem, wodurch ein Volk Staat wird) *peremptorisch* geltend und ein wahrer Friedenszustand werden."

3. Beyond Kant with Kant

Occasionally, being true to a great thinker means going beyond some of the things he or she said in order to be true to the spirit of the thought—thereby perhaps being even truer to that spirit than the author him or herself was in some arguments. Ripstein, a true Kantian, is aware of that. Thus, going beyond what Kant explicitly said, but following him implicitly, Ripstein argues for a set of direct norms that apply to each state by way of an implication of the Postulate of Public Right. While basic human rights follow from the idea of the original contract and thus are merely regulative in Ripstein's view, prohibitions on slavery or genocide are *constitutive* of any public order, and "to violate these prohibitions is an undertaking to abandon a rightful condition and descend into barbarism" (Ripstein 2021: chap. 2, section IV). Ripstein is right, I think, to interpret these norms constructively as being based on the "innate right" to independence of each person, "which is already a robust moral norm" (ibid.). Thus, contrary to Kant's strict prohibition on forcefully interfering in another state's "constitution and government" in the fifth Preliminary Article of *Toward Perpetual Peace* (PP VIII 346), Ripstein argues for the legitimacy of a forcible (although not a unilateral) intervention in cases of massive human rights violations—thus not against "despotic" regimes but only against "barbaric" orders or non-orders that fail to provide a rightful condition for their inhabitants in the first place.

This strikes me as one more example of constructively unpacking Kant's morally grounded legal and political philosophy. We can do that because in the practical realm there is (only) one basic constitutive principle that guides all moral action, including morally grounded political action such as action seeking peace—namely, the constitutive idea of human freedom as the capacity to bring about real change in the world according to the imperatives of morality in the spheres of ethics *and* law. It is the basic idea of practical reason, which grounds all principles of "moral politics" (PP VIII 378). Arthur Ripstein's interpretation of that basic idea is indispensable in helping us to understand Kant. But at the same time, it helps us to move beyond Kant constructively in a Kantian spirit. I take my argument to be a Kantian invitation to Ripstein to make such a move in questioning instead of affirming the necessity of the paradox of peace.[14]

References

Forst, R. (2020). "Kantian Republicanism vs. the Neo-Republican machine: The meaning and practice of political autonomy," in Julia Christ, Kristina Lepold, Daniel Loick, and Titus Stahl (ed.), *Debating Critical Theory. Engagements with Axel Honneth* (Lanham: Rowman and Littlefield, 2020): 17–33.

Huber, J. (2017). "No right to unilaterally claim your territory: On the consistency of Kantian statism," *Critical Review of International Social and Political Philosophy* Vol. 20(1): 677–696.

[14] I thank Lea Ypi, Jakob Huber, Marcus Willaschek, Felix Kämper, Amadeus Ulrich, and Ciaran Cronin for their helpful remarks on an earlier version of this text.

Kant, I. (1996). *Critique of Practical Reason* (ed. and trans. Gregor, M.). Cambridge: Cambridge University Press.

Kant, I. (1996). *The Metaphysics of Morals* (ed. and trans. Gregor, M.). Cambridge: Cambridge University Press.

Kant, I. (1996). *Toward Perpetual Peace* (ed. and trans. Gregor, M.). Cambridge: Cambridge University Press.

Kant, I. (1998) *Critique of Pure Reason* (ed. and trans. Guyer, P., and Wood, A.). Cambridge: Cambridge University Press.

Kant, I. (2000). *Critique of the Power of Judgement* (ed. Guyer, P., and trans. Guyer, P., and Matthews, E.). Cambridge: Cambridge University Press.

Kleingeld, P. (2008). "Kant on historiography and the use of regulative ideas," *Studies in History and Philosophy of Science* Vol. 39(4): 523–528.

Ripstein, A. (2009). *Force and Freedom: Kant's Legal and Political Philosophy*. Cambridge, MA: Harvard University Press.

Ripstein, A. (2021, forthcoming). *Kant and the Law of War.* Oxford University Press.

Willaschek, M. (2018). *Kant on the Sources of Metaphysics: The Dialectic of Pure Reason.* Cambridge: Cambridge University Press.

Ypi, L. (2014). "A permissive theory of territorial rights," *The European Journal of Philosophy* Vol. 22(1): 288–312.

4

Reading Kant's *Rechtslehre*: Some Observations on Ripstein's *Kant and the Law of War*

Thomas Mertens

1. Introduction: Semantic Observations

In retrospect, it was to be expected that Arthur Ripstein's monumental *Force and Freedom* would be complemented by a book on Kant's view on war and peace (Ripstein 2009). True, the earlier book did touch upon matters concerning the law of nations and cosmopolitanism, but not to the extent dedicated to other parts of Kant's doctrine of law. It illuminated in particular Kant's often underestimated views on private law.

Now we do have Ripstein's views on Kant and the law of war in a well-thought through volume, which gives rise to quite some observations and comments, of which I have selected a few. They look in particular at the whole of his Kantian undertaking and at his reading of Kant's views on war and peace.

I start with a very general, semantic, observation which probably reflects not only Ripstein's Kant interpretation, but also that of myself as a long-term reader of Kant and interpretations of Kant, as a modest contributor to that scholarship, and as a translator and editor of some of Kant's texts into Dutch. Since the days I started my Kant studies, much has changed. Kant scholarship has become to a large extent an Anglo-Saxon affair, and Kant is read and interpreted against the background of political and legal problems of that world. This determines to a large extent, I believe, the issues that are now at stake and the linguistic concepts that are now used. History has shown that several readings of Kant are possible, and Ripstein provides us with a new, powerful reading of Kant which is indebted to that Anglo-Saxon background.

Therefore, a first observation concerns translation. Why translate the title of Kant's main work to which Ripstein refers into English the way he and most others nowadays do? The German title reads: *Die Metaphysik der Sitten: Metaphysische Anfangsgründe der Rechtslehre.* It has become customary in the English-speaking world—probably because of the Gregor translation now included in the (1996) Cambridge Edition of Kant's works—to translate this title as "The metaphysics of morals, Part I, Metaphysical first principles of the doctrine of right." This means that *Rechtslehre* becomes "Doctrine of Right." Gregor acknowledges the difficulty in translating Kant's concept of "Recht," because it has multiple meanings. She decided to translate it as "right" for two reasons, namely to uphold the conceptual connection between "das

Thomas Mertens, *Reading Kant's* Rechtslehre*: Some Observations on Ripstein's* Kant and the Law of War In: *The Public Uses of Coercion and Force*. Edited by: Ester Herlin-Karnell and Enzo Rossi, Oxford University Press. © Oxford University Press 2021. DOI: 10.1093/oso/9780197519103.003.0004

Recht" as a system of laws and "ein Recht," as a faculty on the basis of which one is entitled to do something without injustice; the second reason being the implication that all the adjectives derived from "Recht" if translated into "law" should then have to be translated as "legal," which would obscure Kant's distinction between the state of nature and civil society.[1]

Without going into too much detail, I do not find these reasons particularly convincing. A system of "law" indeed contains often a system of (subjective) rights. Bentham would even argue that any substantive right is the "child of law," meaning that there are no such things as rights anterior to the existence of law (Bentham 2001:125). The English language simply makes a distinction between law and right. The German (and the Dutch) language has it differently: "das Recht" as the law or the system of laws and "ein Recht" as a subjective right. Gregor's second reason does not convince me either: what is the problem of understanding the state of nature as a proto-legal system where humans are supposed to "have" an innate right or natural rights (obviously this is what Kant accepts but Bentham does not) and the state (of law) in which humans have legal rights?

Ripstein acknowledges the difficulty as well (Ripstein 2009:14). The German word *Recht* and its cognates have no exact English equivalent, he writes, because it covers both law and the more general idea of legitimate power. He "therefore" joins Gregor in the "right" translation, which has "the merit of preserving some of the ambiguity in a way that neither law nor justice does. In Kant's usage, right refers to the domain of enforceable obligations." I am not convinced by Ripstein's reasons either.

First, *Recht* seems to me as said to refer to law as a system of positive laws and the totality of mechanisms that make law effective: its legislative, administrative, and judicial organs. In this sense, *Recht* has an objective meaning, so that one can study German or ancient Roman "*Recht*," or even Nazi-law, for it is obvious that "Recht" can be just or unjust, efficient or inefficient, defective or complete. As I understand the concept of *Recht*, it does not semantically entail legitimate authority. Even in a broadly "just" or "legitimate" legal system, some citizens will consider particular legal regulations as utterly "unjust." There is also enough to be found in Kant's *Rechtslehre* which we would now consider thoroughly unjust. Just think of Kant's endorsement of capital punishment, not only for the murderer but for his accomplices as well (1914:334). This brings me to Ripstein's second reason. *Recht* may indeed also refer to a subjective element, but is it an "enforceable obligation" or rather a particular claim, entitlement or privilege that may put others under obligations as Kant has it? Everyone who fulfills certain criteria, for example that of age and citizenship as specified by positive law, has "the right" to participate in elections or to get married.

Put slightly differently, *Recht* refers either to the corpus of rules determining external conduct, that is (objective) law; or it refers to a claim, etc., on the basis of a person's faculty or interest, that is, a (subjective) right.[2] Again, according to Bentham all subjective "rights" are derived from "law." Indeed, who is allowed to vote and to

[1] Translator's note on the text of The Metaphysics of Morals. In The Cambridge Edition of the Works of Immanuel Kant (1996:357–358).

[2] Correctly therefore, the academic discipline of *Rechtsphilosophie* is translated "philosophy of law" in order to cover both the objective and the subjective component.

marry in a particular legal order is determined by that order.[3] Some theorists, notably Kelsen, also defend that subjective rights exist only because of positive law, thereby challenging what they call the ideological dualism of law and right. "Rights" only exist independently of "law" if one is willing to accept a form of natural law.[4] *Recht* captures both this objective and subjective meaning, in a sense in which neither law or right does.

One could argue that this semantic exercise leaves unanswered the question on how to translate Kant's concept of "Recht." But it brings us to the core of the matter: Is the *Rechtslehre* a treatise on law (or even a teaching (*Lehre*) of law) or is it a treatise on right(s)? In the latter case, "Doctrine of Right" may sound awkward but it could be correct qua substance. But does Kant's legal philosophy indeed revolve around rights? Something is to be said in its favor. Kant holds that there is an innate right to freedom. There is, however, also a lot to argue in favor of considering it a treatise on law. Important is that the starting position in Kant's practical philosophy in general is not the human being as a bearer of subjective rights on the basis of which a legal framework is to be established, but as a (partly) rational being that finds itself being obligated not merely by the moral law, but as subjected to a system of positive law as well.

This understanding finds at least as much confirmation in the very beginning of the "*Rechtslehre*" as the one focusing on the innate right. Look at how Kant defines the *Metaphysics of Morals* as a whole as a treatise on duties ("Lehre von den Pflichten") and divides it into two parts, into a *Rechtslehre* and a *Tugendlehre*, distinguishing between duties of law[5] and duties of virtue. In that same passage Kant (1914:239) reminds us that "we know our own freedom only through the moral imperative, which is a proposition commanding duty, from which the capacity for putting others under obligation, that is the concept of a right, can afterwards be explicated."[6] My view is that here Kant merely reiterates the position he adopted in the *Kritik der praktischen Vernunft*, namely that freedom may well be the "ratio essendi" of the moral law, but that we only become knowledgeable of ourselves as free beings because we stand under moral laws, as its "ratio cognoscendi." According to this reading, the "Metaphysics of Morals" "merely" adds to this that these moral laws lead to two kinds of duties: those duties the compliance of which can be enforced by external means, that is, duties of law; and the duties that require internal enforcement, that is, duties of virtue.

Is it plausible to read Kant's "Metaphysics of Morals" as spelling out our legal duties and the institutions these require first, and subsequently enumerating the duties of virtue? Some, like O'Neill (O'Neill 1996:140), have indeed vehemently argued in favor of a Kantian primacy of duty over right(s). If this is correct, then one does not find

[3] This is even true for Kant's *Rechtslehre*. It is only within marriage that children "exist" in a legal sense. Those who are born outside wedlock can be treated, Kant writes, as if they were stolen into the commonwealth; see Kant (1914:336).

[4] This is Kelsen's (1992:37–38) position in the *Reine Rechtslehre*.

[5] Does Gregor's (1914:349) translation "duties of right" make sense? What Kant means here are obligations imposed on me by the law.

[6] To rephrase this: the concept of a right (indeed: here Kant's concept of Recht needs to be translated as (a) right) is the capacity to put others under an obligation; this concept (of a subjective right) is thus developed *afterwards*, after the moral imperative. See also Alexy (1992:186–194).

in Kant a modern conception of human rights on the basis of which a system of law should be constructed. His would be a rather classical philosophy of law. This seems to be confirmed by Kant's acknowledgment that legal obligations as derived from (positive) law may at times conflict with moral obligations; Kant even suggests, to say the least, that in such a case priority should be given to legal obligations, which brings him—despite his explicit denial—close to Hobbes's position which is often marked as the starting point of modern legal positivism. It is understandable that Kant's legal philosophy is sometimes characterized as "positivist," as argues Waldron (1996:109, 1535–1566).[7]

If this is a valid interpretation of Kant's text, the translation of the concept of "Rechtslehre" should allow it. Hence: "doctrine of law" instead of "doctrine" of "right." In my view, Kant's "Rechtslehre" is not a modern proto-liberal tract, but a rather classical treatise on legal duties and rights, which contains quite a few passages that we now read with embarrassment and that even seem to contradict the "innate right to freedom."[8]

2. On Relations between States

My second general observation builds on the first semantic one, because the problem of translation returns when Kant uses the concept of *Recht* in connection with the relations between states. Does Kant advocate international "law" or international "right"? Do legal relationships between states consist of a system of agreements—treaties if you like—as a result of what sovereign states have decided on the basis of their rights, or is it a system of obligations? Kant's *Towards Perpetual Peace* suggests the latter, as it formulates six (negative) prohibitions and three (positive) obligations. Ripstein, in chapter 3, suggests the former. According to him, the principle of national independence and national defense—analogously to a person's natural (innate) right to independence—is crucial for Kant, as the "organising principle of the international legal order," prioritizing (again) right over duty. Innovative in Kant's approach is—according to Ripstein (Ripstein 2021: chap. 2, sections II–III)—that the right to national defense is not derived from or based on a more general right of self-preservation. The right to national defense (and to self-defense as well) is based on the right to be independent of others (cf. ibid.). The right to independence of states, and of persons in the case of interpersonal relations, emerges here as the most fundamental right.

This separation of independence and self-preservation, both with regard to persons and to states, can be contested in my view on the basis of Kant's texts. I admit that it is not so easy to find Kant explicitly confirming the right to personal

[7] I am not particularly convinced by those who refer to the appendix to the *Rechtslehre* in which Kant states that one should not obey the legislator in what is in conflict with "inner morality." Who is to decide what violates inner morality? Every citizen?

[8] My "favourite" passage in this regard stems from Kant's view on criminal law in which he writes that "certainly, no human being in a state can be without dignity, . . . The exception is someone who lost it by its own crime, because of which, . . . , he is made a mere tool of another's choice . . . " in Kant (1914:329–330).

self-defense. But from Kant's explicit recognition of the right to national defense in the third preliminary article of *Towards Perpetual Peace* where he writes that citizens can protect themselves and their country against attacks from without, the right to personal defense can be inferred. If citizens of a state have the right to defend themselves collectively against attacks, they surely must have the right to defend themselves individually against an unjustified threat. Is this permission to use force either collectively or individually against attacks derived from a natural right (of independence) or based on a duty? I would think that the latter is the case. According to the account of individual ethical duties that Kant (1914:422) presents in the *Tugendlehre*, it is the first (though not in all circumstances the most compelling) duty of a human being to preserve himself, "as an animal being." And since he has the duty to preserve himself against any inclination to commit suicide (an attack from within, so to speak), would he not also have the duty to preserve himself against an unlawful, external attack?[9]

Therefore, the key in which Kant discusses defense is not independence, but (the duty to) self-preservation. The limits of self-defense are reached when upholding one's life is not possible without violating another, more important moral duty, as in the case in which my prince wants to coerce me into giving a false testimony about an innocent person that he wants to destroy.[10] This also becomes apparent in the situation in which my life is in danger because of what we now call an innocent threat. Kant discusses, relevant here, the case of the plank of Carneades in a part of the *Rechtslehre* which is entitled "ambiguous right."

The ambiguity of the Carneades case lies in my view[11] in the following: Kant insists that no one has the authority to take an innocent life in order to safe his own life; therefore, an agent who kills an innocent person is culpable, even though this culpability is legally of little relevance, because the criminal law on murder has no deterrent effect in the situation of two persons struggling for a plank after a shipwreck: "a threat of an ill that is still uncertain (death by a judicial verdict) cannot outweigh the fear of an ill that is certain (drowning)." The difference between Kant (1914:235–236) and authors such as Pufendorf (1991:chap. 5), who also discusses the "two man on a plank," could then be understood as to either arguing with Pufendorf that the agent had a justification (because of the supposedly absolute right of self-preservation) or arguing with Kant that the agent had merely an excuse (since his duty to self-preservation is not unconditional). Of course, it could also be argued that Kant's rejection of the dictum that "necessity has no law" does not mean much. He still accepts that the killing of innocents out of "necessity" happens and should not be prosecuted. Kant may well insist that in theory all situations of human interaction are covered by law (and that

[9] This disposition to animality is one of the three fundamental human dispositions according to Kant's *Religionsphilosophie* (together with the dispositions to humanity and personality). Under the aspect of animality, humans strive for self-preservation, for propagation of the species and for community. See Kant (1914: 26). Kant's view here seems to be close to that of Thomas Aquinas, *Summa Theologica* (I.II), Question 94, Art. 2.

[10] This is the famous example in *Kritik der praktischen Vernunft*. It shows clearly that upholding one's life is not an unconditional duty according to Kant. Sometimes choosing death is the right thing to do. See also the Balmerino example in the Rechtslehre, in Kant (1914:333).

[11] I discuss Kant's understanding of the plank problem in Mertens (2017).

self-preservation does not trump the duty to respect the life of others), but this means little in practice. In plank situations the law has merely a symbolic meaning.

Kant could, of course, have opted for another solution, namely that perpetrators of killings in plank situations should be prosecuted and convicted,[12] despite the pressure of self-preservation they have experienced: *lex dura sed lex*. The rule is that one has an absolute duty to respect the life of others who do not use violence against us, and this rule can only be upheld by prosecuting its violations and so ensuring its deterrent force in the future.[13] In any case, the fact that Kant considers the killing of an innocent threat excusable, indicates in my view the extent to which he acknowledges the importance of self-preservation and its connection to self-defense. If national defense is indeed analogous to self-defense, it will find its limits when it entails the violation of duties, such as those enumerated in the *ius in bello* rules to which Kant subscribes in his sixth preliminary article.

3. The Public Conception of the State

My third observation regards the specific public conception of the state that Ripstein (Ripstein 2021: chap. 2, section I), ascribes to Kant. Specific for this conception is its rejection of a patrimonial or proprietary conception of the state: states do not own their territory (as if it were private property). Kant states this most clearly, according to Ripstein (ibid.) in the second preliminary article of *Towards Perpetual Peace*: "no independently existing state (whether small or large) shall be acquired by another state through inheritance, exchange, purchase or donation." Should this article, however, not be understood as a rejection of a monarchical understanding of the state in which the state and its inhabitants are the property of the monarch, and as a recommendation of an alternative, republican conception, according to the first definitive article for perpetual state? Is such a republican state a non-proprietary state, or simply a state in which its territory and its natural and economic riches belong to its citizens collectively? Ripstein (ibid., sections I and II) defends the former: Kant rejects all versions of a patrimonial state, since the state is nothing more than (the creating and sustaining of) a "rightful condition" for its citizens; its aim is not to advance some public good or well-being, since it has no external purposes apart from upholding that condition. It seems, however, that the latter interpretation—the citizens have collective ownership in a republican state—is implied in certain passages in the *Rechtslehre*? Look at paragraph 50, in which Kant discusses issues of emigration and immigration. If a "subject" (Kant writes: "Der Untertan, auch als Bürger betrachtet") wishes to emigrate, he has according to Kant the right to do so (because the state does indeed not possess him since he is a person), but he is not entitled to take with him abroad more than his movable belongings. His fixed belongings, even if they are sold and exchanged for money, have to remain within the state. The idea here is that his territory (the "bisher besessenen Boden") ultimately belongs to the

[12] In order to prevent the—in my view not at all so—"strange confusion" on the basis of which jurists take this subjective impunity to be an objective one, contra Kant (2014:236).

[13] This is the position of Justice Keen in the fictitious speluncean explorers case, in Fuller (1949). See also the famous UK legal case *R v. Dudley and Stephens*, (1884) 14 QBD 273 DC.

state. This seems to be confirmed elsewhere in the *Rechtslehre* where Kant (1914:338) discusses the question of whether the sovereign should be considered either as the "supreme proprietor (of the land)" or merely as the supreme commander over the people by law? The answer is this: the sovereign is the lord of the land or supreme proprietor of it (dominus territorii). This all strongly suggests contra Ripstein a kind of "patrimonial" conception of the state in which citizens collectively own the state but are not allowed to harm the patrimony by moving their supposed part broad. In line with this, Kant defends, in the earlier passage, a clear distinction between citizens ("Einsassen") who belong to a specific territory (their fatherland) on the basis of birth and foreigners who can only become citizens through a special legal act. In this sense, the state (or better: "Das Land") belongs to its "native" inhabitants. It is thus understandable that cosmopolitan law does not entitle foreigners, in Kant's (ibid.) view, to residency or citizenship wherever or whenever they want it. On the contrary: a community has the right to turn down such a request for settlement, and to expel the foreigner from its territory, if this can be done without endangering his life.

Is such a "public, non-patrimonial" conception relevant for deciding whether or not Kant subscribes to the just war tradition as Ripstein holds? The problem is that certain passages in *Towards Perpetual Peace* clearly seem to reject that tradition; Kant (ibid.:355) describes its representatives as "sorry comforters" who are "duly cited in justification of an offensive war" but have never moved a state "to desist from its (aggressive) plans." In *Rechtslehre*, however, he seems to subscribe to the right of a state to go to war if it is wronged by another state. And the threshold for such a wrong can be rather low. Merely the menacing increase of another state's power suffices. Many have struggled to reconcile these apparently contradicting positions. According to Hruschka and Byrd (2010), Kant has changed his position (quite drastically and in a short period of time) and the last text, that is, the *Rechtslehre*, should be considered as authoritative. Elsewhere (Mertens 2012), I have challenged this interpretation, defending the *Towards Perpetual Peace* position.

Where does Ripstein stand in this dispute? Key in his interpretation of Kant's view on the law of war is the distinction between the just war tradition and the regular war tradition and Kant's departure from both traditions. This enables Ripstein (2021: chap 2) to carve out what he considers Kant's specific position, which consists in strictly limiting the grounds for a just war to defending the rightful condition of one's state (i.e., national defense as the only legitimate ground for going to war) and in emphasizing and broadly interpreting Kant's view on the law during war (the *ius in bello*). Does this, however, solve the problem of whether or not Kant subscribes to a right to war? I hesitate. Remember that it is, according to Kant, neither conceptually impossible nor historically unrealistic to frame a war of aggression as merely defensive. The right to war can be invoked, Kant writes (1914:346), on the basis of both actively inflicted injuries in which case it would surely count as defensive, but also on the basis of merely imagined threats. Even when a state believes itself to be injured by "an alarming increase" of the power of another state, it can construct its war against this state as defensive. In addition, Kant (1914:349) uses the concept of an "unjust enemy," broadly defined a state that displays a "maxim" expressed either in word (*sic*!) or deed that would make peace among nations impossible. Against such a state, any war would certainly count as defensive.

By interpreting Kant as allowing defensive wars, not narrowly described, Ripstein ultimately seems to bring Kant into the camp of the sorry comforters. This is confirmed by Ripstein's endorsement of humanitarian intervention. Despite Kant's explicit rejection of any intervention in the fifth preliminary article in *Towards Perpetual Peace*—even in the case of a "scandal" that one state gives to another—Ripstein (2016:107) writes that a state is permitted to intervene in another state if it has lapsed into a condition of barbarism, that is, if it fails to provide a rightful condition for its members: "States are entitled to intervene to stop barbarism."[14] But in *Towards Perpetual Peace* one reads that even such a "limited" "right to war" is problematic, if not outright contradictory, because it determines what is lawful not by universally valid laws, but by one-sided physical force. "Right" cannot be decided by "might," that is, military force.

4. Concluding Remarks

Ripstein's *Kant and the Law of War* has not changed my understanding. I still hold that Kant's considerations on the "right to war" and the "unjust enemy" in the *Rechtslehre* are not prescriptive—in the sense that they represent Kant's moral viewpoint—but merely descriptive. In the *Rechtslehre*, Kant merely gives us a description of what the international realm would look like without positive international law. Indeed, in the "state of nature," the "right to war" includes not only the right to self-defense in the case of "active injuries," but also the right to preventive wars against an alarming increase of the power of another state and to humanitarian wars against the barbarisms in other states. In the state of nature, the concept "unjust enemy" is a pleonasm. We shall however not forget that in such a condition moral practical reason pronounces its irresistible veto: there shall be no war.[15]

References

Alexy, R. (1992). *Begriff und Geltung des Rechts*. Freiburg: Alber.

Bentham, J. (2001). "Anarchical Fallacies: A Critical Examination of the Declaration of Rights." In Hayden, P. (ed.), *The Philosophy of Human Rights*. St Paul: Paragon.

Byrd, S., and Hruschka, J. (2010). *Kant's Doctrine of Right. A Commentary*. Cambridge: Cambridge University Press.

Fuller, L. (1949). "The Case of the Speluncean Explores." *Harvard Law Review* Vol. 62(1): 616–645.

Kant, I. (1914). *Kants Gesammelte Schriften. Königlich Preußische Akademie der Wissenschaften*. Vol. 6. Berlin: Reimer.

Kelsen, K. (1992). *Introduction to the Problems of Legal Theory* (Paulson, B., and Paulson, S., trans.). Oxford: Clarendon.

[14] In his *Kant and the Law of War*, Ripstein seems more cautious in his endorsement, see p. 164 (note 20). Be aware that "entitled" means justified, not merely excused.

[15] Cf. Kant (1914:349–350).

Mertens, T. (2012). "Kant and the Just War Tradition." In Justenhoven, H., and Barbieri, W. (eds.), *From Just War to Modern Peace Ethics*. Berlin: De Gruyter Berlin, pp. 231–247.

Mertens, T. (2017). "Emergencies and Criminal Law in Kant's Legal Philosophy." In *ethic@. Revista Internacional de Filosofia da Moral* Vol. 16(1): 459–474.

O'Neill, O. (1989). *Constructions of Reason*. Cambridge: Cambridge University Press.

O'Neill, O. (1996). *Towards Justice and Virtue*. Cambridge: Cambridge University Press.

Pufendorf, S. (1991). *On the Duty of Man and Citizen* (Tully, J., ed.). Cambridge: Cambridge University Press.

Ripstein, A. (2009). *Force and Freedom, Kant's Legal and Political Philosophy*. Cambridge, MA: Harvard University Press.

Ripstein, A. (2016). "Just War, Regular War, and Perpetual War." In *Kantstudien* Vol. 107:187. DOI: https://doi.org/10.1515/kant-2016-0009

Ripstein , A. (2021, forthcoming) Kant and the Law of war, Oxford University Press.

Waldron, J. (1996). "Kant's Legal Positivism." *Harvard Law Review* Vol. 109: 1535–1566.

Other References

The Cambridge Edition of the Works of Immanuel Kant, *Practical Philosophy* (1996). (Guyer, P., and Wood, A.W., eds.). Cambridge: Cambridge University Press, p. 363.

UK legal case *R v. Dudley and Stephens*, (1884) 14 QBD 273 DC.

5

The Moral Basis of State Independence

Anna Stilz

Fundamental to Arthur Ripstein's defense of a Kantian approach to war is the thesis that each state has a right to be independent from the determining choice of other states. As against other states, Ripstein declares that the state is "sovereign—*sui iuris*" (Ripstein 2021: chap 1, section III). The state's right to independence is the basis for its permission to use force in national defense, and also for *in bello* restrictions that limit the permissible means of waging war to those necessary to stop aggression. But what morally justifies the state's right to independence? And can this right be accounted for on Kantian grounds? These are the questions that I pursue in this chapter.

In the chapter's first two sections, I focus on whether the Kantian view, as Ripstein reconstructs it, provides a philosophically satisfying basis for attributing a right to political independence to the state. In the final section, however, I outline an alternative reading of Kant that may provide a more compelling moral foundation for this right.

1. The Analogy between States and Persons

Classical writers on international law represented the state's political independence and territorial integrity as broadly analogous to the rights of individuals to personal independence and bodily integrity (Beitz, 1980: 69–71; Shaw, 2008: 215). Just as individuals should be free to pursue their own ends within some personal sphere of choice, so too, on the classical view, states should be free to pursue their own ends within a collective sphere of choice. This sphere was taken to be the state's territory, which was protected against interference from foreign powers. An implication of this is the *doctrine of non-intervention*: states ought not to intervene in other states' internal affairs. Another implication is that wars of national defense are permissible because they uphold the state's independence from other states.

Yet the analogy between states and persons has been frequently challenged. Why should we think of the sovereignty and autonomy of states as morally significant in the way the sovereignty and autonomy of individuals is usually taken to be? Many authors have objected that states do not have the unity or consciousness that might lead us to respect individuals as ends-in-themselves (cf. Beitz, 1980). Further, as David Rodin notes, few would sanction the use of lethal force to protect the autonomy of groups *other* than the state. If a corporation like Cadbury were to use lethal force to protect itself against a hostile takeover by Kraft, most people would reject this as a blatant

Anna Stilz, *The Moral Basis of State Independence* In: *The Public Uses of Coercion and Force.* Edited by: Ester Herlin-Karnell and Enzo Rossi, Oxford University Press. © Oxford University Press 2021. DOI: 10.1093/oso/9780197519103.003.0005

wrong (cf. Rodin, 2014). So why sanction the use of lethal force to protect the sovereignty and independence of the state?

There are two basic strategies for answering this question. The first theorizes state independence as a *collective* claim: the state as a corporate body is taken to have a fundamental right to govern itself. The second, *reductivist*, strategy for understanding the state's right to independence sees it as *derived* in some way from the rights or claims of that state's members. On this view, state independence is justified insofar as it is necessary to realize the rights, claims, or interests of the individuals who make it up.

2. Kantian Reductivism or Collectivism?

The reductivist strategy would seem the more straightforward one for a Kantian to take. If state independence is necessary to realize the independence of individuals, then national defense can readily be made consistent with Kant's fundamental rationale for the use of force. On Kant's view, force is justified only in order to protect individuals' innate right to external freedom, the right to independence from being constrained by another person's choice, insofar as it can coexist with the freedom of every other under universal law (6: 238). Suppose—as the reductivist suggests—that an aggressor who threatens the state's independence at the same time threatens the independence of its citizens. Then it seems there should be a clear authorization, on Kantian grounds, to repel the attack.

Sometimes Ripstein suggests such a reductivist approach. He argues, for example, that Kant is committed to "an even more robust form of individualism" (Ripstein 2021: chap. 1, section V) than contemporary just war theorists, because Kant holds that all public institutions must be justified based on what is required for individuals to enjoy freedom. Ripstein also holds that "those who are in a rightful condition are entitled to remain in that rightful condition, which means that citizens, considered as a collective body, have a right that outsiders not subvert their legal order" (Ripstein 2021: chap. 2, section IV). So the Kantian argument for state independence might be construed as follows: (1) individuals have a right to external freedom; (2) this right demands interpretation, specification, and enforcement by a state; (3) therefore individuals have a derivative right that outsiders not subvert their state, since its legal order protects their external freedom.

But this Kantian reductivist strategy faces an important difficulty. While Kantian independence requires the existence of some state *in general*, it is not clear how the subordination or destruction of a *particular* state violates the independence of the individuals who make it up.[1] Of course, if the aggressor seeks to impose arbitrary, tyrannical, or exploitative institutions on the former state's territory, individual independence might well be threatened. But this may not always be the case.

Suppose that the United States were to annex Canada, and further suppose that this could be done without any loss of life (perhaps the United States threatens to use nuclear weapons and Canada simply capitulates). The United States then extends its

[1] For this particularity objection to Kantian views, see Simmons (1979, 2016).

body of law and judicial system to the territory of Canada, structuring these in accordance with a presumption of innocence for the accused, and recognizing the private rights previously guaranteed by the Canadian state. The United States further grants equal citizenship rights (including voting rights) to Canada's inhabitants: there are no legal distinctions between "the former Canadians" and other US citizens.

Would this "bloodless" annexation violate the right to independence of individual Canadians, when that right is construed along Ripstein's lines? It is not clear that it would. On Ripstein's view, securing individual independence requires a state that plays two roles. First, it must authoritatively resolve disputes in accordance with public procedures structured by the right to "be beyond reproach," a right that others "overcome the burden of establishing that you have done wrong" (Ripstein 2021: chap. 3, section IV). Second, the state must limit itself to the public purpose of providing a rightful condition to its members, refraining from coercing individuals to serve the private purposes of its officials or of other citizens. As Ripstein often puts it, the state must "rule on behalf" of inhabitants of its territory. (One might think that a state which "rules on behalf" of its inhabitants must, in addition, be based on the will of its people, and that it is just this condition that is violated in the example above. Yet Ripstein does not require that the state reflect the will of its people to be said to guarantee individual independence. I shall return to this point later.)

If these specific procedural guarantees are what protecting individual independence requires, then it is hard to see how any individual's innate right has been violated when the US state takes over the Canadian state in the manner described earlier. While the existence of *some* legal order meeting these procedural standards is an important precondition for securing individual independence, it is not clear why the *particular* Canadian legal order is necessary. Without saying more, then, about why individual independence depends upon rule by a *specific* state, the Kantian view seems unable to account for state independence along the lines sketched by the revisionist strategy.

Indeed, Ripstein makes several remarks that undercut the idea that securing individual independence could depend upon rule by a specific state. He suggests that the state's jurisdiction over the territory it currently controls must be presumed to be rightful, even if that territory has been acquired through past annexation or conquest. On Ripstein's view, each state's claim to rule "does not depend on it having come into being or acquired its territory in the right way" (Ripstein 2021: chap. 1, section III). Instead, he suggests that a state's legal order is "self-certifying" (Ripstein 2021: chap. 2, section V): simply by existing, a state has a claim against other states to noninterference. For Ripstein, then, if the United States actually *does* succeed in annexing Canada and consolidating its rule, it would have a right against other states to jurisdiction over the former Canadian territory and its people.

Here Ripstein is in agreement with international law, which has traditionally grounded states' territorial sovereignty in a principle of effective control. Since the nineteenth century, sovereignty over territory has most often been acquired simply by effectively administering it. Effective control is a relative concept: when two states press competing claims to sovereignty, it is necessary to decide which of the two claimants has done more to exercise sovereign governance over the territory. An organization gains jurisdictional rights when it has been the most effective at physically controlling the territory and exercising governmental functions within it.

Yet this effectiveness standard is morally unsatisfying, since it allows might to make right. Whichever power can maintain physical control for a sufficiently long period of time gains title, even if it conquered the area through bloody conquest or imperial subordination, and even if it maintains its control through continued repression. Such a standard permits territorial disputes to be settled by sheer force.

Ripstein (Ripstein 2021: chap. 2, section V) explicitly denies that a "bloodless annexation," along the lines of the US/Canada example sketched before, would be acceptable on his Kantian view. He holds that citizens whose state is destroyed by such an annexation would have an important moral complaint, since "invasion violates their right to be members of the legal order they are in" (ibid. 51). But what is the basis for the right to be a member of the legal order one is in? It is not—or at least, not obviously—the right to enjoy the generic procedural guarantees required to secure private independence. So what else might it be?

Despite initial appearances, it seems that Ripstein's Kantian view may be better understood as an instance of the *collectivist* strategy for theorizing state independence. As Ripstein notes, Kant draws an "analogy between private law relations and juridical relations among nations" (Ripstein 2021: chap. 2, section II). While this analogy need not rest on the premise that the state is *like* an individual, it does hold that it is equally wrong to subordinate for a nation to subordinate another nation as it is for an individual to subordinate another individual: "each legal order is entitled to be the legal order that it was, consistent with the equivalent entitlement of other legal orders" (ibid., section V).

The entitlement to independence is thus held by the state itself, as a corporate body. This entitlement does not depend on whether or not the subordination of the state would also lead to the subordination of the individuals who make it up. Nor is the state's claim to independence analyzable in terms of the prior rights or interests of individuals: "Each nation's status as a public legal order ... gives it a claim against other nations, over and above the claims its individual members have against outsiders" (ibid., section V). On Ripstein's Kantian view, then, it is simply a fundamental structuring assumption of international right that states are legal equals, with the right to "refuse the jurisdiction of any other legal order" (ibid.).

But this collectivist approach faces problems of its own: it has trouble explaining why groups have rights to be respected as autonomous sources of ends. Why is state independence so morally significant? While collective agency is a metaphysically coherent idea, the thought that states have an irreducible moral status is less attractive. On plausible accounts of value individualism, the importance of state independence should somehow derive from its contribution to individual independence, autonomy, or well-being. Yet the collectivist approach fails to identify any individual interest, right, or claim that is served by the independence of states. If no one's interests, rights, or claims would be set back, then why *shouldn't* we disrespect state independence? When the state as an artificial person is subordinated to another legal order, why should we care?

Further, since groups are themselves composed of individuals, the collectivist approach faces the problem of how to weigh group claims to independence against the claims of individuals, when the two conflict, as they might do in the case of extremely immoral, oppressive, or unjust communities. Ripstein's position is particularly troubling in this regard because it grants a claim to independence to nearly any state, almost no matter how it is internally configured. So long as a state does not commit slavery or genocide, thereby degenerating into "barbarism," it has a right against interference by other states, even if it commits other quite grievous wrongs—including torture; rape; famine; denial of rights to free speech, association, and political representation; or mass repression. Given that the state's independence might in this way override important rights of individuals, we need an extremely persuasive account of why state independence is so important. But Ripstein does little to develop such an account.

Finally, on Ripstein's view, it seems the state has a claim to independence whether or not it came into existence through some representative political process. Since Ripstein holds that a state can gain jurisdiction over a population and their territory simply by effectively governing them, on his view, Israel has a right to jurisdiction over East Jerusalem, Russia can claim jurisdiction over Crimea, and contemporary settler states have rightful jurisdiction over the indigenous peoples they have historically subordinated. But we might balk at the assertion that an organization can gain jurisdictional rights simply by *imposing* itself on a group of people, regardless of whether it reflects their own views or opinions about how they should be governed.

To sum up, then, while Ripstein suggests that there is a fundamental right to state independence, it is difficult to understand the moral basis for this right. While the Kantian reductivist strategy, as he develops it, can show that individual independence requires the general existence of states, it does nothing to explain why individuals have a special claim to be part of any *specific* state. And the Kantian collectivist strategy fails to offer any plausible explanation of why we ought to view state independence as of significant moral value.

3. A Fresh Start?

Can a Kantian view do better? In Chapter 8, on colonialism, Ripstein briefly discusses what strikes me as a more promising Kantian basis for understanding state independence. There, he mentions that the inhabitants of a territory have a right "to rule themselves through their institutions" (Ripstein 2021: chap. 8, section III). Colonialism, he suggests in passing, violates this right, by treating the colonized as merely "passive" in relation to their own governance. One is treated as passive, according to Ripstein, "if someone else determines where you stand in relation to others, if you depend not on your management of your own business but on arrangements made by another" (ibid. 159). Yet while he notes this claim not to be treated as "passive," Ripstein does not develop it into a broader justification for state independence.

That is unfortunate, because the right not to be treated as "passive" provides a promising alternative way for a Kantian to theorize state independence. Kant holds that to have the right to make and enforce law and policy for a population, a state must represent its subjects' *omnilateral will*: "Only the concurring and united will of all, insofar as each decides the same thing for all, and all for each, and so only the general united will of the people, can be legislative" (6:314). Kant also links the omnilateral will to the right not to be treated as passive: "The legislative authority can belong only to the united will of the people. For since all right is to proceed from it, it *cannot* do anyone wrong by its law. Now when someone makes arrangements about *another*, it is always possible for him to do the other wrong; but he can never do wrong in what he decides upon with regard to himself (for *volenti non fit inuria*) (6:314)." One way to understand a community's claim to respect for its omnilateral will, then, is as grounded in the claims of its individual members not to be treated as passive in relation to their own governance. They have a claim to respect for their shared judgments about how, and by whom, they should be governed.

The problem with the Kantian reductivist strategy as reconstructed earlier, then, is that it is too narrow: it focuses solely on the role the state plays in guaranteeing personal independence, by protecting private rights. As institutional "takers," individuals have interest in specification, adjudication, and enforcement of their private rights that only a state can provide. This "taker" dimension of evaluation is concerned with the objective qualities or features of state institutions, independently of anyone's relation to them. But the right not to be treated as "passive" suggests that to be fully independent, individuals must also *relate* to their state in a certain way: they should be authors of its political institutions. Individuals should be not only state "takers," but also state "makers." It may be as important that their institutions reflect their priorities and values (in their role as "makers") as that those institutions be structured in certain way (e.g., by guaranteeing their private rights as "takers").[2]

On my view, in order to have the right to make and enforce law and policy for a population, a government must not only meet certain procedural standards for protecting private rights, it must also reflect the *shared will* of the population it rules. Of course, much here turns on what counts as a shared will. In using the language of "shared will," I do not mean to imply that political groups necessarily share common commitments to shared substantive values. In a modern pluralist society, a shared will is more likely to take the form of a will to associate with one's fellow-citizens in one political community and to support certain institutions. We can interpret a shared will as a joint intention, held by a substantial majority of *cooperators* among the state's population, to support specific institutions and to accept their institutional procedures as a legitimate way of adjudicating and enforcing their private rights. *Cooperators*, as I use the term, are those individuals who are willing to recognize and respect others as bearers of an equal claim to private independence, and who acknowledge the political duties (to set up and defer to a state) that flow from this. On my Kantian reductivist view, then, independence has two dimensions: a legitimate state must not only "rule on behalf" of its inhabitants (by protecting their *private independence*), it must also reflect the shared will of most cooperators on its territory to establish justice through a particular institutional configuration (thus protecting their *political independence*).

Why is it important that the state reflect its citizens' shared will in addition to protecting private rights? The basic thought is that individual independence is jeopardized

[2] I develop this more complex Kantian reductivist account in my recent book *Territorial Sovereignty: A Philosophical Exploration* (Stilz, 2019).

whenever an individual is unilaterally subjected to *alien coercion*. On Kant's view, it is normally wrong to unilaterally use force or coercion to impose one's judgments on other people, including one's judgments about justice, even if one's views are correct. Instead, legitimate lawmaking and enforcement should proceed on the basis of a set of judgments that are *shared*.

To bring out Kant's position, imagine a private individual (say, your next-door neighbor) whose interpretation of justice is substantively correct—that is, she holds justified, true beliefs about everyone's private rights—and who is endowed with sufficient power to successfully bring other people to coordinate around her view (perhaps she has a secret weapon no one else possesses). Even though she could successfully impose a perfectly just scheme of rights by acting unilaterally, Kant claims that she has an important juridical duty not to do that. Her unilateral enforcement would fail to respect her fellows' independence, by treating them as merely "passive."

There are two problems with such unilateral uses of force:

(1) *Inequality*: Unilateral enforcement sets up a hierarchical relationship of subordination. There is the powerful enforcer, on the one hand, and then there are the others, who must obey the decisions that the powerful enforcer makes. These unequal power relations might undermine social equality between the ruler and ruled. If equal social relations are of intrinsic value, then individuals have some *pro tanto* reason to refrain from enforcing their judgments of justice unilaterally (cf. Kolodny, 2014; Viehoff, 2014).

(2) *Lack of Autonomy*: When one person uses coercion or force to impose her judgments of justice on others, she ignores and supplants everyone else's rational agency. An important part of treating others as independent persons involves respecting them as autonomous deliberators, who can reason for themselves how to act.[3] Yet a unilateral enforcer is insensitive to the need to engage her fellows in this way. She does not offer her subjects any reasons that might lead them to share her point of view about what justice requires, nor does she inquire into, or respond to, their reasons for not sharing it. Instead, she simply imposes her judgments, leaving others powerless to do anything but go along.

Treating people as "merely passive" in this way leaves them unable to act independently, on the basis of their own sense of justice, to establish and comply with a political order they can affirm. Instead, they see their activities coordinated by external threats they do not endorse (or perhaps even comprehend). Because of this, they are likely to find life under their powerful neighbor's rule significantly *alienating*. Subject to the neighbor, I imagine I would feel as though a hostile force exercised near-complete control over my life. Further, those treated as passive will not experience their political and social world as a cooperative enterprise, in which they have valuable relationships with other participants, or where they can feel at home.

[3] Famously, Kant's Formula of Humanity, which enjoins us to "so act that you use humanity, whether in your own person, or in the person of any other, always at the same time as an end, never merely as a means" (*G*, 4:429), constrains how we are to interact with others' rational nature, their capacity to set and pursue ends.

For these reasons, I believe treating others as passive objects of unilateral coercion is normally *pro tanto* wrong. There is an important class of exceptions: the requirement to refrain from unilateral coercion is restricted to *cooperators* who are willing to recognize and respect others as possessing an equal private right to independence, and to coordinate in establishing a public authority to specify and enforce this right. Some people—think of murderers or rapists—refuse to recognize others' private independence, even in a minimal way. Suppose I hold you back while you are trying to stab me—thus thwarting your unjust attempt to kill me. It is not reasonable for you to press as an objection to my coercion that it treats you as passive. By attempting to murder me, you have denied me any sphere of independence at all. Here, I agree with Ripstein's important contention that force can permissibly be used in defense of individual independence, even though unilateral uses of punitive, remedial, and administrative force are ruled out.

The essential thought here is that individuals' claim to political independence—their right not to be treated as "passive"—has moral weight only where it is consistent with basic elements of private independence. We are not required, out of respect for their judgments and values, to allow people to determine themselves in ways that clearly violate these moral requirements. However, so long as people *are* willing to recognize others' basic claim to private independence, I believe it is *pro tanto* wrong to coerce them unilaterally. This is so even when their interpretation of justice is mistaken. When other people attempt, in good faith, to respect the claims that, by their lights, I have—even when their interpretation of my claims is incorrect—I respond to them very differently than I do to people who fail to acknowledge that I have any claim to independence at all. Though they are mistaken, they do not treat me with contempt. For that reason, their capacity for making their own political judgments places demands on me, and I owe them some respect for that capacity, by not treating them as passive. I have *pro tanto* reason to interact with them using persuasive means, rather than coercively.

I should note that even in cases where unilateral coercion is wrong—that is, in the case of *cooperators* who respect minimal elements of private independence and are willing to act together to establish a common public authority—sometimes this wrong can be morally outweighed by other important social values. Some people may count as cooperators, in the sense that they are willing to acknowledge others' basic moral claims, and are willing to coordinate under *some* minimally just public authority, yet the particular government they would prefer is simply infeasible, given other people's preferences. If there is no other way to secure the essentials of justice, security, and public order, the wrong of unilaterally coercing these people—though important—may be outweighed. So the constraint on unilateral coercion can be lifted where its recognition would threaten the protection of private independence. While political independence is of very great weight, it is limited by the other requirements of Kantian right, so the imperative to respect it is not absolute.

In sum, on my view, the moral value of *state* independence derives from a distinct worry about the wrong that unilateral coercion inflicts on *individuals*. The principle of individual independence holds that, as rational agents, we ought to be able to direct our lives, to a significant degree, according to our own judgments. This idea grounds important private rights. But I believe that individual independence also has implications for the making and enforcement of political decisions. A scheme of rules about rights might be imposed in a manner that manifests proper respect for the independence of its subjects—treating them as the "active" authors of that scheme, rather than its "passive" objects—or it might not. If a set of rules is forced onto a population of cooperators over their explicit objections, and against their will, then the process of its imposition fails to adequately respect their independence. Surely it would be better to decide and carry out the rules governing our common life through the rational deliberative agency of those subject to them, rather than via a process that circumvents subjects' agential capacities, treating them as merely passive.

How might a state come to be in a position to omnilaterally enforce rules about rights? Individuals who are committed to respecting one another's independence—while at the same time differing about which specific rights and duties this entails—may come to a second-order agreement to associate together in institutions that they collectively accept as a legitimate way to specify and enforce their rights. A state will thus represent the "general united will of the people" insofar as it is authorized by a significant majority of *cooperators* on its territory, who form a joint intention to act together to establish justice through specific institutions. As I elaborate below, a shared will can exist even if not all citizens in a territory unanimously affirm their membership, since there are compelling Kantian reasons for discounting non-unanimity in certain cases.

To be sure, even where this agreement exists, no cooperator's personal judgments will be mirrored by each-and-every political decision. But this diversity of political opinions does not make a shared will among cooperators impossible. There is a second-order sense in which an individual's judgments *are* often reflected in her governing institutions: namely, when she shares a commitment to a political enterprise, and to certain values and procedures by which she believes that enterprise should be structured. Participants can freely accept their group's policies, and the outcomes that result from them, even when those outcomes diverge from their first-order judgments, so long as they share these second-order commitments. A commitment to participating in cooperative political action is therefore very important in enabling individuals' political independence.

Consider a partnership undertaking a joint venture, say, running a coffee shop together. Several philosophers have offered theories of the agency involved in such cases, arguing that it is undergirded by a structure of joint intentions. As they cooperate over time, this coffee shop group will likely develop some shared commitments about how their enterprise should run. This does not mean that the partners will converge in all their first-order judgments. More probably, they will divide on some issues. Yet even when they do not converge, partners are often able to generate second-order shared commitments—not reducible to their own first-order judgments—about how their joint venture should go. Such commitments can unify a group even when they diverge in their views on specific issues. Once such shared commitments have developed, members will generally act on them, and feel entitled to expect the others to act on them. Because a group can arrive at shared commitments, they can develop a group standpoint. This standpoint consists of a set of core values and priorities about how to organize more specific plans of action. The group standpoint will probably not fully correspond to any one person's first-order views. Still, each member can accept the group's standpoint and share in it, if she values the joint venture, and intends to participate in it.

I believe there is a perfectly straightforward sense in which we can speak of this small-scale partnership as sharing a will. Their shared will is to cooperate together in running their coffee shop, and to license the values and priorities that make up their group standpoint. A shared will, then, is nothing more than an interlocking structure of cooperative intentions on the part of each participant, amid common knowledge on the part of all that those intentions obtain. Moreover, the object of a group's shared commitments need not be first-order policy decisions: it can be more abstract ideals and decision procedures.

When each member of a political community participates in the group's shared will, and government imposes laws and policies on the basis of that shared will, then this use of political coercion will not be *alien* to the members. Though she does not personally endorse every political outcome, each member is governed by political institutions that she views as appropriate, since she endorses the higher-order values and procedures by which her joint political venture is conducted. *Omnilateral* coercion differs from *unilateral* coercion, then, because it is based on a joint intention—in which each cooperator shares—to act together to establish justice through specific institutions. Though political independence is ultimately an interest of individuals, it can be furthered by an individual's membership in a self-determining group, if she affirms her participation in the group and accepts the higher-order values and procedures that structure it. In this case, rule by the group's institutions will serve each individual's interest in not being treated as "passive," because the institutions by which she is governed will reflect her own judgments and priorities, as an equal coauthor of these institutions.

Further, unless we suppose that the Kantian state reflects the shared will of its people, I believe it is difficult to see how it could ever solve the problem of unilateral coercion characteristic of the state of nature. The problem of unilateral coercion is a problem of deciding for others what justice requires and compelling them to submit to one's judgments. This is supposed to be a problem even where the coercer's view of justice is *correct.* But if unilateralism is a problem for private individuals in a state of nature, then it is also a problem for the state. What makes the state's enforcement of a system of rights—even an objectively correct system of rights—any less unilateral than the powerful neighbor's enforcement of justice in the state of nature? Only, I submit, the fact that the cooperators among its population endorse its procedures and institutions, judging this an appropriate way to enforce justice among themselves. In that way, they grant their legitimate state the standing to decide on their behalf.

One might object here that it is unrealistic to think that all citizens in a territory will unanimously affirm their membership in the state. My response to this worry is that we should not aim at complete unanimity, since there are compelling Kantian reasons for discounting non-unanimity in cases where (1) dissenters are unwilling to cooperate in a legitimate state that can specify and enforce essential private rights; or (2) where too few dissenters share alternative commitments for a territorial unit reflecting their commitments to be institutionally feasible.

As for (1), the complaints of some alienated groups should be dismissed, on Kantian grounds, because they refuse to recognize the equal independence of others.

Groups that are alienated because they hold aims at odds with others' equal private independence—fascists, racists, theocrats, and so on—are not wronged when their right not to be treated as "passive" is denied. We should also dismiss the complaints of groups who are alienated simply because they refuse to acknowledge others' equivalent claims to political autonomy. Nostalgists for empire have no right to institutions that would uphold their imperial self-image. So the right not to be treated as "passive" is subject to important moral limits: in cases where dissenters' priorities are incompatible with others' independence, there is no moral loss in denying them self-determination.

As for (2), a feasible institutional response to the alienation of dispersed and disaffected minorities is sometimes lacking. While alien coercion is *pro tanto* wrong, this wrong can sometimes be outweighed by other social values, particularly by the need to achieve a stable and functioning legal system. Here I appeal to the importantly *territorial* nature of our Kantian duties: we cannot establish a unitary interpretation of property and contractual rights, enforce those rights, and punish violators, unless people who live in proximity and interact regularly are subject to the same institution. If each person signed up for the jurisdictional organization of his choice, or if all remained in an anarchic condition, then interactions between persons in the same contiguous space would continue to generate conflicts and disputes. So if disaffected individuals refuse participation in *any* feasible institutional configuration, then their dissent should be overridden. Since, on a Kantian view, the state plays an essential role in establishing and enforcing property and other rights, jurisdiction over most matters needs to be territorially defined if states are to carry out morally mandatory tasks. Groups claiming independence must therefore be capable of territorial organization in representative institutions. Since dispersed minorities are unable to fulfill this condition, they cannot claim territorial self-determination, since that would be inconsistent with upholding the private independence of others. Thus, in practice, a state should reflect the commitment of a *significant majority* of cooperators within its territory to associate together politically and to support specific institutions.

Finally, there are some scenarios where I believe we have a duty to allow for the self-determination of those whose shared will does not "fit" with their institutions. Here I have in mind groups who have shared political commitments that are (i) consistent with the provision of basic guarantees of private independence; (ii) who possess or can create a territorially organized structure of representation; and (iii) whose dissent can be feasibly addressed, at reasonable cost, by granting them separate political institutions. In cases of this kind, I believe we compromise the political independence of dissenters—their claim not to be treated as "passive" in relation to their own governance—by coercing them within alien institutions that they reject. It is the violation of this right to political independence that explains what is wrong with colonialism and involuntary annexation. Securing private independence and a functioning legal order for people in the metropole almost never requires coercively imposing political institutions on unwilling subject peoples.

To sum up, then, I believe that the best Kantian strategy for theorizing the moral basis for state independence is indeed a reductivist one. But, as it is usually developed, the reductivist approach is insufficiently complex. We instead need to recognize that the Kantian duty to respect individual independence has two dimensions: on the one hand, we are bound to respect others' *personal independence*, which requires us to cooperate together in a legitimate state that can enforce private rights; but on the other, we are also bound to respect their *political independence*—the claim, held by cooperators, not to be treated as "merely passive" in relation to their own governance. This, in turn, requires that they are ruled in a way that reflects their shared judgments about how, and by whom, they should be governed. It is the violation of individual political independence that in my view best explains what is wrong with colonialism, conquest, and annexation.

Were Ripstein to accept this "two-dimensional" reductivist strategy, he would need to make important changes in his position. First, he could no longer claim that the state's legal order is "self-certifying," and that simply by existing, states have a claim against other states to noninterference. On the view I have sketched, states only have a claim to noninterference where they represent the omnilateral will of their people, and some states (e.g., North Korea) cannot plausibly be said to satisfy this condition. Second, Ripstein would need to abandon the claim that a state can gain rightful jurisdiction over a population and their territory simply by effectively governing them, and thus that *status quo* borders ought not be revised. On the Kantian view I have sketched, the state may lack rightful jurisdiction over (parts of) its population, and there can be reason to revise *status quo* borders to recognize new claims to political independence. My Kantian account will therefore sometimes support claims on the part of minorities to internal autonomy, or even to secession. On my interpretation, then, the Kantian view does not support a statist political morality. Instead, it can justify the prominence that the principle of self-determination has assumed in the wake of decolonization, and may indeed support extending this principle even further.

References

Beitz, C. (1980). *Political Theory and International Relations*. Princeton: Princeton University Press.

Kolodny, K. (2014). "Rule over None II: Social Equality and the Justification of Democracy." *Philosophy and Public Affairs* Vol. 42(4): 287–336.

Rodin, D (2014). "The Myth of National Self-Defence," in Fabre, C., & Lazer, S. (eds.), *The Morality of Defensive War*. Oxford: Oxford University Press, pp. 69–89.

Shaw, M (2003). *International Law, Sixth Edition*. Cambridge: Cambridge University Press.

Simmons, J. (1979). *Moral Principles and Political Obligations*. Princeton: Princeton University Press.

Simmons, J. (2016). *Boundaries of Authority*. Oxford: Oxford University Press.

Shaw, M. (2008). *International Law*. Cambridge: Cambridge University Press.

Stilz, A. (2019). *Territorial Sovereignty*. Oxford: Oxford University Press.

Viehoff, D. (2014). "Democratic Equality and Political Authority." *Philosophy and Public Affairs* Vol. 42(4): 337–375.

Ripstein A (2021, forthcoming) Kant and the Law of war, Oxford University Press.

6

Vulnerability, Space, Communication: Three Conditions of Adequacy for Cosmopolitan Right

Comment on Arthur Ripstein, *Kant and the Law of War (KLW)*

Peter Niesen

In this chapter, I comment on Arthur Ripstein's account of the place of cosmopolitan right in a complete global system of law, that is, a legal system establishing public law on the three levels of state, international, and cosmopolitan order. In a complete global system of law, state law governs the relations between states and their citizens; international law regulates relations between states; and cosmopolitan law governs relations between states, on the one hand, and noncitizen individuals and non-state peoples on the other. Cosmopolitan right (*Weltbürgerrecht*), a body of natural law containing legal obligations as well as subjective entitlements, is a conceptual and substantive innovation introduced by Kant in the aftermath of the French Revolution.[1] It has fascinated interpreters since it first made an appearance in *Toward Perpetual Peace* (1795) and the *Doctrine of Right* (1797). Important debates have revolved round the question of whether cosmopolitan right should be given a narrow or broad reading, but often they have been based on interpreters going over the same few pages. In *Kant and the Law of War* (*KLW*), Arthur Ripstein gives an integrated reading of Kant's ideas on domestic and global law, embedding cosmopolitan law not only into state and international law, but into Kant's private law theory of property appropriation as well. This appears congenial to Kant's thought, since the foundations of cosmopolitan law are laid in the section on Private Right in the *Doctrine of Right*, underlining its intimate relationship with the original appropriation of property and setting out its anti-colonial meaning and purpose.

In the context of *KLW*, it is clear that the function of cosmopolitan law in governing war and peace is to be foregrounded. In the exposition of the work, this is reflected by the fact that Ripstein deals with the anti-colonial function of cosmopolitan law first (chap. 8) and with its place in a stable and lasting global legal order last, in the final chapter of the book (chap. 9). In this, he introduces cosmopolitan law, as formulated by Kant, mainly as an innovation in the law of war.[2] As Ripstein says in the Introduction to his book, "Global institutions and cosmopolitan right are not only the

[1] The best comprehensive guide is still Kleingeld (2012).

[2] Besides *KLW* and Arthur Ripstein's earlier articles leading up to it, this is also the approach pursued in Meckstroth (2018).

Peter Niesen, *Vulnerability, Space, Communication: Three Conditions of Adequacy for Cosmopolitan Right* In: *The Public Uses of Coercion and Force*. Edited by: Ester Herlin-Karnell and Enzo Rossi, Oxford University Press.
DOI: 10.1093/oso/9780197519103.003.0006

path to peace. In their ideal case, they are peace." (*KLW* vi) It is productive to be reminded that cosmopolitan law plays a crucial role in disarming individuals, peoples, and states in their respective encounters, a move on which Kant based his critique of colonial conquest. Ripstein explains how pre-Kantian authors introduced cosmopolitan rights-talk to justify European expansion, under the pretext of the "effective use" of all regions of the earth (*KLW*, chap. 1). Thus, in contrast to the far-reaching and sometimes overblown claims that commentators tried to extract from Kantian cosmopolitan right since the 1990s, it is sobering to be reminded of the negative aims Kant pursued in introducing it: Cosmopolitan right formulates a Janus-headed restriction, combining a ban on annexation with a ban on hostile reactions. Those landing on a foreign shore do not thereby acquire a right to wage war, claim property, settle down, or annex territory. And *vice versa*, innocent new arrivals should not be seen as perpetrating an act of war and therefore not treated with hostility. Cosmopolitan law ensures that "*commercium*—interaction, not just commerce—across borders is subject to law and so not a ground for war" (*KLW*, chap. 9, section IX). Ripstein succeeds in showing that only in Kant, cosmopolitan legal thought is finally completely separated from the doctrine of just war.[3] But at the same time, the relevance of a pure negative reading of cosmopolitan right seems limited, and, in his last chapter, Ripstein advances some more constructive claims. In what follows, I concentrate on the resources cosmopolitan right can draw on to yield anything beyond the absence of an *ius ad bellum* on either side. I argue that Ripstein's elegant and parsimonious reading of cosmopolitan law is not fully explanatorily adequate in terms of its scope and content.

Kant introduces cosmopolitan right as a single unified right to hospitality (8: 358), but from his texts, we can isolate three distinct positive elements. They may or may not flow from a single source, but can jointly be viewed as conditions of adequacy on any comprehensive interpretation of cosmopolitan right. The first is a humanitarian claim about human vulnerability, perhaps most difficult to square with Kant's independence-centered legal philosophy.[4] This feature is triggered when rejection would jeopardize the new arrival's livelihood. Upon arriving on the land of another, a person can only be turned away where "this can be done without destroying him" (8: 358, the German has *Untergang*, i.e., perishing). In a nautical metaphor, his need is to stay afloat and not to drown, in other words: to survive in destitute conditions. The difficulty lies in the fact that nowhere else does Kant seem to derive rights claims from the fact of manifest needfulness. As Ripstein puts it elsewhere: "I do not wrong you if I fail to assist you in your hour of need" (*KLW*, chap. 2, section III).

The second is a spatiotemporal claim, a right to remain where you happen to be on the surface of the earth, where "nature or chance have put" you (6: 262), or a "right to be somewhere" (Huber 2017). This right reflects both our innate right to external freedom and the circumstances of our extended nature, given facts about our natural habitat, planet earth. The combination of innate right, our extended nature, and the

[3] One feature of how current just war theory is "revisionist" is its effort to rejoin the elements of cosmopolitan right and the *ius ad bellum* and *in bello*. See Fabre (2012).

[4] The leading independence-centered approach is Ripstein (2009). But rivaling democratic or obligation-based accounts have an equally hard time in explaining how humanitarian concerns can generate rights claims. See Maus (2015).

spherical geometry of the planet provides us with an equal entitlement to territorial existence, since "originally no one had more right than another to be on a place on the earth" (8: 358). The third claim is encapsulated in the cosmopolitan claimant's "right to visit," as distinguished from any "right to be a guest" (8: 358). Note that this claim is unrestricted by the circumstances of nature or chance. It covers moving about the surface of the earth, offering to engage in commerce (6: 352), and "present[ing] oneself for society" everywhere (8: 358), thereby establishing a communicative right (Niesen 2007). Unfortunately, Kant does not spend much time in justifying hospitality, or its several elements. Hospitality is said to derive from the fact that all human beings share in a "right of possession in common of the earth's surface" (8: 358), but little is said about how original common possession yields either of the three elements. In what follows, I examine how Ripstein's interpretation does with respect to the three elements. My claim is that he does an excellent job with regard to the spatiotemporal element of cosmopolitan right, but that his approach runs into trouble in accounting for the other two, that is, its communicative and humanitarian features. I start with the spatiotemporal claim in section I, move on to the communicative aspect of cosmopolitan right as a right to visit in section II, and conclude with the humanitarian element in section III.

Before I start in earnest, I must point to a fundamental disagreement between Ripstein's account of Kant's cosmopolitan right and mine, which will become relevant only later (in sections II and III), lest it seem ad hoc then. I am aware that the issue cannot be spelled out fully in the course of this chapter, but since it determines the set of options for all attempts to satisfy the three conditions of explanatory adequacy for cosmopolitan right, I need to at least sketchily introduce it here. Kant famously insists, both in his accounts of Private Law and of International Law, on the *provisional* character of under-institutionalized legal orders. All law in a state of nature, in the interpersonal as well as in the international case, is provisionally valid only because in a state of nature, all legal opinion must be unilateral (6: 312, 350; cf. 8: 349). Interpreters agree, therefore, that Kant's natural law theory demands institutional closure in order to make law permanent or "peremptory." In the interpersonal state of nature, states need to be formed and eventually transformed into republics. In the international state of nature, a league of states needs to be founded, all in order to provide the conditions for an impartial resolution of legal conflicts. In the case of the cosmopolitan state of nature between states and foreign individuals, the original predicament is a similar one: where views about entitlement collide between states and foreigners, there exists no impartial authority to resolve them. Kant does not envisage overarching institutions that could decide on cosmopolitan entitlements, in the way that a league of states can resolve international conflicts. But how to resolve the problem of unilateralism, if states continue to coercively reject cosmopolitan claimants? Of course, foreigners will have access to domestic courts to contest any apparent violation and have state action reviewed (Eberl & Niesen 2011). But the domestic jurisdiction, however well-disposed and law-abiding it may be, will still judge the trans-border conflict along the lines of *what seems right and good to it.*[5] It therefore appears likely that, since Kant fails

[5] Cf. the parallelism with the domestic law case, § 44 *Doctrine of Right*, 6: 312.

to discuss a mechanism for adjudicating conflicts under cosmopolitan right, what he is outlining in both *Toward Perpetual Peace* and the *Doctrine of Right* must be *provisional cosmopolitan law*, liable to institutional consolidation and transformation. If the distinction between initial, natural, provisional law on the one hand; and transformed, positive, peremptory law on the other is pertinent for cosmopolitan right, too, then the scope, content and institutional form of *peremptory cosmopolitan law* cannot be reduced to what Kant himself has set out under this heading. I conclude that in Kant's texts, we find an outline only of its "conditions of possibility," so to speak (8: 358).

Arthur Ripstein's account does not commit itself to applying the provisional/peremptory distinction to cosmopolitan right, which makes his account conclusive, and parsimonious. His interpretation is based on the assumption that Kant, in the third Definitive Article of *Toward Perpetual Peace* and in § 62, *Doctrine of Right*, already fixes the scope and content of any future cosmopolitan law, within a permanent architectonic that contains no impartial institutions capable of deciding conflicts between states and foreigners, and in the absence of any positive supranational code of law. My view is that under this completeness assumption, not only do we face a major unresolved systematic issue—all exercise of cosmopolitan right takes place in a state of nature between states and foreigners: a highly structured state of nature, to be sure, but a state of nature nonetheless—but we have too few options for fulfilling the conditions of explanatory adequacy outlined earlier. In *KLW*, no space is reserved for leaving normative issues undecided, to be resolved only through politics and through additional, positive global institutions. This is why I plead for a fuller, less parsimonious account where Kant's provisional cosmopolitan right is to be completed via a process of institutionalization and lawmaking. Of course, cosmopolitan lawmaking would have to chime with Kant's overall strategy of contractual, non-coercive international lawmaking, and cosmopolitan institutions would have to reflect international law's binding, yet nonviolent character.

1. Cosmopolitan Right as the Right to Be Somewhere

Before presenting his own account of cosmopolitan right, Ripstein makes two factual assumptions.

(a) Some people have appropriated some things as private property (*KLW*, chap. 9, section IX)
(b) There are some states (*KLW*, ibid.)

True, it would be interesting to see whether cosmopolitan right claims hold in a possible world in which nobody has ever successfully claimed exclusive ownership over objects, or exclusive jurisdiction over persons, but this would be very far removed from our own world. So why should these relatively uncontroversial empirical assumptions be worth stating? The importance of Ripstein's assumptions can be brought out by comparing his approach with a recent article by Jakob Huber (Huber 2017). Huber starts out without either of the assumptions, and derives cosmopolitan right without reference to property acquisition and the existence of state borders.

I will discuss the divergence between his position and Ripstein's as to statement (a) first.

Ad (a) Huber and Ripstein share the view that Kantian cosmopolitan right has nothing to do with what human beings need. Both reject the traditional understanding of the figure of "original possession in common," according to which any division of the earth ought to be sensitive to distributional constraints, for example, reflecting that people need resources to consume or have materially egalitarian claims to territory. Ripstein paraphrases Kant as saying "I do not wrong you if I fail to assist you in your hour of need" (*KLW* 48), while Huber agrees that "our empirical existence as vulnerable beings with bodily needs cannot have direct rights-grounding justificatory force" (Huber, 2017: 17). Still, their accounts markedly differ. In reconstructing cosmopolitan legal relations as relations among "earth dwellers"[6] in a context of "possible physical interaction" (6: 352), Huber understands them as relations of external interaction between extended beings, and divorces them from their connection to the acquisition or holding of property. For Huber, relations under cosmopolitan law are " 'external' (as located in time and space), but not property-mediated." (Huber, 2017:20) A right to be somewhere relates individuals directly, as holders of innate, not acquired right.[7] It is independent of circumstances in which states can or do meddle, and can be specified in the absence of information about whether states, or more than one state, exist. For Ripstein, in contrast, cosmopolitan right "is not unmediated: it must flow through the idea of a system of public orders." (*KLW*, chap. 9, section IX) Such a system must incorporate private property regimes, territorial states, and rights-bearing individuals. It deals with the facts of past acquisition, with regard to individuals both as *beati possidentes* and as those whose use of their external freedom is limited by others' appropriations of territory.

Is the community of earth dwellers a community of coexistence of extended bodies, as Huber suggests, or is it necessarily a community of exclusive claims over external objects, especially of claims to the possession of territory? In a limited sense, according to Huber, a "right to be somewhere" follows from the fact that people are extended bodies which always already take up some space, and sometimes cannot help being where they are at. They thus need to be immune against charges to vacate any spatiotemporal coordinates they are currently based at. This indeed seems to follow from a right to external freedom given our spatial extendedness, that is, the fact that two people cannot occupy the same spot at the same time. The problem with this reading is that spatiotemporal compatibility is independent of whether or not we assume the earth's territory is finite. Bumping into each other in corridors may violate our right to be somewhere even in non-spherical space. It cannot be the problem Kant addresses in deriving cosmopolitan right, otherwise it would make no difference that we are operating on a limited, spherical surface. But cosmopolitan right is owed to all humans "by virtue of the right of possession in common of the earth's surface on which, as a sphere, they cannot disperse infinitely" (8: 358). The "spherical shape of the place they live in" entails that any ownership of territory "can be thought only as a possession

[6] Huber introduces the notion of earth dwellers for humans living under the predicament of limited territorial space, picking up a formulation by Byrd (2009).

[7] For this distinction, see Byrd & Hruschka (2010).

of a part of a determinate whole, and so as possession of that to which each of them originally has a right" (6: 352). Ripstein, in contrast, explains the relevance of territorial extension by analyzing original possession in common as a "disjunctive" concept, drawing attention to the analogy with the category of "disjunctive judgement" introduced in the *Critique of Pure Reason* (*KLW*, chap. 9, section VII): "Possession in common" covers mutually exclusive options, which are jointly exhaustive of logical space. In other words, cosmopolitan right reacts to a potential scarcity problem—not scarcity of goods or frustration of human needs, as Huber and Ripstein both rightly point out. Kant addresses a scarcity of rightful options, resulting from the historical process of property annexation and the setting of state borders. The actual division of territorial space could, in the absence of a cosmopolitan right to be somewhere, have led to situations in which people are involuntarily and blamelessly, yet illegally present on a patch of land—in the limiting case in which all territory is carved up and divided up, they may have no place to legally stand on at all.[8] The justification of the spatial aspect of cosmopolitan right through the idea of original common ownership is thus intimately connected with the privatization of territory. The initial problem to which cosmopolitan right responds does not even get off the ground in the absence of private processes of appropriation of limited space. In order for the history of unilateral acquisition to unfold, and to enable people to legitimately make exclusive use of parts of the earth, Kant believes we need a normative orientation point, and finds it in the idea of common ownership.[9] Ripstein's property-based account is preferable to Huber's external freedom-based account since the spatial aspect of cosmopolitan right cannot be adequately specified in the absence of all reference to claims to territory on a finite globe. The spatial implications of the idea of an original possession in common of a limited earth flow from acquired, not innate right.

The property-mediated nature of cosmopolitan right is of special relevance in our context, since it can serve as a first step to giving an explanation of its humanitarian, vulnerability-oriented aspect. It is only a first step, since, lacking a right to be somewhere, a newly arrived applicant could be said to be trespassing. However, given that a right to be somewhere is not a right to stay, it is not yet clear why he can only be rejected as long as he need not fear to "perish" (*seinen Untergang besorgen*, 8: 358). It seems easier to see where this latter obligation would flow from in a property-mediated account of cosmopolitan right than in an unmediated account. I will take up this question in section III of this chapter.

Ad (b) The next thing to note is that cosmopolitan right, according to Ripstein, is not only mediated through past and future property acquisition, but also through a system of global public law containing the spheres of state law and a sphere of international law (*KLW*, chap. 9, section VI). There are some states. This is where the motivation for cosmopolitan right comes from, where the problems it reacts to originate, and

[8] The disjunctive property order does not entail that everything needs to be privately owned by someone, but that the amount of ownables is fixed by the claim that I own a piece. What I own is determined by the finite specification of what I don't own.

[9] As Huber (2016:231–249, 238) asserts in a different context: "Original common possession is not an argumentative starting point, but the conclusion of the argument: something like a normative implication of the fact of individuals' acquisition of land under circumstances of spatial constraint constituted by the earth's spherical surface."

also whence the remedy to legal uncertainties and conflicts it proposes derives from. In contrast to interpreters who read cosmopolitan right as referring to the more or less freestanding entitlements of individuals vis-à-vis a global order, abstracting away from their state membership (Habermas, 1998: 165–202, 181),[10] Ripstein insists on its functions in a state-based order: "[C]osmopolitan right coordinates with the right of the state and the right of nations by protecting the independence of each nation's legal order." (*KLW*, chap. 9, section VI) This yields the implication that visitors must defer to the law of the land, underlining Ripstein's reading of the anti-colonial purpose of Kantian cosmopolitan law. However, as I will go on to show, this creates a systematic problem when the law of the land bans interaction with foreigners.[11] In the remainder of this section, I will address a question this view entails about the scope of cosmopolitan right.

To illustrate the compatibility of cosmopolitan law with state law, Ripstein formulates four conditions that can be briefly summarized. *First*, individuals are not to be stripped of their private rights once they cross a state border. *Second*, they have no claim to be treated according to the private law of their home state when abroad. *Third*, they wrong nobody by "interacting with the inhabitants of other legal orders," for example, in engaging in commerce such as buying or selling property or entering into marriage contracts. *Fourth*, other legal orders do not need to permit private transactions between [foreign visitors] and their inhabitants" (all at *KLW*, ibid.). This list of conditions cannot be taken to be exhaustive, since it focuses so clearly on matters of private law only, which have often been taken *pars pro toto* in discussions of cosmopolitan law. But even disregarding public law relations within cosmopolitan law,[12] Ripstein's set of conditions seems inconsistent. On the face of it, conditions three and four appear contradictory, since according to condition four, legal orders could refuse to permit any private transactions between foreign applicants and their inhabitants. In such cases, under condition three, foreigners' attempts to engage in transactions would not be wrongful, but already the attempt to perform a transaction would be deemed illegal. In other words, condition four allows banning non-wrongful exercises of freedom as specified under condition three. While the apparent contradiction can be resolved by distinguishing what is non-wrongful under natural law from what is not illegal under state law, such a resolution could be too quick since the scope of cosmopolitan right to perform private transactions under condition three could then conceivably turn out to be *zero, everywhere*, following the exercise of the prerogatives under condition four.

[10] It is worth noting that in introducing the notion of a *pouvoir constituant mixte*, Habermas changed his position from a monist to a dualist (global *and* state-based) understanding of cosmopolitan law; see Habermas (2012).

[11] In the literature, this is sometimes referred to as the "China-Japan" problem; see Niesen (2007:98–100) and Reinhardt (2019:120–124).

[12] Not all border-crossing exercise of rights, according to Ripstein, is open to such restrictions since not all rights are private rights. Ripstein's view is that although rights to private transactions may be heavily curtailed, rights we enjoy as "members of the public" cannot. These include the "right to stay in a public inn, the right to be accommodated by a ferry across a river with no bridge, and, most fundamentally, the right to use public roads and buy and sell in a public market." (*KLW*, chap. 9) Note, however, that with the reference to the right to buy and sell in a public market, Ripstein's view can no longer account for the China-Japan problem.

This apparent contradiction is no coincidence, nor does it arise with Ripstein's account exclusively. On the contrary, it seems to reflect Kant's own doubts about the universality of private law claims and their capacity to travel. When Kant discusses the cases of China and Japan, he says they have "wisely" placed restrictions on commerce, "the former allowing them access but not entry, the latter even allowing access to only a single European people, the Dutch, but excluding them, like prisoners, from community with the natives" (8: 359). He may have had something in mind like legitimate unilateral restrictions on cosmopolitan interaction in commerce, in the narrow sense of economic blockades.[13] Or he may have had a pragmatic historical justification in mind for comprehensive closures, allowing China and Japan to learn from their grievous experiences with European nations, as argues Williams (2007).[14] But take Ripstein's example, marriage across borders. Engaging in marriage, according to Kant, is one form of private transaction (covered in the private law section of the *Doctrine of Right*, 6: 277ff.) which could, according to Ripstein, be both non-wrongful and still be open to being declared illegal by unilateral state action. This would make a situation perfectly consistent where everybody had the right to propose to anybody else abroad, but states would have an absolute prerogative to frustrate all those proposals without thereby violating foreigners' cosmopolitan rights (cf. Kleingeld, 2012: 81). While it is tempting to say that the non-wrongfulness of interaction is one that persists in natural law, whereas its contingent illegality only enters via restrictions in positive law, the question is whether this can be the last word on global rightful intercourse. Can positive law place restrictions on natural rights to engage in intercourse across borders? This leads to a general question about the transformative potential of positive cosmopolitan law that I will take up in closing.

2. A Right to Travel or a Right to Visit?

I now turn to the topic of voluntary movement under cosmopolitan right and leave behind the arguments from nature or chance, on which the spatiotemporal claims of cosmopolitan right were founded. The third element that needs to be explained as a condition of adequacy on all interpretations concerns cosmopolitan right as a right to visit. A visit involves an intentional movement in space, but is not identical with that. In this section, I want to suggest that Ripstein treats cosmopolitan right as an innate mobility right, paying too little attention to its raison d'être, communication with all others, that is definitive for the right to visit. To bring out Ripstein's position, consider first the difference between a right to travel and right to visit. "Travel" means getting from A to B, while "visit" means getting from A to B to contact C, where C is an individual or group sharing a symmetrical entitlement. Intuitively, the difference can be

[13] A narrowly economic reading of the China-Japan problem, as proposed in Niesen (2007), will not work either since China and Japan simultaneously closed their borders to cultural and religious exchange. See also Kleingeld (2012) and Reinhardt (2019). For a nuanced broad reading of commerce, see Dilek Huseyinzadegan (2019).

[14] See also Karoline Reinhardt (2019:123) and Arthur Ripstein, "Bringing Rights and Citizenship under Law on a Globus Terraqueus" (unpublished).

brought out with reference to recent anti-tourism protests in Barcelona, Venice, by ecological NGOs in Tibet or elsewhere. While travelers claim a right to enter a country to look at natural or cultural monuments, not a right to visit, says Hughes (2018), the local populations are contemplating a *Trexit*, that is, an exit from tourism.

Ripstein treats both claims to travel and claims to visit under cosmopolitan law, but the dominant perspective is that a right to travel has universal scope, can be justified independently of whether it serves as a means to visit, and at the same time yields an entitlement to visit in all cases. It is questionable whether this entailment does in fact hold, given Ripstein's austere premises. In introducing the provision of public goods as a condition of legitimacy of the global legal order, he uses the example of public roads as freestanding requirements, independently of the uses they serve. Public roads, "understood as public rights-of-way on which all members of the public are free to pass" are "to accomodate all travelers": "[T]he possibility of getting from one location to another within a rightful condition without requiring the permission of any other private person is itself a formal requirement of right. There must be a public right-of-way between any two locations even if many people had no real interest in going beyond their immediate neighbours" (all *KLW*, chap. 9, section IX). Free passage covers a universal right to get from A to B, but it leaves C out of the picture. Ripstein argues that the right to visit is a corollary of the more fundamental right to travel: "Disjunctive possession generates a right to travel and with it a right to visit, but nothing more." (*KLW*, chap. 9, section IX). The focus on traveling in contrast to visiting is further strengthened through his reliance on a recent paper by Aravind Ganesh. Ganesh (2018) suggests to explain cosmopolitan right through the interaction between innkeeper and traveler, treating the innkeeper as a public functionary who has no right to reject (short-term) patrons. In this, "[t]he right to use public roads" is again said to be "paradigmatic" (*KLW*, chap. 9, section IX). In other passages, Ripstein seems to say that mobility rights are not freestanding, but teleologically oriented to visits and communication: "[I]f all land is privately owned, then in order for the resident of one plot of land to get to any non-adjacent plot of land *to interact with its residents*, he or she—the 'traveler'—would need the permission of everyone owning land in between those two plots in order to travel." (*KLW*, ibid., my emphasis). But finally, according to Ripstein, "the specific purposes of the traveler" are irrelevant, and they may or may not contain any contact with the resident population (*KLW*, ibid.).

Cosmopolitan right, understood as a right to travel, yields less than what is needed for purposes of explanatory adequacy. Remember that for Kant a right to visit is a right to access (though not a right to enter). A universal mobility right, even involving the crossing of borders and the use of public roads everywhere, for example as a right to safe passage,[15] would not generate such access since it would not ensure opportunities for world citizens offering to engage in commerce, in the broad sense, or "present[ing] themselves for society" (8: 358). Remember the case of the anti-tourism protests. What they contest is a universal spectators' right, in the absence of any interest the spectators may have in contacting the "old inhabitants." A right to visit is the exact opposite, a claim to enter into a communicative situation in which speech

[15] For a discussion, see Huber (2020).

act offers can be made and accepted or rejected. A visit is a constellation in which the hearer can be said to owe the speaker at least a rejection.[16] This is not automatically the case for travelers, Kleingeld would claim, even if they come equipped with a human right to free speech on top of their universal mobility right (Kleingeld 2012).

My hunch is that its explanatory weakness regarding the communicative aspect of cosmopolitan right is grounded in a one-sidedness in Ripstein's account of the idea of possession in common of the surface of the earth. In deriving the scope of cosmopolitan right from the conditions of global mobility, Ripstein draws on Kant's use of the idea of a permanent possession in common only in the disjunctive sense mentioned before: Given that people have divided up the globe into states and private territory, they have to provide for each others' border-crossing use of roads, etc., in a *remedial* fashion. Otherwise their patchwork territories would effectively immobilize each other. But there is another sense of original possession in common in Kant that plays no role in Ripstein's argument, yet may be crucial for deriving the right to visit. This is the non-disjunctive sense of the idea of original possession in common. In Kant's view, not the entire spherical surface of the earth is up for grabs for privatization. Ships and camels can transport people toward each other across uninhabitable surfaces such as oceans and deserts. They "make it possible to approach one another over these regions belonging to no one and to make use of the right to the earth's surface, which belongs to the human race in common, for possible commerce" (8: 358). No matter whether we explicate such non-disjunctive common possession in the language of no-ownership, of persisting common ownership or of communal property,[17] it seems clear that Kant plans to reserve some natural global infrastructure for approaching one another by *protecting* it from unilateral acquisition, not by compensating for its personal or official annexation. In this he mirrors an argument in Grotius's *Mare Liberum*, which says that the necessary conditions of mobility must not be privatized:[18]

> [A]ll those things which are so ordained by nature that anyone using them they may nevertheless suffice others whomsoever for the common use are at this day (and perpetually ought to be) of the same condition whereof they were when nature first discovered them. Grotius (2004:24)

Kant does not use the *Mare Liberum* reference in his writings on cosmopolitan right, although he substantively agrees with it and rejects the *mare clausum* alternative for the open seas in the Private Right section of his *Doctrine of Right* (6: 269). He goes beyond Grotius, who argues that common things must remain in use for common

[16] Although not a justification. Here, Kantian communicative entitlement and contemporary intersubjectivist conceptions of communicative rights and obligations, such as Jürgen Habermas's account of 'communicative freedom' and Rainer Forst's 'right to justification', part way.

[17] See Stilz (2014).

[18] In Ripstein, ownership of roads, whether public or private, makes no difference for the right to travel. He rejects Grotius's idea of harmless use, referring to the legitimate interference with private property rights, not the perpetual usufruct of what was originally in common and has never been privatized. On the one hand, this is a stronger account of the right to travel than the one hinted at here, since it covers artificial, privately owned infrastructure. On the other hand, as pointed out earlier, mobility is not the same as the opportunity to communicate.

purposes, by specifying that these purposes are genuinely joint purposes. The seas belong to humankind not in order to harvest fish and then privatize one's catch, but "to approach one another ... for possible commerce" (8: 358). Grotius's scholastic label for this distinction is that some things are common "affirmatively," and some "privatively" (ibid. 26). For Kant, engaging with one another via ships and camels across the seas or across deserts uses an affirmatively common resource. It does not just take place under *the idea* of original possession in common, but makes actual use of a joint resource that mankind share. The fact that we share a common infrastructure to approach, visit, and engage with each other provides for a more comprehensive source for global communicative rights than the idea of compensation for disjunctive ownership.

3. Whence the Humanitarian Element in Cosmopolitan Right?

Perhaps all right-based conceptions of global order find it difficult to recognize humanitarian claims, if by humanitarian concerns we mean those reacting to human vulnerability to various kinds of dangers to life and limb.[19] Yet Kant is adamant that the humanitarian component in cosmopolitan right is not "a philanthropic principle, but a principle having to do with rights" (6: 352), and that the responsibility for not letting a visitor perish is shifted to the community she is visiting (8: 358). Ripstein refers, in justification, to what he calls a human being's "right of humanity in his or her own person" (*KLW*, chap. 9, section VII), a notion that plays a founding role already in *Force and Freedom* (30–56). It is unclear, however, how the derivation works exactly, especially if it is to bypass need. Sending back a refugee to a situation of mortal danger would obviously amount to a violation of cosmopolitan law, since it would lead to letting a visitor "perish juridically" (*KLW*, chap. 9, section X). But new arrivals will require more than just non-refoulement in order not to perish juridically. Among other things, those obligations seem to demand, contra Ripstein, that I should not "fail to assist you in your hour of need" (*KLW*, chap. 2), although need can, as we agreed earlier, only play an indicative, not a foundational role. Kant's conception of social welfare is of no help here, since he believes that welfare obligations not to let others perish are generated by the civic bond to our fellow citizens (6: 326). This means that we are still in need of a justification for the *domestic* obligation not to let *non-members* perish.[20]

If we cannot account for humanitarian claims under cosmopolitan law by referring to need itself, nor to innate or acquired entitlement, what can those obligations result

[19] In what follows, I work with an intuitive, undefined understanding of what 'humanitarian' or 'vulnerability' mean in this context. Often, 'humanitarian law' is understood as international law governing warfare. I do not use the term in this way. Sometimes, 'humanitarian' justifications are contrasted to rights-based justifications, as in Forst (2012). I do not rely on this definition either.

[20] It is a red herring to distinguish, as has become customary, between the duties toward voluntary and involuntary arrivals since the crucial distinction is that between their voluntary or involuntary, and blameless, sojourn. Kant does not help in this case by toying, in the *Vorarbeiten*, with the idea that visitors in a desert sandstorm may stay if their visit (not their failing to leave) came about involuntarily (23: 173). Reliance on this over-used passage should therefore be taken with a grain of salt.

from? As with the two other elements already discussed, the spatiotemporal and communicative claims under cosmopolitan right, Kant invites us to look for a justification in the idea of original common ownership. In the case of humanitarian claims, a look at the *provisional character* of dominion over property in land and territorial borders seems promising. This reading is not necessarily incompatible with Ripstein's interpretation, since the argument derives from the disjunctive nature of original possession in common, and therefore from the right to be somewhere, not from the non-disjunctive nature of continuous possession in common of the globe's communicative infrastructure, as under the right to visit. The normative force of humanitarian concerns triggered by non-voluntary sojourn can perhaps best be explained by reference to the provisionality of the global property distribution and location of state borders, an issue that does not feature prominently in *KLW*, and is not applied there to the cosmopolitan predicament (but see *KLW*, chap. 2, section I for the domestic case and chap. 9, section VII for the international case). Necessarily, Kant argues, the claiming of property rights came about under the law of the strongest, and it was followed by the unilateral, belligerent manipulations of state borders (6: 312, 350). It appears therefore not implausible to suggest that whatever a peremptory division of property and territorial jurisdiction will look like, it should compensate for the vulnerabilities of those who are stuck somewhere, and who may have lost out in the global race for appropriation.

4. Conclusion

Explaining cosmopolitan right in the sense of giving an account of its spatiotemporal, communicative, and humanitarian elements must perhaps draw on features that cannot be extracted from provisional cosmopolitan right, and only come into being in the efforts of mankind in realizing peremptory cosmopolitan right. I cannot go into the extended debate here on how we should think provisional relates to peremptory law and dominion in Kant.[21] Ripstein understandably steers a skeptical course regarding new institutions and new entitlements that could compromise non-domination among states. He reads Kant's assertion that public law does not contain anything that could not have been thought of in private law (*KLW*, chap. 2, section I), and that therefore peremptory law must build on the legal relations already in place in provisional law, not only in a prescriptive but also in a limiting sense. In parallel to the case of international law, Ripstein insists that no form of world government would be compatible with Kant's architecture of public law. He is therefore committed to cosmopolitan law (i) being more or less completely captured in a description from the theorist's point of view and (ii) being more or less perfectly realized institutionally already, in the absence of an international regime that could decide cosmopolitan right claims, if necessary, against the will of nation-states.

[21] Rafeeq Hasan (2018:850–876) does not address cosmopolitan provisionality in his useful taxonomy of different types of provisionality, but it seems to involve both what he calls 'strong' (i.e., rectificatory) and what he calls 'anticipatory' (i.e., institutionally minded) provisionality. For a discussion of the senses of provisionality required for cosmopolitan law, see Niesen (2017).

This is not the place to resolve the issue of *what* cosmopolitan claimants should receive under the humanitarian, or under the other two aspects of cosmopolitan law, once a complete transformation from provisional to peremptory cosmopolitan right has been achieved. But both of Ripstein's presumptions (i) and (ii) seem to me not to hold. As to (i), the question of how to heal unjustified unilateral historical annexations of property and territory may be something that those advantaged and those disadvantaged by them should jointly resolve. As to (ii), cosmopolitan right, although public law in name (6: 311), in *Kant and the Law of War* still pits the parties concerned in potential conflicts, that is, states and non-members, horizontally against each other. This cannot be the last word for cosmopolitan right. We are familiar with such horizontal constellations from the sections on private right and on pre-league of nations *ius ad bellum* in the *Doctrine of Right* (§ 55–59, 6: 344–349), where they constitute the problem, not the solution. Kant's demand is to transcend them, in order for them to be regulated in a peremptory public legal order. So peremptory, institutional cosmopolitan law is necessary from a normative point of view.

But is it also permissible? In commenting on the standing of nations within an overarching system of global public law, Arthur Ripstein argues that their independence is not to be compromised and they are not to be "subject to compulsion from without" (*KLW*, chap. 9, section IV). In this, he insists on Kantian "sovereignty" against the authority of omnilateral institutions, especially for the case of war. I do not think nation-states' independence will be violated in the case of the formation of cosmopolitan law. In my reading, states and their peoples are to have a say, and not be subject to compulsion, in the creation of conflict-resolving cosmopolitan authorities, the formation of a cosmopolitan code, and in the administration of positive cosmopolitan law. What states do need to give up in a complete global system of public law is their unilateral normative authority in deciding which cosmopolitan claims hold against them, even if no coercive mechanisms are being provided to enforce those.

References

Byrd, S. (2009). "Intelligible Possession of Objects of Choice." In Denis, L. (ed.), *Kant's Metaphysics of Morals: A Critical* Guide. Cambridge: Cambridge University Press, pp. 93–111.

Byrd, S., & Hruschka, J. (2010). *Kant's Doctrine of Right. A Commentary.* Cambridge: Cambridge University Press.

Eberl, O., & Niesen, P. (2011). Kommentar. In *Immanuel Kant. Zum ewigen Frieden und Auszüge aus der Rechtslehre.* Berlin: Suhrkamp, pp. 265–266.

Fabre, C. (2012). *Cosmopolitan War.* Oxford: Oxford University Press.

Forst, R. (2012). "Justice, Morality, and Power in the Global Context." In Forst, R. (ed.), *The Right to Justification.* New York: Columbia University Press, pp. 241–250.

Flikschuh, K. (2000). *Kant and Modern Political Philosophy.* Cambridge: Cambridge University Press.

Ganesh, A. (2018). "Wirtbarkeit. Cosmopolitan Right and Innkeeping." *Legal Theory* Vol. 24(3): 159–190.

Grotius, H. (2004). *The Free Sea* (Armitage, D., ed.). Indianapolis: Liberty Fund.

Habermas, J. (1998). Kant's Idea of Perpetual Peace with 200 Years' Hindsight. In Habermas (ed.), *The Inclusion of the Other*. Cambridge, MA: MIT Press, pp. 165–202, 181.
Habermas, J. (2012). *The Crisis of the European Union*. Cambridge: Polity, pp. 53–70.
Hasan, R. (2018). "The Provisionality of Property Rights in Kant's Doctrine of Right." *Canadian Journal of Philosophy* Vol. 48(1): 850–876.
Huber, J. (2016). "Theorising from a Global Standpoint. Kant and Grotius on Original Common Possession of the Earth." *European Journal of Philosophy* Vol. 25(2): 231–249.
Huber, J. (2017). "Cosmopolitanism for Earth Dwellers: Kant on the Right to Be Somewhere." *Kantian Review* Vol. 22(1): 1–23.
Huber, J. (2020). "The Underrated Premium of Territorial Arrival." In Shachar, A. (ed.)., *The Shifting Border*. Manchester: Manchester University Press.
Hughes, N. (2018). "'Tourists Go Home': Anti-tourism Industry Protest in Barcelona." *Journal of Destination Marketing & Management* Vol. 9(1): 374–376.
Huseyinzadegan, D. (2019). "Rereading Kantian Hospitality for the Present." In Brown, G., & Telegdi-Csetri, Á. (eds.), *Kant's Cosmopolitics. Contemporary Issues and Global Debates.* Edinburgh: Edinburgh University Press, pp. 151–176.
Kleingeld, P. (2012). *Kant and Cosmopolitanism. The Philosophical Ideal of World Citizenship.* Cambridge: Cambridge University Press.
Maus, I. (2015). *Menschenrechte, Demokratie und Frieden.* Berlin: Suhrkamp.
Meckstroth, C. (2018). "Hospitality, or Kant's Critique of Cosmopolitanism and Human Rights." *Political Theory* Vol. 46(4): 537–559.
Niesen, P. (2007). "Colonialism and Hospitality." *Politics and Ethics Review* Vol. 3(1): 90–107.
Niesen, P. (2017). "'What Kant Would Have Said in the Refugee Crisis.'" *Danish Yearbook of Philosophy* Vol. 50: 83–106.
Ripstein, A. (2009). *Force and Freedom. Kant's Legal and Political Philosophy.* Cambridge, MA: Harvard University Press.
Ripstein, A (2021, forthcoming), Kant and the Law of War, Oxford: Oxford University Press.
Reinhardt, K. (2019). *Migration und Weltbürgerrecht. Zur Aktualität eines Theoriestücks der politischen Philosophie Kants.* Freiburg/München: Karl Alber.
Stilz, A. (2014). "Provisional Right and Non-State People." In Flikschuh, K., & Ypi, L. (eds.), *Kant and Colonialism.* Oxford: Oxford University Press, pp. 197–219.
Williams, H. (2007). "Kantian Cosmopolitan Right." *Politics and Ethics Review* Vol. 3(1): pp. 57–76.
Waligore, T. (2009). "Cosmopolitan Rights, Indigenous Peoples, and the Risks of Cultural Interaction." *Public Reason* Vol. 1(1): 27–56.

7

Three Models of Territory

Arthur Ripstein on the Territorial Rights of States

Alice Pinheiro Walla

Introduction

In *Kant and the Law of War*, Arthur Ripstein rejects two models of understanding the relation of states to their territory. According to the first model defended by natural law theorists, national territory is conceived as the private property of states. A second model, conceived as an alternative to the property model, regards national territory as constituting the *body* of the state. In this view, a state is connected to its territory in an indissoluble way, analogously to the relation between a person and her body.

Ripstein criticizes both models and proposes a third account, which links territory to the state's provision of a civil condition to its citizens. In this model, territorial integrity is closely connected to juridical independence from other public legal orders. States are not wronged by the existence of other, independent public law systems; therefore, they must respect the territorial boundaries of other states.

Ripstein's model is indeed more promising than the two previous models and, in some aspects, more in line with the spirit of Kant's theory. However, I will argue that it fails to recognize that possession of land that constitutes the territory has normative implications that are not eliminated by the fact that an independent legal order has been established over the territory.

I will argue that regardless of whether a legal order has been established over the territory, possession of land *itself* already imposes obligations on persons outside the territory to respect it. However, possession of land also imposes duties on the holders of territory that are global in scope. It is therefore not possible to reduce territorial rights to claims of juridical independence in virtue of a state's internal civil condition. The reason is a fundamental asymmetry between forms of private right that involve external objects, including land (*Sachenrecht*), and forms of private right involving only the wills of the parties. I argue that Kant recognized this asymmetry in his legal theory. As a result, rights to external objects require a form of consent that can only be achieved under a global rule of law. Therefore, the problem of territory (the fact that possession of land must impose an obligation on all others to respect that territory) is not yet solved in a world of "monadic" public legal orders coexisting in an international state of nature. A state that enforces a public legal order over a specific area solves the problem of consent only in regard to its own citizens. It does not yet solve the problem of possession of land and the obligations it gives rise to in regard to other

Alice Pinheiro Walla, *Three Models of Territory* In: *The Public Uses of Coercion and Force*. Edited by: Ester Herlin-Karnell and Enzo Rossi, Oxford University Press. © Oxford University Press 2021. DOI: 10.1093/oso/9780197519103.003.0007

states and foreign individuals. However, recognizing this problem does not commit us to adopting a conception of territory as the property of states, as Ripstein would object.

1. *Par in parem non habet imperium*

Ripstein defines states as *public legal orders.*[1] They provide a condition of public justice to their members. This means that when conflicts about rights arise, individuals no longer need to enforce their claims unilaterally against each other. A public legal order provides not only the means but is in fact the only entity with the *authority* to resolve rights conflicts in a way that can be *binding* to the parties, due to its omnilateral, public character.

Because several states exist side by side on the surface of the earth, we also have a plurality of public legal orders, whose coexistence requires external regulation. The law of peoples or *Völkerrecht*, which according to Kant should be called *Staatenrecht* (the right of states) (cf. MS 6:343), has to do with the external relations of states *qua* public legal orders regarding each other. The fact that a state is an internal civil condition is relevant for the way states can treat each other in the international domain. This is compatible with the claim that the jurisdictional authority of a state does not extend to other states. States are thus juridically independent and yet their internal civil condition imposes constraints on their external relations to each other. For instance, Kant argues that no state shall be permitted to acquire another state as if it was a thing. Instead, states are to be regarded as *moral persons*:

> For a state is not (like the land on which it resides) a belonging (*Patrimonium*). It is a society of human beings that no one other than itself can command or dispose of. Like a trunk, it has its own roots; and to annex it to another state as a graft is to do away with its existence as a moral person and to make a moral person into a thing, and so to contradict the idea of the original contract, apart from which no right over a people can be thought." ZeF VIII:344

Ripstein interprets the moral personhood of states with an analogy to individuals' right to be *sui iuris* or their own master (cf. MS 6:238, 270). This is a liberty right: there is no way another person can *bind* me to give up my authority to decide in my own affairs if I am not infringing the equal liberty of another to decide in *their* own affairs. This is of course a normative claim, not a descriptive one. As Ripstein stresses, "each nation is *sui iuris*, rather than *alieni iuris*." Their independence, including territorial integrity, follows from the fact that "no nation is subject to another nation's public law." Further, the mere existence of a state as a public legal order does not constitute

[1] Parenthetical references to Kant's writings give the volume and page number(s) of the Royal Prussian Academy edition (*Kants gesammelte Schriften*). I use the following abbreviations: MS = The Doctrine of Right; TP = "On the Common Saying: That May Be Correct in Theory, But it is of No Use in Practice"; ZeF = Zum ewigen Frieden. Unless stated otherwise, all English translations are from the *Cambridge Edition of the Works of Immanuel Kant*.

in itself a *wrong* to other states. States have thus a liberty right (no duty not to) to be independent from each other in the international domain, analogous to the right of individuals to be *sui iuris*.

The status of states as independent civil conditions is not conditional upon the recognition of other states. Therefore, it is not a voluntary choice of states to recognize international law. No permission from the international community is required for a state to be constituted as a juridical equal among equals. *Par in parem non habet imperium*: equals among equals have no power (over each other). This principle of international law is the foundation of the immunity of a sovereign power from interference from other sovereign powers. However, this is not to say that all states are equally just. Sovereign immunity only means that "no optional affirmative act" by other states is required for the validity of a state's juridical status (cf. Ripstein 2021: chap. 2, section V) The public powers of a given state remain restricted to the territory under its jurisdictional authority and have "no extra-territorial application" (ibid.).

Although states are public orders, they stand in *private-like* relations in regard to each other. If states are essentially private entities, to which private right claims apply, war can be understood as *a means of acquisition* in the international arena. Ripstein argues that Kant's rejection of the patrimonial state as a proprietary of territory amounts to rejecting the possibility of *acquiring* territory (Ripstein 2021: chap. 1, section III).[2] States' only acceptable external purpose for action is the preservation of its existing civil condition (ibid.19). States are not allowed to use its citizens as tools of war either (MS 6:344). Kant identifies three titles of private right: (1) property, which governs the ways in which separate persons interact independently; (2) contract, which governs the way they interact interdependently and consensually; and (3) fiduciary type relations, in which persons interact interdependently but non-consensually (Ripstein 2021: chap. 2, section V). These private relations are then transferred to the right of nations, although Ripstein denies that they are straightforward applications of private right. National territory is an analogue of property, international treaties are analogues of contracts, and specific relations between states are analogues of status relations between persons.

Further, Ripstein argues that the basic private law analogies must be understood *relationally*. This disqualifies the property model and the body model as adequate accounts of territory (ibid.). Territory is the place where a state "exercises its jurisdiction as a legal order" (ibid.). Its legitimacy does not depend upon it having followed a procedure or having been acquired in the appropriate way. This is because acquisition of territory precedes the institution of the state and its procedures. Therefore, Kant's legal theory preserves the *status quo* when it comes to territory: whatever territory a given state happens to have, will be considered its rightful territory.

[2] Unless one understands "territory" as a geographical area under the jurisdiction of a legal order (and not merely as land), Ripstein's conclusion does not follow. Kant's claim is that it is impermissible to acquire another *state*; the principle of sovereign immunity does not exclude acquiring a territory over which there is no state authority, although, as we will see, Kant recognized indigenous, non-state peoples as having genuine title to their lands. Kant's point is thus not against acquisition of territory per se, but acquisition of another public order.

2. Why Territory Cannot Be Reduced to Juridical Independence

Ripstein's account of territorial rights is indeed distinctive among recent Kantian accounts of territorial rights, even statist ones, such as Anna Stilz's (this volume). According to Stilz, states are justified in holding territorial rights if they succeed in implementing justice over a certain area, including upholding minimal human rights standards. Although failed states would not be justified in holding territorial rights in Stilz's account, she does not want to argue that they should lose their territory and includes provisions which would exclude loss of territory. I argued elsewhere that this account is far more demanding than Kant's own understanding of territorial rights (cf. Pinheiro Walla, 2014). Ripstein's account does not fall prey to this objection. As he observes, "Kant's answer sets a very low bar for satisfying what he calls 'the postulate of public right,' that is, for solving the problem of the state of nature; all that is required are institutions for making, applying and enforcing law in a way that allows each person to have private claims against the others." (Ripstein 2021: chap. 2, section III). His account is more in line with Kant's own theory by being fairly minimal in this regard. States do not have to satisfy liberal standards of justice in order to be entitled to their territory. In contrast, the idea of an original contract, which functions as a regulative ideal for the state is far more demanding than satisfying the postulate of public right (cf. Stilz, 2011; Stilz, this volume).

In contrast to Lea Ypi's (Ypi, 2014) account, there is no emphasis on the permissive law of practical reason as a principle permitting territorial control under the condition that states are also committed to implement rightful relations internationally (cf. Stilz, this volume). The problem with Ypi's account is that if the permission to control territory is conditional upon a commitment to implement international law, then we must be able to determine (1) when we can safely assume that a state has adopted such a commitment, and (2) what are the consequences for states of which are clearly *not committed* to the implementation of international law. In contrast, Kant suggests that control of territory imposes *a duty* on states to implement an international rule of law; whether they are committed to the promotion of this end or not is irrelevant, since the obligation arises from the fact that they control territory.[3] Ripstein adopts a different strategy. Territorial rights and political independence go hand in hand in his account. If each state is *sui iuris*, states do not have the *authority* to interfere in the internal affairs of other states, which are equally *sui iuris*. The citizens of a given polity "have a

[3] "Since a state of nature among nations, like a state of nature among individual human beings, is a condition that one ought to leave in order to enter a lawful condition, before this happens any rights of nations, and anything external that is mine or yours which states can acquire or retain by war, are merely provisional. Only in a universal association of states (analogous to that by which a people become a state) can rights come to hold conclusively and a true condition of peace come about. But if such a state made up of nations were to extend too far over vast regions, governing it and so too protecting each of its members would finally have to become impossible, while several such corporations would again bring on a state of war. So perpetual peace, the ultimate goal of the whole right of nations, is indeed an unachievable idea. Still, the political principles directed toward perpetual peace, of entering such alliances of states, which serve for continual approximation to it, are not unachievable. Instead, since continual approximation to it is a task based on duty and therefore on the right of human beings and of states, this can certainly be achieved." (MS VI:350).

right that outsiders not subvert their legal order" (Ripstein 2021: chap. 2, section V). This claim is based on the public nature of a state, as a civil condition. Interference in the territorial integrity of a state amounts to violating the equal juridical standing of all states in regard to each other in the international domain.

While other states may not interfere with another jurisdiction, Ripstein's account tells us little about other incidents of territorial rights, such as the right to control borders and immigration and control of natural resources. One may wonder if Ripstein's theory does offer a sufficient account of territorial rights to cover those prerogatives as well. All the theory provides is an account of the *wrong* constituted by interference in another state's public legal order. Although Ripstein discusses migration, the cosmopolitan right to visit, and the wrongness of colonialism, these are not derived directly from the territorial claims of states. They follow from the right of states to be free from external interference or imposition of the public law of one nation over another. Further, colonialist conquest also leads to humiliation and power asymmetry (Ripstein 2021; chap. 8).

The gist of Ripstein's argument (Ripstein 2021: chap. 2, section IV) is that a state does not wrong other states by existing. The creation of a state does not change the rights of others. I agree with Ripstein that a state does not wrong other states in virtue of its existence. However, it is noteworthy that Kant already prohibits coercion of and external interference with non-state groups which occupy land (to whom he often refers to as "inhabitants," *Einwohner*).[4]

> Lastly, it can still be asked whether, when neither nature nor chance but just our own will brings us into the neighborhood of a people **that holds out no prospect of a civil union with it**, we should not be authorized to found colonies, by force if need be, **in order to establish a civil union with them** and bring these human beings (savages) into a rightful condition (as with the American Indians, the Hottentots and the inhabitants of New Holland); or (which is not much better), to found colonies by fraudulent purchase of their land, and so become owners of their land, making use of our superiority **without regard for their first possession.** Should we not be authorized to do this, especially since nature itself (which abhors a vacuum) seems to demand it, and great expanses of land in other parts of the world, which are now splendidly populated, would have otherwise remained uninhabited by civilized people or, indeed, would have to remain forever uninhabited, so that the end of creation would have been frustrated? But it is easy to see through this veil of injustice (Jesuitism), which would sanction any means to good ends. Such a way of acquiring land is therefore to be repudiated. (cf. MS VI:266, my emphasis)

Kant recognizes a duty to respect territory *independently* of the existence of a condition of public law over a specific territory. Founding colonies without the consent of

[4] Kant justifies the right to coerce individuals to enter a civil condition together with their possessions to a prerogative of right. Although no individual in the state of nature has the authority to coerce another unilaterally, they have "practical reason on their side" since it is a duty to enter a civil condition with all others (MS VI:312). However, once a group occupies an area, this prerogative no longer applies, even if the group is not a state yet. Kant defines a state (*civitas*) as the union of a multitude of human beings under laws of right (*die Vereinigung einer Menge von Menschen unter Rechtsgesetzen*) (MS VI:313).

the inhabitants is an impermissible way to acquire land. The first acquisition of the inhabitants of a certain area imposes constraints on how others can settle in the area. Depending on how close settlers come to their territory, they may require the express consent of the natives via a contract that is not fraudulent. If there is a duty to respect first acquisition regardless of whether the people in question constitute a state, we are left with two possibilities: either Kant regards the inhabitants as constituting a civil condition *of some sort*, in which case Ripstein's argument still holds; or the first acquisition of a group is itself *sufficient* for imposing a duty on others to respect the occupation of land, regardless of whether they can be said to be in a civil condition. If so, this suggests that *acquisition of land* is what is doing the fundamental work in Kant's theory when it comes to the right to exclude from the territory.

The case of non-state peoples shows that acquisition of land and collective occupation can impose an obligation on other nations independently from the public legal order argument defended by Ripstein ("one must not interfere in nations' internal civil condition and therefore must respect their territory"). While the public legal order argument for territorial rights needs not be ruled out, I argue that it provides an *additional* argument in support to the first acquisition argument.[5] Possession of land is based on a principle of private right (*Privatrecht*), which is normative already in the state of nature, despite its so-called "provisional" character (cf. MS VI:264). In contrast, a public legal order is binding only to those under the scope of its jurisdiction. The social contract is *internally binding* (it determines among others the duties of the sovereign in regard to its own citizens) while acquisition and ongoing occupation of an area are also *externally binding* (they impose a duty on other nations and outsiders in an international state of nature to respect that group or nations' territory).

If this is correct, Kant's account of territorial rights is even less statist than Ripstein suggests. The existence of an internal will of the people does indeed constrain the way sovereigns may dispose of the territory, but this has to do with the sovereign's obligation *towards her own citizens*. The internal will of the people is itself not binding to foreign states; its jurisdiction does not extend to other legal orders and is confined to its own territory and people. However, their occupation of space ought to be respected by all states even in the international state of nature, that is, *even in the lack of a* common jurisdiction. Territorial rights *protect* the internal legal order, not the other way around. If there is a duty to enter a legal condition with others it would be contradictory to base the right to exclude individual outsiders from *joining* that civil condition on the legal order argument itself. The expansion of the rule of law is the aim of Kant's legal theory. The right to exclude from land and territory is based instead on a duty to respect first acquisition (or the oldest ongoing occupation of territory). Therefore, uniting into a larger public legal order requires the informed *consent* of the inhabitants (and is not ruled out if based on genuine consent).

One may object by reminding us that states have a right to decline voluntary visitors in Kant's theory. He famously restricted cosmopolitan right to a minimal right to

[5] My position is therefore not to reject the public legal order argument as false, but to show that acquisition of land is a sufficient argument against colonialism and other practices that would interfere with territorial integrity and control. Thanks to Enzo Rossi for pressing me to make my position on this point more explicit.

visit and endorsed the protectionist practices of China and Japan to exclude European "guests" (cf. ZeF VIII:359). A duty to join a civil condition, as expressed by the postulate of public right, does not mean that anyone wishing to enter a jurisdiction has a right to admission. Visitors have no right to settle and can be denied entry by the host country, unless rejection would mean their own demise.[6] But the reason for this is, as explicitly mentioned in both *Perpetual Peace* and the *Doctrine of Right*, is *common possession of the earth* (ZeF VIII:358; MS VI:352). Kant's theory of acquisition and possession is thus at the core of Kant's arguments against colonialism, an aspect of his legal theory that is often ignored by commentators who are dismissive of or reluctant to engage with his theory of acquisition and possession of land. Note that respect for first acquisition would rule out annexation of a foreign territory, but it does not exclude the possibility of a group or state to unite *voluntarily* with another civil condition (as Kant suggests it should be the case with international public institutions).[7]

3. Legal Orders and Territory

We tend to assume that a legal order is necessarily territorially bounded. This is not the case. In fact, territorially bound jurisdictions are relatively recent in European history. For instance, in the early Middle Ages, Lombards and Latins had their own respective legal codes. There would be special rules for resolving conflicts between individuals subjected to different laws.[8] In principle, different jurisdictions could coexist without being separated by hard borders, if the international rule of law would regulate global mobility. This is because hard borders demarcate not only the limits of a state's jurisdiction, but also signalize that there is a state of nature between states; it sets a limit not only to the scope of a state's jurisdiction, but also to the state of nature *without*.

Imagine a distant future in which international juridical institutions have developed to such an extent that national territory has *become obsolete*.[9] Territorial borders exist merely *pro forma* or as memory of times past, such as the checkpoints of the former German Democratic Republic in reunified Germany, or for local governance administrative purposes. Let us assume that all possession of land on the globe is now regulated by an international rule of law, leaving no state or people on the globe outside the scope of its jurisdiction. Under those circumstances, national territory would no longer perform its primary function, namely, to determine the scope of a specific legal order in regard to outsiders as well as to establish the boundaries between itself

[6] I have argued elsewhere that the right to be admitted in a situation where one's situation is involuntary is Kant's own revised version of the right of necessity (Pinheiro Walla, 2016). The alleged right of necessity discussed in the *Doctrine of Right* (the right to kill an innocent to save one's own life) is not a real right (cf. MS VI:235–236). On the right of necessity in modern political philosophy, see Olsthoorn (this volume).

[7] I will analyze common possession of the earth in more detail in the last chapter of the chapter.

[8] I would like to thank Enzo Rossi for pointing this out to me.

[9] By this I do not mean that we can do away with the fact that our interactions take place in space and that Right *itself* is space bound. The idea is that if all former territorial states came to be under a common public legal order, all claims to land would be fully regulated and conclusively binding to all living on the face of the earth. There is therefore no need to *assert* territorial claims regarding each other. The status of all land distribution on the earth would be considered juridically determined (peremptory). In other words: the international state of nature would have been fully overcome.

and other legal orders. Although this may sound as a truism, borders can only play this function if there is someone or something to *exclude*. In contrast, if no one can be considered a non-member or outsider, borders are merely nominal. Borders are no longer *excluding* anyone.

The scenario I am proposing would not entail the *dissolution* of internal public orders, but higher-order regulations coordinating previously isolated public legal orders, similarly to the federalist model of the European Union. Instead of swimming side by side in an international state of nature, each legal order for itself, these states have voluntarily agreed to join an international rule of law which imposes constraints on their external relations to each other (in this regard, this international rule of law could be fairly minimal compared to the juridical integration imposed by the institutions of the European Union on member states).

Now imagine that a few generations later, aided by stability and coordination provided by the international legal order, internal legal orders have evolved to such an extent as to decouple membership from territory. We now have public law order red, public law order blue, and public law order green (for the sake of simplicity, I will assume there are only three internal legal orders). The jurisdiction of red applies to the citizens of red, that of blue to the citizens of blue, and that of green to the citizens of green. Citizens live scattered in different parts of the planet alongside each other. Territory is no longer relevant for identifying membership to a specific public legal order; each legal order has its own passport, as the only criteria for membership. There are no hard-national borders in a significant sense, since all persons can move freely on the globe and will belong to one of those legal orders, which are all regulated and integrated under international law. The relations between citizens of different public legal orders are regulated by international rules and principles specifying which jurisdiction has the right to coerce whom and in which case. In this scenario, it is still possible to interfere into a specific legal order. This would amount to overstepping one's jurisdiction or hindering justice as specified by the higher-order legal order. Interference in a public legal order would not entail interference in the territory. Instead, it would be a mere question of *authority*, namely who has the right to judge and coerce in regard to the members of legal order red, blue and green. We learn from this scenario that the right of public legal order to noninterference does not necessarily require territorial rights.[10] Territorial disputes would no longer arise because the international rule of law would be developed enough to settle, with due legal authority, any disagreements about land. The more developed the international rule of law and the greater its scope, the less important territoriality becomes for particular jurisdictions within that scope. Territorial rights are required when there is *lawlessness* between states, that is, in the absence of an international legal order that regulates the relations between two previously isolated legal orders.

This brings us back to our more familiar international order made up of sovereign states whose external relations remain largely unregulated. We tend to think that the

[10] The spatial character of Right itself is not made obsolete; this would be impossible. Conflicts between individuals and the impact their external actions have on each other are always in space. The difference is that the division of the earth into distinct *national territories with hard borders* is no longer relevant. My intent is to show that it is possible to think of jurisdiction without conflating it with a specific territory.

rationale for territorial rights is the protection of a public legal order. But the reason why this seems necessary in the first place is that possession of land is still unregulated and therefore can be *contested* in interstate relations. If the internal constitution of a state is itself not binding to other states and is confined to its own territory and people; and if no state can bind another not to decide in their own affairs (which is necessarily the case in an international state of nature), it follows that *only the acquisition of land* of each state is normatively binding for other states. As Kant notes, although there is no binding authority to decide in matters of possession in the international state of nature, provisional acquisition must be treated as "true acquisition," even in the state of nature (cf. MS VI:264).

4. Possession of Land and International Wars

Since taking possession of things[11] in Kant's legal theory gives rise to a duty to enter the civil condition, David Miller noted that this duty can be applied to states in regard to their claim to jurisdiction over a territory. In response to Miller, Ripstein argues that the parallel between individual acquisition of property and a state's territory needs to be drawn "at the right place" (Ripstein 2021: chap. 2, section VI). By possessing territory, it is true that states change the context in which both other states and individuals act. But decisive for Ripstein is the fact that no states or persons are thereby *wronged.*

"The acquisition of property by individuals puts other individuals under new obligations of non-interference; dealing with the object of another person's property right is a wrong, even though dealing with the same object is non-wrongful if nobody owns it. No such analogy applies to states; the birth of a state does not stop another state from doing anything it was previously entitled to do, because the only way in which a state can rightfully deal with territory is by exercising jurisdiction over it. But if no other state has jurisdiction in that location, no other state is wronged." (ibid.)

Ripstein's argument relies on the distinction between wronging someone and merely changing the context in which someone acts. If a public legal order has established itself over a specific geographical area, it relates to territory "in the only way a state can rightfully relate" to land; therefore, it precludes any other competing claims to that territory. A state which did not respect territory that is under another state's jurisdiction would be wrongfully interfering with the public legal order of that state. This is because the existence of the latter does not wrong anyone who is not part of it. Although the existence of that state will exclude the possibility of appropriating the land on which it is located and of setting up an alternative state in that territory, no one has an innate or acquired right to have a specific set of options available to them: "I wrong you if I interfere with your means, but I do not wrong you if I do not use my

[11] These include any external objects that can become the possession of a person such as immovables (e.g., land) and movable objects. Land is the external object par excellence, since acquiring a piece of land for the first time entails acquiring all the "accidents" (movables) inherent on that land (cf. MS 6 260–261). The first proprietor can then alienate those movables, thereby dissociating possession of these objects from possession of the land.

means in ways that are best suited to your successful use of yours" (ibid.). The relevant aspect of individuals' innate right remains untouched, according to Ripstein.

While a private person's claim to property depends on how that property was acquired, Ripstein argues that there is no backward-looking procedure that qualifies as the rightful way to acquire territory. I will argue that Kant's theory allows a more nuanced approach to territorial disputes that does not require contesting the existing rule of law on that territory. It is worth revisiting the relationship between his theory of acquisition and the constraints imposed by the duty to leave the state of nature, the so-called the postulate of public law (cf. MS VI: 307). Indeed, Kant's account of territory is conservative and future oriented: its main priority is not to correct past injustices but to make possible a future rule of law.[12] For pragmatic reasons, we must assume that a territory was acquired in good faith. Even if territory was acquired by violent means, its legitimacy must not be questioned if appropriation of the territory was ultimately *successful*, lest we allow instability to threaten an emerging and fragile rule of law. However, *future* acquisition must be rightful and certain ways of acquiring territory must be ruled out in the future. This principle should not be applied backward since it could endanger the establishment of a rule of law. But in the case of *contested territory*, the history of acquisition may well be relevant; a people could have a stronger title to land than another group, although the group with the stronger title may nevertheless lack the authority to impose its (correct) judgment about rights on the group with lesser or no title to the territory. As stressed before, only the verdict of a *common* public institution would be binding *to both parties*. Therefore, Kant's theory of acquisition and the history of acquisition may be required for determining reparation or compensation claims, although the *legitimacy* of an existing public law order does not depend on the history of acquisition satisfying specific criteria.

Extensive inquiries into the history of each act of acquisition would make a public condition ultimately impossible; therefore, unless a dispute arises, acquisition must be presumed to be lawful. The same applies to transfers of property according to the rules of the market. Kant discusses the case of a stolen horse that was bought by a good faith purchaser in accordance with the rules of the market (MS VI:302). There are two types of legal reasoning one can identify in such a scenario: legal reasoning as in the state of nature (in accordance with the principles of *commutative justice*, which Kant identifies with *right in itself*), and the legal reasoning of a court of justice (*distributive justice*, which is right according to *empirical principles*).[13] If we follow commutative justice, the horse should go back to the original owner, from whom it was stolen. Distributive justice, in contrast, would permit the good faith purchaser to keep the horse, with some compensation to the first owner. Interestingly, even though the verdict of a court of justice is deemed subjective, public justice must prevail (*in favorem iustitiae distributivae*, MS VI:303). The verdict of the court is thus *binding*. The reason is not that there is no principle for determining whether acquisition was rightful in a

[12] This is not to say that past injustice does not matter or should be forgotten; the point is that when implementing the rule of law, we must give priority to its *realization* over compensating the past.

[13] Note that distributive justice (*iustitia distributiva*) means for Kant a condition of *public justice*, and not what we nowadays understand under that concept (i.e., justice in the distribution of goods and resources).

state of nature, but the need to maintain a well-functioning market and maintain the stability of the legal system.

Ripstein dismisses the analogy between territory and property, but one could argue with the above example that what is at stake is in fact not a "disanalogy" between individual property and territory, but rather the normative priority of the postulate of public law over the normative implications of acquisition in the international state of nature. In a civil condition, *legal certainty* (justice as defined by positive public laws) takes precedence over *substantive justice* (what common human reason would consider just). Where a condition of public justice already exists, public law must take priority over the history of acquisition and commutative justice, although not without residue, as I argued somewhere else (cf. Pinheiro Walla, 2019).[14]

Although within a particular state the problem of distributive justice has been solved (there is now a public institution with the authority to decide and coerce), its title to its territory may still be contested by other states at international level. This is why Kant argues that the *exeundum* must also apply at the international level, although states must not be *coerced* to leave the state of nature.[15] Kant's claim that states must regard each other as moral persons is a *precondition* for the possibility of states moving voluntarily toward an international rule of law, and not where the development of legal orders should inevitably stop (cf. Pinheiro Walla, 2018). Until a civil condition between states has been created, all possession on the surface of the globe is *provisional* (*provisorisch*). Unlike what is often suggested in the literature, "provisional" does not mean that possession is merely "temporary" but refers to the *modality* of a possession claim: while in the state of nature possession is simply a matter of fact (*facto*) and is therefore *contingent*, possession acquires the status of *necessity* (*lege*) in a civil condition because the decision of a public court of justice on the matter is omnilaterally *binding*. What about territory? Crucial for the modality of a possession claim is whether or not there is a common public institution with the *authority* to impose an obligation on all others to respect that claim. From the perspective of a state in regard to its own citizens (within its civil condition), the sovereign is to be regarded as the *symbolic proprietor of the land.* (MS VI:323) Because of her public character, the sovereign is in a position to issue binding verdicts concerning disputes over property

[14] Kant has a great deal to offer to contemporary global justice debates. However, one must be careful when trying to translate aspects of Kant's theory into these debates (see previous footnote). Would it be permissible to wage a war of subsistence against another country on (contemporary) distributive justice grounds, say, either as a remedial right for past wrongs or as a "right of necessity"? Since distributive justice means for Kant not the just distribution of resources but the implementation of a public legal order, this is clearly ruled out, although there are possible ways of arguing for rights to welfare from the perspective of Kant's legal philosophy. For a defense of a right to wage subsistence wars as a remedy for past wrongs, see Fabre (2008). For a discussion of Fabre's position as well as of a right of necessity to wage wars of subsistence, see Olsthoorn (this volume).

[15] "In accordance with reason there is only one way that states in relation with one another can leave the lawless condition, which involves nothing but war; it is that, like individual human beings, they give up their savage (lawless) freedom, accommodate themselves to public coercive laws, and so form an (always growing) *state of nations* (*civitas gentium*) that would finally encompass all the nations of the earth. But, in accordance with their idea of the right of nations, they do not at all want this, thus rejecting *in hypothesi* what is correct *in thesi*; so (if all is not to be lost) in place of the positive idea *of a world republic* only the *negative* surrogate of a *league* that averts war, endures, and always expands can hold back the stream of hostile inclination that shies away from right, though with constant danger of its breaking out." (ZeF VIII:357)

within the boundaries of that territory. In contrast, from the perspective of the international state of nature, the claims of states to their territory are provisional (i.e., contingent with regard to their bindingness). Despite the existence of isolated, internal public law orders at international level, we have what Yitzhak Benbaji called "regulated anarchy" between states. (cf. Benbaji, this volume) Since there is no common public institution with the authority to bind all others, a state is in a situation where it needs to defend its territory against others even though her possession is valid in the state of nature. In case of rights disputes, no state has the authority to issue a binding verdict to all conflict parties and to enforce such a verdict; unilateral interpretation and enforcement of rights is therefore inevitable.

While there are important affinities between territory and possession of land, there are also significant dissimilarities. Territory is not the private property of states; even though states may claim title to territory in a way that is similar to the way individuals claim land, states are not allowed to dispose of the territory like a thing (for instance, by selling parts of the territory to other states) because of their representative function in regard to their people; further, states' right to control the territory may be limited by common possession of the earth: the equal original right of all human beings to be somewhere in the planet. States have a *duty* to admit foreigners into their territory when refusing them would lead to their destruction (cf. ZeF VIII:358). Acquisition by conquest is also ruled out, but not simply because Kant rejects a "patrimonial" conception of national territory. It is ruled out because even in the state of nature it would be impermissible to dispossess a group who already lived on that land.

Kant introduces the idea of a *lex permissiva* in order to enable possessors to impose an obligation on all others to respect an act of unilateral appropriation, an authority they would otherwise not have as private persons.[16] In the case of acquisition through sale or marriage, states would be indeed acting as private persons, disposing of territory as their private property, without respect for the people they are supposed to represent. However, it is still the case that from an international perspective states occupy, that is, possess land. One could argue that states merely claim jurisdictional rights over an area and not possession of land; however, precisely *because* they do not share a common public legal order, the only thing they can possibly claim in relation to other nations that could bind them is a title to the territory, their possession of land.

The difference between property (*Eigentum*) and possession (*Besitz*) in Kant's legal theory is merely a *modal* one. While possession is *contingent* (*facto*; a matter of who arrived first and who controls what), property is *necessary* (*lege*). Possession and property do not differ in content. This is why Kant says that private rights are imported into the civil condition. (cf. MS VI:312–313) There is a duty to respect possession already in the state of nature, even though possession only becomes property in a civil condition binding all individuals and nations on the globe.

[16] "This postulate can be called a permissive law (*lex permissiva*) of practical reason, which gives us an authorization that could not be got from mere concepts of right as such, namely to put all others under an obligation, which they would not otherwise have, to refrain from using certain objects of our choice because we have been the first to take them into our possession. Reason wills that this holds as a principle, and it does this as practical reason, which extends itself a priori by this postulate of reason." (MS VI:247)

5. Common Possession of the Earth and the Cosmopolitan Duties of States

Ripstein devotes a section to the discussion of common possession of the earth in chapter 8 (Ripstein 2021). However, the account does not consider the full implications of common possession for the territorial claims of states and for the rights of individuals as "citizens of the world." Although Ripstein correctly highlights the fundamental role of *space* in external, rights relations and consequently for the problem of right (*Recht*), the role of *possession of land* for cosmopolitan right does not receive due attention. Further, since Ripstein's interpretation of public right is too statist it is unable to identify a distinctive role for cosmopolitan law that is not conflated with the law of nations.

As Ripstein explains, Kant understood community of land as *disjunctive possession*. Disjunctive judgment in the *Critique of Pure Reason* conceives two or more things as excluding each other but at the same time constituting a *community* through this relation (cf. Longuenesse, 2011). Community is thus a relational idea which presupposes "reciprocal determination with respect to location." If we apply this logical structure to territorial states, it is possible to understand their relation both as excluding each other from their specific locations and as standing in regard to each other as a community of all nations on the earth (as the distinctive parts of a determinate whole). The mutual exclusion regarding their specific locations nevertheless *unites* territorial states into a community of the earth.

However, what Ripstein (Ripstein 2021: chap. 9, section VII) sees in disjunctive possession is "precisely the formal structure of innate right: independence from being bound by others to more than one can in turn bind them." Ripstein's strong analogy between states' independence and individuals' right to be *sui iuris* leads him to regard common possession of the earth as just another, large-scale version of the individual innate right, this time involving nations instead of individuals. But this parallelism between innate right and common possession of the earth is problematic. There is a significant difference between occupying space *simpliciter*, as in the case of an individual with innate right; and *occupation of land*, as in the case of territorial states. A body can never cease to occupy space due to its impenetrability; it can move around or be displaced from a specific location without ceasing to be extended in space. In contrast, territory means occupying a specific part of the planet on a permanent basis, to the exclusion of others *from that specific location of the earth*, not simply from the space a body must necessarily occupy as extended matter. Mere occupation of space is consistent with being moved away and placed somewhere else. In contrast, settling down presupposes ongoing possession of an area. Since land is immovable, in order to occupy an area of the earth on a permanent basis, one must also be able to *claim title* to that location in case of disputes. Further, we also assume that occupying that area also entitles the occupants to the natural resources within that area. Since common possession of the earth involves something external to persons (i.e., land and its resources), it necessarily goes beyond the scope of innate right itself. (cf. MS VI:247–248, 250)

Although Kant identifies common possession of the earth with the logical structure of disjunctive judgments, this notion is not merely about a right of states to be independent from other states. Ripstein's interpretation neglects the *community* constituted by the disjunction. As Gottfried Achenwall (1995:91) pointed out, "common" is that to which many have an equal right. All persons must have an equal right to all external things *before* they can acquire anything (that is, *separate* something for themselves from the pool of things to which they are in principle equally entitled). Although Kant's conception of common possession of the earth is different from his natural law predecessors, including the above-mentioned Achenwall, the notion plays nevertheless the same general function in his theory of acquisition: it accounts for the fact that all persons must *originally* stand in a relevant juridical relation to each other in regard to all land on the planet (*ex ante*), so that all subsequent acquisition of land can *count* as a legally relevant deed of persons, capable of binding others to respect that acquisition already in the state of nature.[17] Kant's disjunctive community of the land accounts for why *priority in time* is a valid principle of distribution when it comes to the division of land while avoiding the idea that the normative starting point was a communism of land use. The latter could not be original, but only *adventitious* (derived from the will of another). The community of the earth *reminds us* that while states are permitted to possess and control territory to the exclusion of others, all groups and individuals have an original equal right to *acquire* any location in that common pool of possible locations that is planet earth. And that is something that continues to impose normative constraints on the territorial rights of states, even after the distribution of all land on the planet has taken place. Cosmopolitan right is none other than that *normative residue of common possession of the earth.*

According to Ripstein (Ripstein 2021: chap. 9, section VIII), the right to political independence and territorial integrity provides the content of international law. Cosmopolitan right, in turn, presupposes the idea "that a visitor is already a member of some state" (ibid.). Ripstein understands the task of cosmopolitan right as ensuring that no jurisdiction is dominated by another when a member of one state happens to be present in a foreign territory. This is because individuals can carry their private rights into another jurisdiction (e.g., their movable property and their marital status). While the host jurisdiction must not be subject to "alien" regulations within its territory, we must still ensure that the private rights of the visitors are protected in a foreign jurisdiction. Cosmopolitan right ensures that the private rights of travelers are consistent with the independence of the states they visit. We also need cosmopolitan right in case some persons are left without membership in a state. According to Ripstein (ibid.), every person must belong in a state: "[Cosmopolitan right's] organizing idea is that every visitor is always already [*sic*] a member of some state and every place visited to always already be in a legal order."

Instead of understanding the three levels of public law (the right of states, law of nations, and cosmopolitan right) as three specific levels of *juridification* or expansion of the civil condition, as to overcome all external relations in which a state of nature still obtains, in Ripstein's interpretation the right of states constitutes the *core* of public law. International and cosmopolitan law are meant to *supplement* the right of states. The role of international law is to exclude arbitrariness in the relations between states.

[17] Original stands for "prior to any acts of choice"; in regard to acquisition, original means "not derived." Byrd, S., and Hruschka, J. (2010), *Kant's Doctrine of Right. A Commentary*. Cambridge University Press, p. 4, fn. 8.

Cosmopolitan Right is supposed to enable commerce between nations. (cf. ibid. VII) Ripstein (ibid) also says that global movement needs to be regulated by states: "If people and things can move from one legal order to another, that movement must itself be brought under public law." However, it is not clear why the specific function Ripstein assigns to cosmopolitan right cannot be a task of the law of nations. Since states can agree on how to regulate property across borders, recognize the marital status of foreigners, and regulate and control global movement, it is difficult to see why we need cosmopolitan right *additionally* to international law in order to perform this task. There is no distinctive function for cosmopolitan right

"So understood cosmopolitan right coordinates with the right of the state and the right of nations by protecting the independence of legal orders from each nation's legal order. Cosmopolitan right is required to ensure that the independence of legal orders from each other does not become a means through which individual human beings are dominated; it is limited to the right to visit to ensure that relations between individual human beings do not become a means of subordinating one legal order to another." (ibid.).

In contrast, I will argue that cosmopolitan right must be understood as protecting freedom of global movement from being completely reduced to the decisions of states (in other words, from becoming the matter of the right of states alone). The right to visit without being treated with hostility remains a prerogative of all persons as world citizens regardless of their citizenship or of the host state in question. In other words, cosmopolitan right becomes a safeguard that the possibility of seeking interaction with others will not be fully reduced to the control of states.

I called cosmopolitan right the "residue" of the original common possession of the earth, which continues to impose certain normative constraints on territorial states even after all land of the planet has been divided. The view that individuals have an original (i.e., non-acquired) right to seek interaction with a foreign people without being treated with hostility if their intent is peaceful is the ultimate expression of this "residue." Unless cosmopolitan right is institutionalized in the form of a cosmopolitan constitution binding all states on earth, cosmopolitan right remains restricted to the non-positive, minimal "right to visit." Anything less than a cosmopolitan legal order, that is, a public legal order *above* the law of nations, would necessarily be no more than an attempt to translate the content of cosmopolitan right (the right to visit) into international law. There is of course nothing wrong with that. But in order to understand what is *distinctive* about cosmopolitan right as opposed to the law of nations we must not lose sight of the reason why cosmopolitan right is strictly needed as a third form of public right. Cosmopolitan right must not be reducible to the right of states (domestically) and the right of states in their external relations regarding each other (internationally). My view is that cosmopolitan right is strictly needed in order to safeguard the original right of all persons to *minimal freedom of global movement* in a world in which global mobility is no longer free. This minimal right must be safeguarded regardless of one's nationality or lack thereof. Commerce in this sense entails much less than trade or the international regulation of private rights. It means the universal possibility of *establishing* those relations in the first place. Restrictions on global mobility mean ultimately constraints on the very possibility of communication for all persons as cosmopolitan citizens.

Because planet earth is now divided into territorial states, each with the authority to control their own territory and movement across borders, the mere *possibility* of seeking interaction with others is severely restricted by territorial states. States now regulate global movement via international agreements, most notably via the international passport regime. It is nearly impossible to travel nowadays without a valid passport and consequently without being the citizen of some state. While states cooperate with each other to restrict the movement of the globe's poorest, citizens of the Global North and some wealthy individuals can travel cheaply, fast and safely around the world or even to buy citizenship of countries which would allow them greater mobility privileges. Access to global mobility infrastructure remains the privilege of specific nationalities while others are barred from the same opportunities (cf. Spijkerboer, 2018).[18] Is this somehow a realization of cosmopolitan right? By no means. We could not be further away from the cosmopolitan ideal.

Ripstein is correct that, all things equal, merely reducing the scope of choices available to other citizens of the world does not amount to a violation of their freedom. However, implementing a civil condition over a specific territory, while other areas are still under a state of nature, has normative implications; only a global civil condition could reach closure in regard to the territorial claims of states. Ripstein's view of territory as a legal order does not eliminate those normative implications.

Territory continues to impose obligations on states which require them not to stop the expansion juridification at the level of national territories. States are therefore required to limit their claim to control territory in case person or groups are deprived of a place to live in safety; they also have a duty to move voluntarily beyond their own domestic legal orders toward an international rule of law, where this can be done safely (without risk of domination or tyranny).

The fact that a territorial state controls territory to the exclusion of others who even may want to join that civil condition is normatively significant. A challenge to the idea that territorial rights are absolute is the current climate crisis. We know that large amounts of people will be displaced by virtue of climate change. The question is thus what states owe to each other and foreign individuals in face of this challenge. I believe that Kant's legal theory offers a normative guide to addressing our current global challenges whose potential is only starting to be fully explored. The idea that we originally share the land of the planet with all others and therefore have duties toward them that arise from the very fact that we control territory is a good place to start.

References

Achenwald, A., & Pütter, J. (1995). *Anfänge des Naturrechts*. Insel Verlag.

Byrd, S., & Hruschka, J. (2010). *Kant's Doctrine of Right. A Commentary*. Cambridge University Press.

[18] Spijkerboer (2018:452) defines global mobility infrastructure as the physical structures, services, and laws that enable some persons to move across the globe with high speed, low risk, and low cost. In contrast, those excluded from the global mobility infrastructure travel slowly, with high risk and high costs.

Fabre, C. (2008). "Cosmopolitanism, Just War Theory and Legitimate Authority." *International Affairs*, Vol. 84(5): 963–976.

Kant, I. (1793). "Über den Gemeinspruch: Das mag in der Theorie richtig sein, taugt aber nicht für die Praxis / On the Common Saying: That may be correct in theory, but it is of no use in practice. Academy edition Vol. 8." In Gregor, M. (ed., trans.) (1996). *The Cambridge Edition of the Works of Immanuel Kant*. Practical Philosophy. Cambridge University Press.

Kant, I. (1795). "Zum ewigen Frieden / Toward Perpetual Peace. Academy edition Vol. 8." In Gregor, M. (ed., trans.) (1986). The Cambridge Edition of the Works of Immanuel Kant. Practical Philosophy. Cambridge University Press.

Kant, I. (1797). "Die Metaphysik der Sitten / The Metaphysics of Morals. Academy edition, Vol. 6." In Gregor, M. (ed., trans.) (1986). The Cambridge Edition of the Works of Immanuel Kant. Practical Philosophy. Cambridge University Press.

Longuenesse, B. (2011). "Kant's Standpoint on the Whole: Disjunction Judgment, Community, and the Third Analogy of Experience." In Payne, C., & Thorpe, L. (eds.), *Kant and the Concept of Community*. Boydell & Brewer.

Miller, D. (2016). "Neo-Kantian Theories of Self-Determination: A Critique." Review of International Studies, Vol. 42(5): 862.

Pinheiro Walla, A. (2014). "A Commentary on Anna Stilz, 'Nations, States, and Territory' and Lea Ypi, 'A Permissive Theory of Territorial Rights.'" In *Territory and Justice Symposia 3: Kantian Theories of Territorial Rights*. Retrieved at: https://seis.bristol.ac.uk/~plcdib/territory/papers/AlicePW.pdf

Pinheiro Walla, A. (2016). "Common Possession of the Earth and Cosmopolitan Right." *Kant-Studien*, Vol. 107(1): 160–178.

Pinheiro Walla, A. (2018). "Global Government or Global Governance? Realism and Idealism in Kant's Legal Theory." *Journal of Global Ethics*, Vol. 13(3): 312–325.

Pinheiro Walla, A. (2020). "A Kantian Foundation for Welfare Rights." *Jurisprudence*, Vol. 11(1): 76–91 (Published online: 18 Dec 2019).

Ripstein A, (2021, forthcoming) Kant and the Law of War, Oxford. Oxford University Press.

Spijkerboer, T. (2018). "The Global Mobility Infrastructure: Reconceptualizing the Externalization of Migration Control." *European Journal of Migration and Law*, Vol. 20(1): 452–469.

Stilz, A. (2011). "Nations, States, and Territory." *Ethics*, Vol. 121(3): 572–601.

Ypi, L. (2014). "A Permissive Theory of Territorial Rights." *European Journal of Philosophy*, Vol. 22: 288–312.

8
A Kantian Defense of Remedial Wars

Alon Harel

1. Introduction

Ripstein's *Kant and the Law of War* presents, develops, and defends a reconceptualization of Kant's just war theory. While it is presented as an interpretation of Kant's political theory, Ripstein also wishes to defend the Kantian position against alternative views. Hence, the book can be described as both a book in the history of political theory as well as a book in contemporary political philosophy. In its later incarnation, it is the best response I am familiar with to the recent inclination of many theorists to reduce political philosophy to ethics and explain the structure of the state and state institutions as a mere means for the promotion of individual well-being.[1] Ripstein defends the less intuitive (but, in my view, the sound) claim that political institutions have their own distinct logic which cannot be understood purely in terms of individual ethics. It is this understanding which guides also his account of wars. To understand what is permissible or impermissible in wars, one has to first account for what states are about, what their status vis-à-vis each other is, and what the independence of states entails.

I share Ripstein's conviction that the reductionist-instrumentalist approach to wars (and, more generally, to political theory) is fundamentally mistaken (Harel 2014:3). I am also sympathetic to the boldness of his enterprise. There are three dimensions in which Ripstein's enterprise is exceptionally bold: it is anti-consequentialist, anti-intuitionist, and anti-conventionalist.

The temptation to yield to consequentialist reasoning, intuitionism, and conventionalism is particularly great in the context of wars. The grave consequences of wars and the horrific costs in human life almost inevitably lead just war theorists to design rules which would minimize these costs. Humanitarian concerns weigh particularly heavily in this context and this inclination on the part of political theorists (too) often comes at the expense of an understanding of what wars are, when they can legitimately be fought, and what the proper constraints on the conduct of soldiers in wars are.

Ripstein resists successfully these three temptations. He rejects any argument that rest on measuring the well-being effects of the rules of war on the right to conduct war, or on the permissible means that can be used in wars. The question of whether a state has a right to conduct a war and what the permissible means to use in wars are, rest not on the likely consequences of having particular rules but on an understanding of the status of states vis-à-vis each other. Ripstein also rejects the intuitionist

[1] In the context of just war theory, the leading advocate of this "revisionist" view is McMahan (2011).

Alon Harel, *A Kantian Defense of Remedial Wars* In: *The Public Uses of Coercion and Force*. Edited by: Ester Herlin-Karnell and Enzo Rossi, Oxford University Press. © Oxford University Press 2021. DOI: 10.1093/oso/9780197519103.003.0008

approach which dominates much of the literature. He believes that to understand the norms which govern wars, one needs to provide a theory and such a theory may, at times, radically differ from the pre-theoretical intuitions that people may have. Last, Ripstein is also anti-conventionalist in that while he often discusses existing legal conventions governing the conduct of wars, he does not regard those conventions as definitive. This rejection of consequentialism, intuitionism, and conventionalism follows inevitably from Ripstein's ambition to take seriously "*Kant's claim to provide a metaphysics rather than an anthropology of morals*" (ibid.20).

Yet, there are instances in which Ripstein himself yielded to the seductive force of consequentialism, intuitionism, and conventionalism. In particular, I shall argue that Ripstein's reasons for rejecting a right to remedial wars fail. The underlying Kantian principles guiding Ripstein's own account dictate that remedial wars are permissible. There are very powerful consequentialist, intuitionist, and conventionalist arguments against recognizing a right to remedial wars. But the logic provided by Ripstein's account of Kant cannot justify the prohibition on such wars. Precisely as the right to conduct defensive wars follows inevitably from our understanding of states as independent of each other, so I shall argue, the right to conduct remedial wars follows from this very same principle.

2. On the Kantian Justification of Remedial Wars

Ripstein differentiates among three types of wars: self-defense, remedial, and punitive. In his view, the only legitimate wars are defensive wars: namely wars resisting the present (or, at times, future) aggression of another state; these are wars that are designed to prevent a present or a future wrong. The right to conduct such wars follows inevitably from the recognition that nations are independent of each other and have the right to makes choices of their own.[2] If no such right to a defensive war exists, states are not truly independent of each other as their independence can be compromised by the aggression by other states. Self-defense wars involve preventive force. Preventive force is necessary to uphold the independence of the state against present or future aggression. In contrast, punitive and remedial wars are wars designed to punish states for past wrongdoing or to undo past wrongs. Those wars are not necessary for preserving the independence of states and are consequently impermissible.

As the main purpose of this chapter is to establish the permissibility of remedial wars, let me present Ripstein's characterization of remedial wars:

> Remedial wars are wars, the purpose of which "*is to reclaim things that have been wrongfully seized, those who wrongfully kept things that were unjustly acquired would be appropriate targets of force, even if they had acquired those things innocently, perhaps by mistake, or believing those from whom they had acquired them that they were entitled to them.*" (Ripstein 2021: chap. 3, section 1 and chap. 4, section 1)[3]

[2] For a discussion of the independence of states and its normative implications, see also Gur-Arye, M., & Harel, A.

[3] Note that this characterization rests on the intention of the state conducting the war. I believe the exclusive reliance on intentions may raise some difficulties, but those are irrelevant for my purposes here.

As stated earlier, Ripstein believes that while states typically have a right to conduct defensive wars (although not always[4]), they have no rights to conduct remedial wars. Hence, there is an asymmetry between the past and the future. A state has a right to defend itself against present and future aggression, but it has no right to undo the products of past aggression (namely, to engage in remedial wars), or to punish states for past wrongs (namely, to engage in punitive wars). I shall challenge this claim and argue that the very same considerations that justify *self-defense wars* also justify *remedial wars*. To do so, let me first examine in greater detail the arguments provided by Ripstein to justify the right to conduct self-defense wars.

Ripstein maintains that right to self-defense follows inevitably from the premise that states are independent of each other. Hence:

> *If that aggression cannot even be resisted, then the state that is attacked is subject to the other state's choice after all. Preventive force simply upholds the independence of the state that is attacked.* (ibid.71)

The absence of a right to self-defense implies inevitably compromising the independence of the state and its ability to make choices. If a state A violates the independence of another state B and B had no right to self-defense, it follows that state B is not fully independent, as it is subject to the will or choices of state B. Hence, the right to self-defense can be described as the normative embodiment of the principle which dictates that states are independent of each other.

Arguably, this very same logic should apply to remedial wars. To see it, assume that France commits an aggression against Germany. Given the right of self-defense, Germany has a right to defend itself. In the absence of such a right, we cannot really uphold the position that Germany is independent. Assume now that France succeeded in violating Germany's independence and has annexed parts of Germany. Let us assume also that it imposed a peace treaty which was signed by both parties. How can the success of its wrongful aggression annul the right to Germany *to undo the wrong*? Note that being faithful to Ripstein's methodology I do not maintain that Germany's remedial right to undo the wrong is justified on the grounds that it deters future violations of the right to independence or because it is conducive to peace. I only argue that the right to conduct remedial wars follows inevitably from the principle of the independence of nations. How can we say that Germany is a fully independent member of the community of nations if it cannot act to undo the wrong that was committed to it and regain its territory?

Another way to defend this conclusion rests on the conviction that a state has a duty to undo wrongs that it has committed in the past and that, as a general rule, a failure to undo these wrongs is in itself a violation of the independence of the state which is the victim of the past aggression. In the case of defensive wars, France invades Germany and the right of Germany to independence dictates that Germany has a right to self-defense, namely a right to prevent the infringement of its independence. In the case of remedial rights, France has already annexed wrongfully a territory of Germany and it

[4] Self-defense wars can affect the interests of third parties and, at times, those interests are so powerful that the state has no right to engage even in self-defense. See Ripstein (Ripstein 2021: chap. 3).

fails to undo the wrong. By failing to undo the wrong, it violates the right of Germany to territorial integrity and thereby infringes the independence of Germany. If this is so, Germany's right to conduct a remedial war is dictated by Germany's right to independence--a right which requires France to undo the wrong of holding the territory which it wrongfully annexed. Precisely as the right to self-defense is legitimated on the grounds that it is a right to prevent the wrong of *present (or future)* aggression threatening the independence of states, so the remedial right can be legitimated on the grounds that it is a right to prevent the wrong of *present (or future)* failure to undo past aggression.

Perhaps, it could be argued, that while present (or future) aggression is a wrong as it compromises the independence of the states, upholding the products of past aggression is not a present wrong. The past wrong of aggression does not entail that the holding of the products of the past aggression is a wrong. Or perhaps, even if it is a wrong, it is perhaps a different type of a wrong: not a wrong that compromises the independence of the state.

I find this conclusion implausible. I do not deny that there are circumstances in which upholding the products of past aggression is not wrongful. Perhaps circumstances have changed so radically that it would be wrong to undo the past wrong of aggression.[5] But in the absence of special circumstances, holding a territory that is the product of past aggression is a wrong. Further it is the same type of wrong as the past aggression because the refusal to undo the wrong compromise the territorial integrity and consequently also the independence of the state.

Ripstein does not raise this particular concern but he devotes much attention to the question of why punitive and remedial wars are impermissible. Unfortunately, I believe that punitive and remedial wars are fundamentally different and that while Ripstein's counterargument addresses successfully the case of punitive wars, it fails to address the case of remedial wars. I shall argue that those two cases should be discussed separately and that the case for a right to remedial wars is much stronger than the case for a right to punitive wars.

3. Remedial and Punitive Wars

Ripstein presents the case for recognizing a right to punitive and remedial wars as follows:

> *On this view, reclaiming lost territory, seeking compensation for historical wrongs, or punishing wrongdoing might also provide grounds for using force, since each is a response to wrong, and, it might further be thought that responses meant to remedy such wrongs might ... be modelled on extended or ongoing versions of active violations ... Rather than preventing a wrong, such a conception of war would be ... remedial or punitive. Its remedial aspect would consist in reclaiming something wrongfully*

[5] This is the theme defended by Waldron (1992) in a different context.

> *taken, and its punitive aspect as responsive to the wrongfulness of the acts of the enemy.* (Ripstein 2021: chap. 3, section IV)

Ripstein's response to this argument rests on the absence of authoritative institutions that can settle disputes concerning punitive and remedial wars. While in the domestic context we use adjudicatory institutions to make determinations of guilt and wrongdoing and to render decisions concerning remedial rights, we have no such institutions in the international arena. Hence, to recognize a right to punitive or remedial wars would inevitably entail that each state becomes "a judge in its own case" and therefore, in effect, this would imply the establishment of a system in which each nation can "*unilaterally act on its own judgments about the rights of another*" and that such a system is "*necessarily defective as a system of law.*" (ibid.)

But can this conclusion also affect the right to self-defense? After all, the question of whether a state acted in self-defense can in itself be subject to dispute and consequently, arguably recognizing a right to self-defense also implies the establishment of a defective system in which each state can make decisions which bear on the right of other states. For instance, in the *Six-Day War*, Israel attacked Egypt and justified its attack on the grounds of self-defense. It maintained that the concentration of Egyptian military forces on its borders was designed to violate its independence and that it simply defended itself against the anticipated Egyptian aggression. This claim on the part of Israel is controversial on both factual and normative grounds.[6] Factual concerns include concerns such as whether Egypt indeed intended to attack Israel; how many soldiers and what size of army are sufficient to pose a threat to the independence of Israel; can the decision of Egypt to concentrate the forces be reasonably interpreted as an act of self-defense, or does it indicate an intention to commit aggression? The normative concerns include questions such as these: Can Israel conduct a war of self-defense only when the anticipated aggression endangers its independence, namely is likely to succeed, or can it do so even in cases it is able to resist the aggression at a later stage (although, perhaps at a higher cost)?[7] Arguably, settling these controversies require adjudicative institutions and the absence of such institutions in the international arena dictates that ultimately there is no right to self-defense.

Ripstein is fully aware of the complexities of the judgments required to substantiate the claim that states have a right to conduct defensive wars. Nevertheless, Ripstein maintains that there is a fundamental difference between cases involving judgments required to conduct defensive wars, on the one hand, and judgments required to conduct remedial or punitive wars, on the other. He maintains that only the former can be justified, even in the absence of adjudicative institutions. This is because:

> *Still the case of defensive force is different in one important respect, the defender is protecting what it already has—its political independence and territorial integrity, that is, itself—and so preventing the aggressor from interfering with it . . . Therefore, although there is one sense in which the defender acts as judge in its own case—it makes the*

[6] For a discussion, see Quigley (2013).

[7] And indeed, under the conventional view, Section 51 to the UN Charter allows states to conduct preventive wars under certain circumstances.

> *judgment of fact that the aggressor is attacking it—the defensive force serves to prevent a unilateral violation by the aggressor. . . Given the absence of any such procedure in the international case . . . the defender's right to be beyond reproach takes the form of its entitlement to retain what it already has."* (Ripstein 2021: chap. 3, section IV)

To show the difficulty of differentiating between the case of wars to self-defense and remedial wars, let me reiterate the passage with the necessary changes and examine whether these considerations can also apply in the case of remedial rights:

> *Still the case of [remedial] force is different in one important respect [than in the case of punitive force], the defender is protecting what it already has—its political independence and territorial integrity, that is, itself—and so preventing the aggressor from interfering with it [by failing to undo the wrong it has committed] . . . Therefore, although there is one sense in which the defender acts as judge in its own case—it makes the judgment of fact that the aggressor [was] attacking it—the defensive force serves to prevent a unilateral violation by the aggressor . . . Given the absence of any such procedure in the international case . . . the defender's right to be beyond reproach takes the form of its entitlement to retain what it already has [namely to take back the territory that was wrongfully taken from it].* (ibid.)

To support the observation concerning the similarity between the case of defensive and remedial wars, let me point out two fundamental differences between the right to conduct remedial wars and the right to conduct punitive wars: With respect to both of these issues, remedial wars resemble defensive wars and not punitive wars. First, I shall reiterate my earlier claim that the right to conduct remedial wars is an inevitable byproduct of the conviction that states are independent of each other. Second, I shall argue that the type of judgments required to enforce the right to conduct remedial wars is similar in nature to the type of judgments required in order to conduct wars of self-defense. Hence, if by having such a remedial right the state becomes "a judge of its own case," or that it "can unilaterally act on its own judgments," it follows that the same considerations would also apply with respect to the right of self-defense and, consequently, self-defense wars would be impermissible.

First, as argued before, like the right to self-defense, the right to remedial wars is an inevitable byproduct of the principle that nations are independent of each other. There is no fundamental difference between the case in which the right of a nation is being compromised, and the case in which it was compromised, and the products of aggression remain subject to the control of the aggressor. Or, to use the alternative description provided earlier, there is no difference between the case in which the right of a nation to independence is being presently compromised by aggression and the case in which it is being presently compromised by failing to undo the products of past aggression. In the latter case, the state that acts on the basis of its remedial right is also acting to prevent *a present wrong*, the wrong of failing to undo the previous wrong of aggression.[8] In contrast, the right to punish wrongdoers does not follow from

[8] This of course is subject to the proviso that circumstances have not changed such that the holding of the territory is not a wrong anymore. See Waldron (ibid.).

the independence of states. A state can be independent when it is able to resist present aggression (by conducting a war of self-defense), or when it is able to resist present failure to undo the products of past aggression (by conducting a remedial war), without having a right to punish the aggressor. Thus, the right to conduct remedial wars is grounded in considerations which it shares with the right to self-defense and both of these types of wars differ from punitive wars.

Second, Ripstein believes that recognizing a right to punitive or remedial wars "*makes war conditional on some idea of fault or past wrong, permitting each nation to be . . . judge in its own case*" (Ripstein 2021: chap. 3, section IV). In his view, such a system "*in which one party can unilaterally act on its own judgments about the rights of another is necessarily defective as a system of law*" (ibid.). But, while this observation is applicable to punitive wars, it is not applicable to remedial wars.

It is evident that Ripstein is right when he claims that the judgment required to establish the justifiability of punitive wars is one that can only be made by adjudicative institutions and, in the absence of such institutions, no state has a right to conduct punitive wars. This follows from the fact that no state is superior or else can be presumed to be superior in its moral powers to any other state. Hence, even if a particular state operates in good faith, and even ifs judgments are likely to be correct, it has no right to make determinations of guilt or to determine the appropriate punishment for a past wrong as such a judgment makes it "a judge in its own case."

In contrast, Ripstein believes that the state can make a judgment that another state commits an aggression (and, consequently, that it has a right to conduct a defensive war). Admittedly, it would be better if public institutions could make such a judgment. But, despite this deficiency, it is permissible on the part of the state to act on the basis of such a judgment. It seems, however, that the judgment that *at present* another state commits an aggression is similar in type to the judgment that *in the past* another state committed an aggression and failed to undo the products of its aggression. To the extent that it is illegitimate on the part of the state to make such a judgment concerning past aggression, it should also be illegitimate on the part of the state to make a judgment concerning present or future aggression.

Note that I do not deny that sometimes the question of whether there was a wrongful aggression in the past and the products of this wrong are wrongly upheld by the aggressor at present is more difficult to settle than the question of whether there is a wrongful aggression at present. It is often the case that knowing facts about the past is more difficult than knowing facts about the present (although, at times, it could be the other way around). The issue is not the prospects of a mistaken judgment. As Ripstein says, even if the states operate on the basis of a careful and wise judgment, they are not entitled to conduct punitive wars (ibid.). The issue is not whether the judgments are correct (or likely to be correct) but whether they are the type of judgments that cannot be made unilaterally. While the judgment that a state has committed a wrong and that it should therefore be punished cannot be made by another state without undermining the independence of the former state, a judgment that is made by one state that its right to independence is violated by a present aggressor (in the case of defensive wars), or is violated by failing to return a territory that was wrongly acquired in the past (in the case of remedial wars), can be made without violating its right to independence.

To conclude while there are fundamental differences between self-defense and punitive wars, there are no fundamental differences between self-defense wars and remedial wars. In particular, like the right to conduct a defensive war, the right to conduct remedial wars follows inevitably from the observation that states are independent of each other.

4. Conclusion

The conclusion of this chapter is one I wish could be avoided as it conflicts with humanitarian intuitions. Very powerful consequentialist arguments justify endorsing a rule prohibiting remedial wars. It is also one that conflicts with our moral intuitions and, to a large extent, it also conflicts with current deeply established conventions. But, Ripstein rightly wishes to develop a theory of just wars that resists the temptation to yield to consequentialist considerations, intuitions, and conventions. Hence, he is bound by his own theoretical convictions to concede that states have a right to conduct remedial wars.

References

Gur-Arye, M., & Harel, A. (forthcoming). Taking Internationalism Seriously: Why International Criminal Law Matters. In Kevin Jon Heller et al. (eds.), *Oxford Handbook of International Criminal Law*. Oxford: Oxford University Press. Chap. 9.

Harel, A. (2011). *Why Law Matters*. Oxford: Oxford University Press.

McMahan, J. (2011). *Killing in War*. Oxford: Oxford University Press

Quingley, J. (2013). *The Six-Day War and Self Defense: Questioning the Legal Basis for Preventive War.* Cambridge: Cambridge University Press.

Ripstein A. (2021, forthcoming) *Kant and the Law of War*, Oxford. Oxford University Press.

Waldron, J. (1992). Superseding Historic Injustice. *Ethics* 103(4): pp. 4–28.

9

National Defense and the Value of Independence

Massimo Renzo

Two models are traditionally invoked in discussions of the morality of war.[1] The most popular, at least in philosophical circles, is the "just war model," which goes back to the writings of Aquinas and Augustin. This model offers a highly moralized understanding of the principles that regulate the use of force in war, in that it constructs war as an effort aimed at enforcing preexisting moral rights. Only by determining who holds such rights, and who is responsible for violating them, can we determine who has a just cause for taking part in hostilities and who hasn't, which in turn determines who is protected by belligerent rights and can comply with *jus in bello* principles. Anyone fighting on the unjust side cannot possibly respect *jus in bello* principles, according to the just war model, as *jus in bello* principles set constraints on how wars may be permissibly fought, and there is no permissible way to fight a war with an unjust cause.[2]

The just war theory model is traditionally contrasted with the so-called "regular war model," which instead treats *jus in bello* constraints as independent from *jus ad bellum* considerations. Within the regular war model (defended most notably by Grotius, Pufendorf, and Vattel), the same sets of rights and duties apply to all belligerents, independently of the justice of their cause. This is because, absent a superior authority with the power to adjudicate a dispute between states, there is no meaningful sense in which any of them can be said to have a just cause. Thus, war becomes the only way to decide the justice of the matter at hand.

This is a broad classification, of course, and like all classifications should not be interpreted too rigidly. But it has largely served its purpose, and with the caveat that the theories offered by each of the thinkers listed earlier are inevitably more nuanced than this dichotomy suggests, there is a general agreement about how each of them should be classified.

Kant's theory is the exception, Arthur Ripstein suggests. Kant's views cannot be easily classified within the traditional dichotomy because Kant offers a genuine alternative both to the just war model and the regular model—one that, according to Ripstein, avoids many of the problems each of them encounters. *Kant and the Laws of War* (Ripstein 2021) is a meticulous reconstruction of this model.[3]

[1] I'm grateful to Ester Herlin-Karnell, Enzo Rossi and, especially, Arthur Ripstein for helpful comments and stimulating discussions.

[2] For the "just war model," see McMahan (2009).

[3] References to this work are in brackets in the text.

Massimo Renzo, *National Defense and the Value of Independence* In: *The Public Uses of Coercion and Force.* Edited by: Ester Herlin-Karnell and Enzo Rossi, Oxford University Press. © Oxford University Press 2021. DOI: 10.1093/oso/9780197519103.003.0009

The book bears the marks of Ripstein's previous work on Kant: a rigorous, illuminating presentation of his ideas, whose aim is not merely philological or exegetical, but rather to show why and how those ideas can help us answer a number of central questions in legal and political philosophy. To achieve this, Ripstein successfully puts Kant's views in conversation with some of the most sophisticated positions that have emerged in the contemporary debate on war, from the vexed question of the relationship between *jus ad bellum* and *jus in bello* to the permissibility of humanitarian intervention to issues of jus post bellum.

In this chapter I will consider only one of these issues, namely Ripstein's treatment of wars of national defense. I choose this topic because, despite being treated both in common-sense morality and in international law, as the paradigm of wars that states are permitted to fight, wars of national defense are surprisingly difficult to justify within the predominant just war theory approach. Attempts by some of the most prominent philosophers working within this paradigm have either concluded that wars of national defense are impermissible (Norman 1995: chap. 5; Rodin 2014), or that they are permissible only in a restricted number of circumstances. If they are right, and if Ripstein's account does a better job in dealing with these cases, that would be a reason to favor the Kantian approach over the traditional just war model (assuming that we are reluctant to abandon the verdict of common-sense morality and international law on this issue).

I will argue that neither of these claims is correct. Ripstein's view ultimately lacks the resources to justify a right to wage wars of national defense, and contemporary just war theorists are wrong in believing that their approach can at best ground a weak version of such right. This gives us some reasons to prefer the just war approach over the Kantian one championed by Ripstein.

1. Wars of National Defense

A war of national defense is a war waged by a victim state (V) to defend itself against an aggressing state (A), whose aim is to acquire control over V's political institutions and subject V's members to its own authority.[4] As in the case of other wars, wars of national defense typically end up involving a certain amount of killing. But unlike other wars, they need not. This is because, ex hypothesis, A's goal in attacking V is only to subject V's members to its authority. Although A is ready to kill any members of V willing to resist, no one would be killed if V's members simply surrendered. If met with surrender, A's aggression would be bloodless.[5]

Why does the just war theory paradigm struggle with justifying wars of national defense? The reason is that defensive harm is subject to a proportionality constraint, which just war theory typically articulates in terms of the contrast between the harm

[4] Some include the preservation of other so-called "purely political goods," such as territory and natural resources or existing cultural values, in the justification for wars of national defense. See, for example, Moore (2014). For reasons of space, I will not discuss these other political goods here.

[5] Contrast these with what some call "genocidal wars" in which A's aim in attacking V is to kill V's members. See, for instance, Rodin (2014).

inflicted and the good consequences produced by a given defensive conduct. Roughly, resorting to defensive lethal force is permissible only in order to protect vital interests whose fulfillment is necessary to have a minimally decent life (cf. Rodin 2014: chap. 3; Frowe 2014). But on the face of it, A's aggression does not aim to threaten such interests. This is because insofar as A can secure its goal of subjecting V's members to its authority, the vital interests of such members, such as the interest in not being killed, tortured, or enslaved, are adequately protected. True, the interest in being enfranchised or in being part of one's own political community would be wrongfully set back, but these are deemed insufficiently important to justify resorting to lethal defensive force.[6]

The standard example presented in support of this view is one in which you are stopping me from voting in a general election, and the only way in which I can reach the polling station is by killing you (Hurka 2011; Frowe 2014: chap. 6; Lazar 2014; Fabre 2014: 15). Clearly, in this case I am not permitted to kill you to defend my interests in exercising my right to political participation or my interest in acting as a member of my political community. Thus, many conclude, such interests must be insufficiently weighty to justify resorting to lethal violence. And if these interests are not sufficiently weighty to justify killing the wrongdoer in the voting case, a fortiori they cannot be sufficiently weighty to justify V in waging a defensive war against A, given that wars always involve inflicting a significant amount of collateral harm on civilians and other innocent parties, in addition to the harm inflicted on those responsible for the unjust attack.

In other words, waging a defensive war against A is impermissible for the same reason why it would be impermissible for me to kill you to prevent you from getting me to vote. Although you do wrong me by doing so, the defensive harm required to prevent this wrong would be disproportionate to the good achieved by it.

Much ingenuity has gone into trying to resist this counterintuitive conclusion, but these attempts have fallen short. Existing accounts vindicate only heavily qualified versions of the right to wage wars of national defense, often limiting it to cases where other interests, such as personal bodily integrity or international security are threatened by it (Lazar 2014; Stilz 2014; Frowe 2014: ch. 5; Fabre 2014). Can the Kantian model articulated by Ripstein do better?

2. The Kantian Model

Ripstein identifies two pillars of the Kantian model. The first is Kant's view of the state: the only purpose for which states are permitted (indeed, obligated) to act, according to Kant, is providing a "rightful condition" for their citizens. What justifies the right of the state to make and enforce its laws on a given territory is that this is required for those living in that territory to be able to secure their independence from each other.

[6] An excellent collection of essays attempting to solve this problem is Lazar & Fabre (2014).

Each of us is his or her own master, and as such has the right to set his or her own plans. This right is innate, but only by living in a state can we enjoy the range of enforceable rights required to protect it (Ripstein 2021: chap. 2, section IV). This is why the moral power of the state to impose and enforce its laws over us is not incompatible with any of us being "sui iuris" (ibid.).

The second pillar is that absent a public international authority capable of adjudicating disputes among states, states lack the right to enforce their own judgments about the wrongs perpetrated by other states. This is not to say that anything goes and force can be used to determine who is right. Rather, the point is that disputes can be resolved on their merits only in a condition of peace. But peace can exist only if states accept the results of past disputes independently of their merits, refraining from reopening those disputes even if they genuinely believe that they were wrongly settled. "Peace precludes future wars by accepting the results of past ones" (Ripstein 2021: chap. 2, section VI).

In this sense, peace is a "constitutive principle" of war: peace is the guiding idea around which our theory of war should be built. Since peace cannot be achieved unless past disputes are accepted by all as settled, we're required to do so, even when they are settled by force, and independently of their merits. This is why remedial or punitive wars are impermissible. Only defensive wars are compatible with peace as a constitutive principle of war.

My focus will be on the first pillar, so let me elaborate on that. In particular, it's worth stressing that the notion of self-mastery (or autonomy) invoked by Kant is to be understood in "relational and contrastive terms" (Ripstein: chap. 2, section III). As Ripstein puts it, "the entire content of the idea that you are your own master is exhausted by the thought that no other person is your master, that you are entitled to be independent of another person's determining choice" (ibid).[7]

This notion (call it "independence") can be contrasted with the romantic notion of self-mastery (call it "personal self-determination"), according to which what matters is our capacity to act as self-determining agents, i.e., as agents in charge of shaping our life in accordance to plans we have chosen for ourselves. Here the focus is on the importance of being the authors of our life, rather than on not being subjected to someone else's choices. These two ideas are importantly related of course, since being subjected to someone else's choices normally prevents us from being the authors of our life. But one might not be subject to anyone else's choices without being the author of one's life. This might be the case if someone lacks the capacity to exercise her self-determining agency, or if someone has the capacity and fails to exercise it, by acting on whims or on the bases of impulses and inclinations that are somehow alien to him (Frankfurt 1988a, 1988b; Watson 1975). Such person would enjoy independence without enjoying personal self-determination.

[7] See also Ripstein (2009).

3. Ripstein's Account

How can the Kantian model justify the permissibility of waging wars of national defense? The reason why aggression is morally impermissible, according to Kant, is that "rightful relations between nations are like rightful relations between persons" (Ripstein 2021: chap. 2, section I). Just like individuals have a right to be independent, and thus free from interference in pursuing their own ends, so states have a right to be independent, and thus free from interference, in exercising their functions. Importantly, this is not because states are in any way similar to individuals, Ripstein observes. Kant does not subscribe to Michael Walzer's "domestic analogy" between individuals and states (ibid.). Rather, it's because the relationship between states as independent legal orders is similar to the relationship between individuals as independent agents (ibid.). "The relevant analogy [between human beings and states] concerns the relations between them, rather than the relata understood apart from them" (ibid.).

The reason why V is permitted to wage war of national defense against A, is thus that A has violated the terms of the rightful relation existing between A and V. V's permission mirrors the one that I have to use defensive force against you when you wrongfully attack me, which in turn is grounded in your violation of the terms of the rightful relation existing between you and me.

On the face of it, this approach has a significant advantage over the just war theory model with respect to the justification of defensive wars: insofar as Kant does not ground the right to wage these wars in the value of protecting certain interests, his view is not left with problem of explaining how infliction of lethal harm may ever by justified to defend allegedly minor interests, such as the interest in being enfranchised or the interest in being part of one's own political community. The claim that relationships between states are relevantly similar to those between individuals is far from obvious. Why should we accept it?

After all, that individuals have a right to pursue their projects without interference, and that this right may be protected by lethal force is intuitively plausible. Different moral theories disagree about what justifies such right, but the existence of the right is rarely challenged. By contrast, the claim that states have such right, and the claim that, even assuming that they do, the right may be protected by lethal force, are less obvious (as confirmed by the fact that many are inclined to question them). In light of this, why should we believe that there is analogy between the relationships between states and those between individuals? The easiest way to be persuaded that the relationship between *x* and *y* is similar to the relationship between *w* and *z* is that the pairs {*x*, *w*} and {*y*, *z*} are relevantly similar. Ripstein rejects this argument but I'm unsure as to what is meant to replace it.

One way to see the worry here is to consider who exactly is wronged when the right of states to be independent is interfered with. In one passage, Ripstein writes that "one legal order wrongs another by subordinating the other legal order to its own" (Ripstein 2021: chap. 2, section V). But it is not clear what it is for a legal order to be wronged. If by "legal order" we mean a set of institutions and norms, then the claim that a legal order can be wronged sounds puzzling. The claim is more plausible if it's understood as being about the members of the legal order, as some other passages in

Ripstein's book seem to suggest. But although this claim is more plausible, it raises the question of why the members of V would be wronged by A. Why should it matter to them whether the civic condition required to protect their independence is provided by a particular legal system rather than another?

Remember the first pillar of the Kantian model: although I have a claim against your interference with my independence, I lack the same claim against the state's interference because being subject to a political authority is necessary for each of us to enjoy our rights to independence. If this is correct,[8] why should it matter whether the political authority in question is *this one* rather than *that one*? On what grounds could the current members of V object if V's legal system, to which they are currently subject, is replaced by a different one that performs exactly the same task? Of course, they might object if the new system is not as effective as the previous one, or if the transition costs from one to the other are too high. But once we go down this road, we are discussing costs and benefits, and thus the answer to the question of whether interference is permissible can only be "it depends on the costs and benefits." This is emphatically not how Ripstein thinks we should frame the problem.

In defending Kant's view, Ripstein writes that "those who are in a rightful condition are entitled to remain in that rightful condition, which means that the citizens, considered as a collective body, have a right that outsiders not subvert their legal order" (ibid.). But my worry is that the Kantian framework is not robust enough to ground such right. Clearly the right to independence—the only pre-institutional right that individuals are said to have within that framework—cannot do the job. If it's in virtue of the vitally important functions performed by V that the right to independence of V's members is not threatened by their subjection to V's political authority, why should it be threatened by their subjection to A's authority, once we assume that A would be performing the same functions? As far as I can see, the Kantian approach lacks the resources to vindicate this conclusion. Insofar as A provides a rightful condition for V's members, it does not seem to violate their pre-institutional right any more than V does.

At the same time, A does not seem to violate any of the other rights possessed by V's members, i.e., any of those rights that can be fully enjoyed only within a civil condition (and whose existence is ultimately required for the right to independence to be protected). This is because, ex hypothesis, A would do an equally good job in implementing such rights in V. So, for example, instead of being protected by V's property law, the right to independence of V's members would be protected by A's property law. What complaint could they have against this change, provided that the bundle of rights typically associated with the existence of the rule of law is enforced?

Ripstein writes that "[V's] entitlement to remain in a rightful condition cannot be a question for any outside agency, because that would subordinate the rights of its inhabitants to outsiders" (ibid.). But notice that in the case we are discussing, A would not be deciding whether V's members remain in a rightful condition. A would only decide whether they remain in their current rightful condition rather than a different one. The question is: Why should that matter? This is what we need to explain

[8] I have some sympathy for this view, though I do not subscribe myself to the full Kantian picture. Cf. Renzo (2011).

to vindicate the view that wars of aggression that aim to replace an existing state with a new one are impermissible and the claim that lethal force may be used to prevent this wrong.

Ripstein's strongest argument is probably this: if A's aggression was successful, V would lack the capacity to provide closure with respect to disputes arising within its members, because any attempt to do would so be at the mercy of A's approval (ibid. 49, 61). Effectively, this would leave V's members without the capacity to securely enjoy their private rights. This is a genuine worry, but one that arises only if V becomes A's colony, in the sense that V persists as a subaltern political authority, whose laws and decisions are ultimately subject to A's approval. Importantly, the worry does not arise if A simply incorporates the members of V and treats them as its own citizens. In the latter case—call it "forcible annexation," to distinguish it from colonial domination[9]—the capacity of V's former members to securely enjoy their private rights is not threatened, because it would be protected by A. Their private rights would be protected in the same way in which the rights of A's members are protected by A.

So, I do think that Ripstein is right in claiming that states could not fulfill their public functions if they were not be able to provide closure, as well as in claiming that states would not be able to provide closure if they were subject to the authority of a different state. But while this would give V the right to use defensive force to resist colonial domination, it would not give V the right to do so in order to resist forcible annexation. If A annexes V, there is no insecurity among V's members in how they enjoy private rights (Ripstein 2021: chap. 2, section V). They do so in exactly the same way in which A's members do. Yet, intuitively V would have a right to defend itself against aggressions aiming at forcible annexation, and not simply against those aiming at colonial domination.

If I am right, the Kantian framework, built as it is on the idea that defensive force can only be justified by appealing to the value of independence (Ripstein 2021: chap. 2, section VI), is unable to vindicate the permissibility of waging wars of national defense in cases of forcible annexation (though it might be able to do so in cases of colonial aggression). A different framework, centered around a more robust notion of autonomy, is called for.

4. An Alternative

We have seen that the Kantian notion of independence can be contrasted with a more robust notion of autonomy (or "self-mastery," as Ripstein often refers to it). Instead of following Kant in articulating the value of autonomous agency in negative terms, as requiring merely non-subjection to someone else's will,[10] we can follow the romantics and conceive of autonomous agency as involving the choice and pursuit of a plan of

[9] The formulation of wars of national defense offered above is intentionally ambiguous between these two forms of domination. They both raise the question of whether a lethal defensive response is permissible. On the structural differences between the two, see Renzo (2019b).

[10] An alternative formulation of the Kantian view to the one favored by Ripstein is offered by Annie Stilz in this volume (Stilz 2021).

life we have chosen for ourselves. According to this view, we enjoy the value of autonomy when we get to shape our own life as we wish. Only when we successfully do that, can we rightfully see ourselves as the authors of our own life.

I believe this notion of autonomy as personal self-determination is more promising as a starting point to vindicate the permissibility of waging war of national defense. Here's a sketch of how the account would go.[11] To begin with, notice that our interest in choosing how to shape our own life is not limited to the choices we make when we act by ourselves,[12] but extends to the choices we make when we act together with others we are in a valuable relationship with. In the same way in which it matters to me that I get to pursue the career I choose for myself, as opposed to one that is forced on me, it matters to me that my wife and I get to raise our children as we choose. If you interfere with the latter choice, you wrong me for the same reason why you wrong me if you interfere with the former.

Of course, the two choices are importantly different. When I act by myself, my choices normally (though not always) consist in identifying specific courses of actions I favor: I choose that career, that lifestyle, or that companion. Things are different when I act together with others. In that case, often what matters to me is that the choices I make are such that they can be attributed to the whole group, and not just to me. So, for example, it matters to me that my wife and I decide together how to raise our children, i.e., that the final decision is not simply my decision, but the decision of the family I'm part of. Similarly, it matters to me that the decisions of my department are ones that my colleagues and I can see ourselves as making together, rather than as a decisions I make.

Of course, families, departments, and other groups make decisions in different ways. The point here is that in cases of collective action, what matters to me, at least insofar as autonomy goes, is not that the group acts as I wish, but rather that it acts in ways that allow its members (including me) to have an adequate input in a process of collective deliberation, so that they can see whatever the group does as an expression of their collective agency.[13]

Importantly, this means that you would wrong me if you were to interfere with the result of this process of collective deliberation, even if in doing so you would get the group to act as I personally desire. For example, you would wrong me if, knowing that my wife and I disagree about how to raise our children, you intervened to force us to act in the way I personally favor. Even if your interference leads my family to act in the way I prefer, I am wronged by you because my interest in deciding together with my wife how to raise our children has been undermined.

I believe that, despite important differences between political communities and the other groups I have mentioned, the right to make decisions together with the members of one's own political community can be justified along these lines. For those who value membership in their political community, one valuable way of exercising their

[11] I develop the account in Renzo (2018b). This section draws on that paper.

[12] Yes, we have switched to the language of interests. More on this below.

[13] This idea can be articulated in different ways. Probably, the most influential model is the one defended in List & Pettit (2011), but see also recent work on the agency of political communities, including Stilz (2016) and Moore (2015).

self-determining agency consists in making decisions together with the other members of the community in question. When this is the case, preventing them from doing so is wrong because it interferes with their interest in shaping their life as they wish. Thus, the demand to respect the capacity of the group to make decisions together (call this "political self-determination") is ultimately grounded in the demand to respect the self-determining agency of its members, which in turn is grounded in the value of their being able to shape their life as they wish.

Obviously here I am painting with a broad brush, but suppose we can get the account to work.[14] We would then be able to avoid the problem faced by the Kantian view. The problem, remember, was that the view seems unable to vindicate the conclusion that V is permitted to use lethal defensive force against A because, at least when A aims at annexing V and treating V's members as its own citizens (as opposed to colonizing V), V's rights to independence would be protected equally well under A's authority as it is under V's authority. If this is correct, independence cannot be what justifies the use of defensive force against A. The notion of personal self-determination, by contrast, seems well placed to do that. For once we accept that a component of what it takes to shape our life as we wish is that we get to make decisions together with others we are in a valuable relationship with, we can use this idea to explain why the interest that V's members have in shaping their life as they wish is compromised by A, whether V is colonized or forcibly annexed by it.

In cases of colonialism, the interest of V's members in making decisions together as a group is set back because, even if we assume that V's members do retain some capacity to determine their own laws and policies (as happens in cases of indirect colonialism), A has a significant input, through the imposition of its own laws and political institutions, in determining what those laws and policies should be. But the interest of V's members in making decisions together as a group is also set back in cases of forcible annexation because, even if there is a sense in which, once incorporated in A's political community, the former members of V would continue to make decisions together, clearly this is not the relevant sense of "making decisions together." For what we value is the capacity to make decisions together with the members of certain groups so that the decisions reflect the inputs of the members of the relevant group *and only of those members*.

Clearly, for example, my interests in making decisions together with the members of my department would be compromised (though perhaps justifiably so) if we are incorporated in your department, so that members of your department now have a say in what we should do. This is true even if, strictly speaking, in this scenario I would still be making decisions together with the members of my department; just not *only* with them.[15]

The advantage of appealing to personal self-determination, rather than independence, is that our interest in acting together with others we are in a valuable relationship with identifies specific individuals as non-fungible members of the relevant group within which we value exercising our self-determining agency. We don't value making decisions together with the members of *a* family; we value making decisions

[14] I consider the main problems of the account in Renzo (2018b).
[15] I develop this point in Renzo (2019b) and Renzo (2018a).

with the members of *our* family. We don't value making decisions together with the members of *a* department; we value making decisions together with the members of *our* department. Similarly, we don't value making decisions together with the members of *a* political community; we value making decisions together with the members of *our* political community. Thus, while the value of independence can be realized equally well whether one is subject to A's authority or V's, the value of personal self-determination cannot.

Notice that in abandoning the thinner notion of independence in favor of the richer notion of personal self-determination, we have also abandoned the idea that the wrongfulness of aggression is best understood by focusing on how certain rightful relations between states are undermined. The focus of the competing approach I am offering is rather on a particular interest that individuals have: the interest in shaping their life as they wish. My claim is that this interest is sufficiently important to generate a pre-institutional right against interference with valuable exercises of our self-determining agency, and that this right may be coercively enforced. Thus, we are back within the just war theory framework: we possess moral rights whose existence does not depend on the existence of special authorities or particular juridical relationships, and these rights determine who has a just cause for using lethal force and how these rights may be permissibly enforced.

But what about the problems encountered by just war theorists in justifying wars of national defense? The main issue was that the right to be part of a given state does not seem sufficiently weighty to justify resort to war. However, once we see this right as grounded in our interest in shaping our life as we wish, the idea that we may use lethal violence to enforce it doesn't seem implausible. After all, it seems clear that I may resort to defensive force to protect my right to act together with the members of certain groups, such as my family. If so, there is no reason why the right could not extend to cover the right to act together with the members of other valuable groups, such as the political community.

Of course, this is not to say that the two groups are equally valuable (they aren't). Nor is it to say that resort to force is always permissible to protect exercises of our self-determining agency, no matter how flimsy or insignificant. The permission to resort to lethal force is restricted in two ways. First, it is limited to exercises of our self-determining agency within spheres of conduct that are central to our capacity to live the life we have chosen for ourselves. For example, I may not resort to lethal force to prevent you from forcing me to dress more formally, even if in doing so you do interfere with my capacity to exercise my self-determining agency. Given the relative insignificance that matters of fashion have within the life I have chosen for myself, resort to lethal defensive force would be disproportionate in this case. For the same reason, the right to fight in a war of national defense does not extend to cover those for whom membership in a given particular community is not central to the life they have chosen for ourselves.

Second, the permission is limited in response to attacks that will significantly set back our capacity to exercise our self-determining agency within the relevant sphere of conduct. For example, although I may resort to lethal force to prevent you from keeping me apart from my family for years (given how important spending time with them is for the plan of life I have chosen for myself), I may not do so to prevent you from keeping me apart from me for an hour. This explains why resorting to lethal defensive force is not permissible in the voting case that has exercised just war theorists so much. Preventing me from voting has a negligible impact on my capacity to make decisions together with the members of my political community, insofar as I can engage in a range of other relevant activities such as taking part in public debates, assemble and associate with them, run for election, etc.

With these two conditions in place, the view I am offering avoids being unpalatably permissive and justifies resorting to lethal force only in what intuitively are the right cases. At the same time, the view provides a suitably robust account of the right to wage wars of national defense, in the sense that the right is not conditional on further vital interests being threatened. If I am right, this gives us reasons to stick with the traditional just war theory paradigm instead of following Ripstein down Kant's path.

References

Fabre, C. (2014). "Cosmopolitanism and Wars of Self-Defence." In Fabre, C., & Lazar, S. (eds), *The Morality of Defensive War*. Oxford: Oxford University Press.

Frankfurt, H. (1988a). "Freedom of the Will and the Concept of the Person." In *The Importance of What We Care About: Philosophical Essays*. Cambridge: Cambridge University Press.

Frankfurt, H. (1988b) "Identification and Wholeheartedness." In *The Importance of What We Care About: Philosophical Essays*. Cambridge: Cambridge University Press.

Frowe, H. (2014), *Defensive Killing*. Oxford: Oxford University Press.

Hurka, T. (2011). "Proportionality in the Morality of War." In *Drawing Morals: Essays in Ethical Theory*. Oxford: Oxford University Press, pp. 251–254.

Lazar, S. (2014) "National Defence, Self Defence, and the Problem of Political Aggression." In Lazar, S., & Fabre, C. (eds), *The Morality of Defensive War*. Oxford: Oxford University Press, pp. 10–38.

Lazar, S., & Fabre, C. (eds) (2014). *The Morality of Defensive War*. Oxford: Oxford University Press, pp. 10–38.

List, C., & Pettit, P. (2011). *Group Agency: The Possibility, Design, and Status of Corporate Agents*. Oxford: Oxford University Press.

McMahan, J. (2009). *Killing in War*. Oxford: Oxford University Press.

McMahan, J. (2014). "What Rights May Be Defended by Means of War?" In Lazar, S., & Fabre, C. (eds), *The Morality of Defensive War*. Oxford: Oxford University Press, pp. 116–156.

Moore, M. (2014). "Collective Self-Determination, Institutions of Justice, and Wars of National Defence." In Lazar, S., & Fabre, C. (eds), *The Morality of Defensive War*. Oxford: Oxford University Press, pp. 186–199.

Moore, M. (2015). *A Political Theory of Territory*. Oxford: Oxford University Press.

Norman, R. (1995). *Ethics, Killing, and War*. Cambridge: Cambridge University Press.

Renzo, M. (2011). "State Legitimacy and Self-Defence." *Law and Philosophy* Vol. 30(1): 575–601.

Renzo, M. (2018a). "Helping the Rebels." *Journal of Ethics and Social Philosophy* Vol. 13(3): 222–239.

Renzo, M. (2018b). "Political Self-Determination and Wars of National Defense." *Journal of Moral Philosophy* Vol. 15(6): 706–730.

Renzo, M. (2019a). "Revolution and Intervention." *Noûs* Vol. 54(1): 233–253.

Renzo, M. (2019b). "Why Colonialism Is Wrong." *Current Legal Problems* Vol. 72(1): 347–373.

Ripstein, A. (2009). *Force and Freedom: Kant's Legal and Political Philosophy*. Cambridge, MA: Harvard University Press.
Ripstein A. (2021). *Kant and the Law of war*. Manuscript.
Rodin, D. (2014). "The Myth of National Self-Defence." In Lazar, S., & Fabre, C. (eds), *The Morality of Defensive War*. Oxford: Oxford University Press, pp.70–89.
Stilz, A. (2014). "Territorial Rights and National Defence." In Lazar, S., & Fabre, C. (eds), *The Morality of Defensive War*. Oxford: Oxford University Press, pp. 204–227.
Stilz, A. (2016). "The Value of Self-Determination." In Sobel, D., Vallentyne, P., & Wall, S. (eds.), *Oxford Studies in Political Philosophy Vol. 2*. Oxford: Oxford University Press.
Stilz, A. (2021). "The Moral Basis of State Independence." This volume.
Watson, G. (1975). "Free Agency." *The Journal of Philosophy* Vol. 72 (April): 205–2020.

PART TWO

KANTIAN THEORIES OF WAR AND RELATED QUESTIONS

10

Exactitude and Indemonstrability in Kant's *Doctrine of Right*

Katrin Flikschuh

> A doctrine of right wants it to be determined (with mathematical exactitude) what belongs to each. (6:233)

> No one need be surprised that theoretical principles of outer mine and thine lose themselves in intelligible grounds. (6:252)

1. Introduction

It has become common to refer to Kant's *Doctrine of Right* as his legal theory, or his philosophy of law. I take it that a philosophy of law concerns itself with the origins, justification, and codification of law as an internally consistent, comprehensive, and complete (though dynamically developing) body of coercively enforceable rules designed to create and regulate permitted and prohibited forms of interaction between members within its antecedently specified domain of application. By "origins," I here mean the metaphysical and/or normative bases of positive law; by "codification," I mean the contextually informed specification of abstract metaphysical and normative principles into a system of positive laws. This is no doubt an inadequate specification of law and of the philosophy of law, but it will have to do for my purposes here. I am no expert in the field, and the following discussion of Kant's *Doctrine of Right* comes from a non-legal perspective. What most strikes me about approaches to the *Doctrine of Right* that speak of it as Kant's *legal* philosophy is their emphasis on clarity, precision and, perhaps most importantly, the determinability of the metaphysical and/or normative grounds of proposed positive laws. From the perspective of law, this may well be appropriate: given law's essentially self-referential coercive authority, its normative bases as well as the scope and conditions of its application had better be determinable "with mathematical exactitude." Even then, such demands for exactitude presumably refer to a regulative ideal that simply applies with special stringency: after all, being man-made, not even "the law" is infallible—indeed, perhaps one should say

Katrin Flikschuh, *Exactitude and Indemonstrability in Kant's* Doctrine of Right In: *The Public Uses of Coercion and Force*. Edited by: Ester Herlin-Karnell and Enzo Rossi, Oxford University Press. © Oxford University Press 2021.
DOI: 10.1093/oso/9780197519103.003.0010

that precisely *because* of its claim to preeminent authority, the law, being man-made, should be treated as *the least* infallible of all regulatory systems.[1]

I do not myself read the *Doctrine of Right* as primarily a work in legal philosophy. For one thing, the text is just as much about the origins (in the metaphysical sense) of statehood, and while the idea of the state comprises that of law it is also wider than that. I am thus more inclined to read the *Doctrine of Right* as a work of *political* philosophy, hence as a work that includes a concern with law but is not confined to that concern alone. I also believe, however, that the state is a less comprehensive idea for Kant than it is for some of his immediate predecessors and successors on the one hand and for most contemporary political philosophers on the other hand. The structure of the *Metaphysics of Morals* bears this out: whereas for Hobbes—and more so for Rousseau and Hegel—the state constitutes a comprehensive moral order, for Kant, it occupies but half the moral universe: the domain of virtue is conceivable outside the bounds of the state. Indeed, religious community arguably constitutes a third domain of essential human experience that is reducible to neither ethics nor law. The *Doctrine of Right* thus has a remit that is both wider than the philosophy of law and narrower than more comprehensive conceptions of political philosophy: it is concerned with the idea of the state and, correspondingly, with right as a distinctly associative political morality, but it is also the companion piece to the *Doctrine of Virtue*, which means that not everything of moral significance falls within the domain of law or that of political morality more generally (Brandt 1990; O'Neill 1996).

It is perhaps in any case inaccurate to suggest that current readers reduce the text to a philosophy of law. It may be more accurate to say that many bring philosophy of law desiderata to bear on their political readings of the work. There is, in other words, a tendency to assume that the degree of precision, clarity, and determinability that applies to the study of law extends to the whole of Kant's political thinking. In my view, this goes against the spirit of the text, which is often a good deal less determinate than recent interpretive trends suggest. A specific example may illustrate this. A number of years ago, when contemporary interest in the *Doctrine of Right* was still in its infancy, Reinhard Brandt (1882) cut an interpretive Gordian knot when he announced that it was plain for all to see (so long as they were sufficiently attentive) that the heart of the argument lay in the property chapters; more specifically, in paragraphs 1–6 of Chapter One, Part I. Brandt's immediate interest was the permissive law, or the "juridical postulate of practical reason," stated in paragraph 2 of the originally published version and later moved into paragraph 6 by Bernd Ludwig (1988) to replace the "missing"

[1] In "Towards Perpetual Peace" Kant says the following: "The exalted epithets often bestowed upon a ruler ("the divinely anointed," "the administrator of the divine will on earth") have frequently been censured as gross and dizzying flattery but, it seems to me, without grounds. Far from making the ruler of a country arrogant, they should rather humble him in his soul, if he is intelligent (as must be assumed) and considers that he takes on an office far too demanding for any human being—namely, the most sacred office that God has on earth, that of a trustee of the right of mankind—such that he must always be concerned about having in some ways offended against this "apple of God's eye." (8:353) Kant rather brilliantly recasts the divine right of rule as serving the ruler as a reminder of his own fallibility. It is the office of ruling that is "sacred" not its incumbent. Given human fallibility, any incumbent will necessarily be inadequate to the task of lawmaking. The wider point is to suggest that we cannot ultimately comprehend the a priori grounds of law which "lose themselves in the intelligible realm" (verlieren sich im Intelligiblen) (6:252).

deduction of intelligible possession as a synthetic a priori proposition of right. The details of Brandt's interpretation need not detain us here; the point is simply that Brandt identified as central to the entire text—correctly, in my view (cf. Flinkschuh 2000)—what are arguably the most obscure and confounding passages within it.

If one accepts Brandt's thesis about the centrality of the *lex permissiva*, the metaphysical grounds of property claims, and with them the grounds of the state, "lose themselves in intelligible grounds" (6:252): these grounds are ultimately indemonstrable. On reflection, this is not surprising, given that the idea on which these claims are based—freedom—is itself incapable of a theoretical deduction. It turns out, then, that we have only very limited insight into the grounds of our rights claims against one another: strictly speaking, we cannot *know* whether we are rights-holding agents.[2] Recent readings of the *Doctrine of Right* evince little interest in Kant's metaphysical caution. To the contrary, the current trend is to start from Kant's—strangely positioned—statement of innate right as unambiguously affirming a kind of original or natural right to external freedom of each and then to derive the entire system of right from this supposedly self-evident rights premise.[3] It seems to me that much of the philosophical nuance in Kant's political thinking is lost in this way.

In what follows, I shall consider neither the innate right nor the *lex permissiva*. Instead, I shall examine two recent approaches to Kant's position on the law of nations, where his ambivalence between a world state, a coercive federation, and a free league presents another long-standing interpretive puzzle. The first approach which I shall consider is that of Joachim Hruschka, the second that of Arthur Ripstein. Both interpret Kant from a philosophy of law perspective; both argue that Kant's position on the right of nations is clear and unambiguous, at least in the *Doctrine of Right*. Yet both also reach conflicting conclusions over Kant's final position on the matter. It is not my intention here to berate Hruschka or Ripstein for their philosophy of law perspective; to the contrary, I engage with them because I think them among the philosophically most astute and systematically rigorous exponents of Kant's legal and political thought. Nonetheless, I want to suggest that even in the *Doctrine of Right*—Kant's most mature political work—his position on the right of nations is less settled than either account claims. In one sense, this is of course borne out by the very fact *that* Hruschka and Ripstein reach textually well-supported yet conflicting conclusions: given that they do, claims that the law's precision and exactitude may be sustainable only against a background acknowledgment of human finitude and fallibility (O'Neill 1992). As we shall see, my general impression is that Ripstein's account in particular does in fact recognize the ultimate indemonstrability of the metaphysical grounds of law—or comes at any rate very close to doing so.

[2] Much then depends, of course, on how one goes on to interpret the idea of freedom as a "fact of reason"—whether one takes it to have the status of a fact from the perspective of practical reason, hence as "as good as provable" practically speaking, or whether one takes it to an inscrutable practical given whose validity for us we have to accept on good faith. See Beck (1965:200–214).

[3] Prominent recent examples include Hodgson (2010) and Pallikkathayil (2010).

2. Kant's Sovereignty Dilemma

Kant's dithering over the requisite institutional form of rights relations between states is well known: between *Universal History* and *Doctrine of Right* he entertains various conflicting proposals that range from the idea of a world state to a coercive federation of states to a free league and, finally, to a permanent yet always dissolvable congress of states. The tension between some of these alternatives is brought to a head in *Perpetual Peace* in the context of the discussion of the second definitive article: "The right of nations shall be based on a federation of free states" (8:354). In the back and forth between two chief options on the table—coercive federation or free league—Kant initially rejects the former as contrary to the idea of a right of nations: a supranational coercive authority would entail the "melting together" ("Zusammenschmelzen") of the plurality of states presupposed by the very idea of a right of *nations*. Besides, and unlike individual persons in the natural condition, states "already possess an internally rightful constitution, so have outgrown coercion by others" (8:355–856). The free league—a voluntary federation without overarching coercive authority—is thus proposed as alternative analogue to the coercive civil condition for individuals. Yet in the final paragraphs of the section, Kant changes tack: practical reason is said to demand a world state (Völkerstaat) after all, and it is only because states won't do as they ought, that is, submit to a coercive public authority, that the free league of states remains as "negative surrogate" to the theoretically correct idea of the world state.

Kant's ambivalence in *Perpetual Peace* has generated a large literature on what he says and should have said.[4] His position in the *Doctrine of Right* is different. In stark contrast to *Perpetual Peace*, he there appears reconciled to the idea of a right of war—at least on the presumption, as both Hruschka and Ripstein emphasize, that wars are so many attempts to establish a lasting peace.[5] In the *Doctrine of Right*, discussion of the institutional form which a right of nations should take is conducted against background acceptance of the principle that states use war as a means to obtain what they believe to be theirs by right. Hruschka and Ripstein both treat the *Doctrine of Right* as Kant's final word on the right of nations—both take the text to resolve any outstanding ambiguities in *Perpetual Peace*. However, they disagree about what Kant's final positions amounts to. While Hruschka envisages the gradual establishment of a coercive federation, Ripstein endorses a free league. In one sense, then, the interpretive disagreement simply continues. As noted, however, what interests me here is not the interpretive disagreement itself so much as the claim on both sides that, from a *systematic* perspective, Kant's position in the *Doctrine of Right* is unambiguous. This leaves three possibilities: either Hruschka is wrong about the coercive federation, or Ripstein is wrong about the free league, or both are wrong in thinking Kant's position systematically unambiguous. My own money is on the third possibility. This then raises two further questions: Why is Kant finding it so difficult to settle the matter decisively, and

[4] See the collection of essays in Hoeffe (1995) and more recently Niesen & Eberle (2011).

[5] I am not sure that I do in fact share this reading. The chief reason why Kant endorses rights of war in the *Doctrine of Right* seems to me to be the fact that he believes this to be the only possible way in which states, as juridically sovereign entities, can settle rights disputes between them.

why do we find it so difficult in turn to accept that he does not, and perhaps cannot settle it?

3. The Right of Nations as a Coercive Federation

Let me turn first to Hruschka's strikingly imaginative attempt to resolve Kant's ambivalence between a free league or coercive federation. According to Hruschka, Kant's position on the right of nations is in fact settled from "Theory and Practice" onward. There, Kant demands "a right of nations, based on public laws accompanied by power to which each state would have to submit" (8:312). Hruschka claims that "Kant holds fast to this demand in 'Perpetual Peace' as well as the *Doctrine of Right*" (Hruschka 2015).[6] The only difference between the two texts is that while *Perpetual Peace* rejects the *compulsion* of states into a coercive federation, the *Doctrine of Right* endorses it. This difference is explained by Hruschka as a merely temporary departure on Kant's part from a thoroughgoing analogy between individual persons and states in the natural condition. Thus, while *Perpetual Peace* temporarily diagnoses a constitutive difference between individual persons who can be externally compelled and states who cannot, the *Doctrine of Right* reverts to Kant's position in "Theory and Practice" by endorsing war as a means to peace. Hruschka believes this endorsement to include the right of one state to compel another into a coercive federation with it. The problem of rightful compulsion that beset *Perpetual Peace* is thus resolved in the *Doctrine of Right* by way of reverting to a right to war.

In his reading of *Perpetual Peace*, Hruschka introduces an important terminological distinction on which he draws in his subsequent analysis of the *Doctrine of Right*. I shall spend some time on this, as it strikes me as a highly significant move. Hruschka works with the German terms, "Völkerstaat" and "Völkerbund." These translate into English as "state of states" and "league of states" respectively. The first of these—a state of states—is often equivalently referred to as a "coercive federation." A "state of states" envisages the imposition of a coercive authority over and above that of states—this distinguishes it from a non-coercive "league of states," which represents an association of states with joint decision-making procedures absent a coercive, supra-state authority. It is common in the literature to treat a coercive federation and a free league as opposites. Against this, Hruschka avers that Kant treats "league of states" as a higher-order concept (Gattunsbegriff) of which a free league and a coercive federation are two possible variants. The two concepts are thus not opposites but different species under the same genus. On Hruschka's account, therefore, when Kant speaks of a league of states in general, he could have in mind either a coercive federation or what Hruschka calls a "mere" league of nations (blosser Völkerbund)—where "mere" indicates the absence of a coercive authority.

Hruschka concedes that the terminological distinction between "mere league" and "coercive league" is his own. He claims, however, that unless we introduce the distinction, the following sentence in *Perpetual Peace* is unintelligible: "This would

[6] All translations of quotes from Hruschka's article are mine.

be a league of nations, which need not, however, be a state of nations" (8:354 /130) (Dies wäre ein Voelkerbund, der aber gleichwohl kein Völkerstaat sein müsste.) Now I happen to think that, while it is open to different interpretations, the original sentence poses no *intelligibility* problem: Kant is either saying that a league of nations does not *have* to take the form of a state of nations—this is Hruschka's preferred reading. Alternatively, Kant is saying that a league of nations is not to be *confused with* a state of nations—this is the reading I myself endorse. In short, there may be ambiguity, but there is no unintelligibility.[7]

As noted, according to Hruschka, to say that a league of nations *need not* be a state of nations is to say that a league of nations *can be* a state of nations. A state of nations is then not the opposite of a league of nations but simply one possible variant of it.[8] This not the first time Hruschka has introduced a terminological distinction of this type. In an earlier article on the permissive law, Hruschka similarly distinguishes between a "permitted action" and a "merely permitted action" (Hruschka 2004). There, the claim is that while a permitted action may also be required—the mother is both permitted to feed her child and morally and legally required to do so—a "merely permitted action" is permitted without being required: take it or leave it. This distinction allows Hruschka to class the act of acquisition as authorized by the *lex permiss*iva as merely permitted, hence as not required. Acquisition is then a liberty or a morally indifferent act (which does of course raise the question as to why a law is required at all in relation to it). Again, the terminological distinction is his own, not Kant's.

Although, in the context of the right of nations, Hruschka ostensibly introduces the notion of a "mere league" in order to clarify the meaning of a particular sentence in *Perpetual Peace*, he redeploys the distinction between "league" and "mere league" when he turns from *Perpetual Peace* to the *Doctrine of Right*. There, in paragraph 54, Kant speaks of the demand for "a league of nations in accordance with the idea of an original contract." (6:344) The envisaged league is said to have this peculiarity that it "involves no sovereign authority (as in the civil constitution), but only an association (federation)." This non-coercive league can be both renounced and renewed at any time; Kant explains that its optional dissolution "is a right in subsidium of another

[7] As Enzo Rossi has helpfully pointed out to me, the source of ambiguity lies in the possibility of interpreting "müsste" in terms of either deontic or conceptual necessity. Immediately following the sentences quoted by Hruschka, Kant in fact goes on to say that "in this there would be a contradiction: because each state contains within it the relation of a superior (legislator) to an inferior (those of obey, namely the people), whereas many nations in one state would make only one people—which contradicts the presupposition [of a right of nations]" (8:354). At this point in the argument Kant thus in fact rejects the idea of a state of states as contradicting the idea of a right of nations; so he is not in fact treating a coercive federation as one possible variant of a league for states.

[8] One question that arises is whether, for Hruschka, a coercive league, i.e., a state of nations, is equivalent to a world state. The latter is entertained by Kant in some of his early writings though in "Perpetual Peace" where he rejects the idea of a world state on the grounds that it would issue in a "soulless despotism" (8:367). Given Hruschka's view of "Perpetual Peace" as a temporary departure from Kant's settled view, he might indeed regard world state and a state of nations as equivalent. I think it more likely, however, that Hruschka has in mind the US model of divided sovereignty, according to which federated states cede some sovereign powers to the federal government while retaining sovereign powers with regard to state-internal issues. My own view is that for Kant sovereignty is indivisible so that for him a state of nations is in fact equivalent to a world state.

and original right, to avoid getting involved in a state of actual war among the other members." (6:345) It looks then, that the right to withdraw from the league—thereby dissolving its original constitution—tracks an original right of each state to decide for itself when to go to war. The "subsidium" of the original right—the right of withdrawal from the league—thus protects a state from being drawn into a war by associated states against its own choosing.

Under Hruschka's distinction, however, the mentioned "right held in subsidium of another, original right" takes on a rather different meaning. According to Hruschka, a federation without coercive authority is a "mere league." It is this "mere league" that is introduced in "subsidium of" the "original right." Hruschka somewhat oddly takes the "original right" to refer to the right of a state to be in a civil condition with other states. More specifically, for Hruschka, the original right to war of each state is a right to compel all other states into a coercive federation with it. To say that entrance into a free league is a right held "in subsidium" of the original right to be in a civil condition, (Kleingeld 2004), is thus to treat a free or "mere" league as a deficient form of a "proper," that is, coercive league.

As with Hruschka's original distinction between "permission" and "mere permission" in relation to the *lex permissiva*, the word "mere" does much of the heavy lifting. "Mere" (bloss) is a normatively loaded term that invokes an absence or a deficiency. A "mere federation" is one that is characterized by an absence of coercive authority, implying that a more fully developed federation would not possess this deficiency. For Hruschka (2015:134), therefore, Kant's reference to the alliance as "a subsidium" of an original right indicates that the non-coercive alliance is a deficient version of a coercive one. The free league functions as a stop-gap until the fully coercive version of a league is realized: "If the right of the mere league has a merely subsidiary function, this means that the primary or original right is the right of a state of states [i.e., a coercive federation]."

Why all the hair-splitting between "mere" and "proper" league? Well, it allows Hruschka to claim that any interpreter who is puzzled by Kant's apparent dithering between either a free league or a coercive federation simply fails to notice the forensic precision with which Kant specifies, lawyer-like, his terminology. On Hruschka's account, there is no equivocation on Kant's part: a free league is a mere league and a coercive federation is a proper league. Kant could not have been clearer—the impression of ambivalence or equivocation on the part of the reader is a consequence of the reader's inattentiveness to Kant's precise terminological distinctions. (Again, though, bear in mind, that the distinction between "coercive league" and "mere league" is in fact introduced by Hruschka himself.)

4. The Right of Nations as a Free League

Ripstein, too, believes Kant's final position in the *Doctrine of Right* to be unambiguous. In contrast to Hruschka, however, Ripstein holds that position to be broadly consonant with that of *Perpetual Peace*. This is not surprising, given that Ripstein takes the *Doctrine of Right* to affirm a free league and, indeed, takes Kant explicitly to *reject* a coercive federation on principled grounds. Of course, Ripstein faces the potential difficulty of explaining why, having categorically ruled out a *right* to war in *Perpetual Peace*, Kant reinstates such a right in the *Doctrine of Right*. That said, transitional wars of self-defense are not ruled out in *Perpetual Peace* either. Still, the *Doctrine of Right* explicitly acknowledges a state's right to war as the only possible *juridical* means to settle disputes among sovereign entities. It is doubtful that *Perpetual Peace* similarly regards war as a juridical means rather than as a permission to violate laws of peace during the transitional phase from the *status quo* to a condition of international right. Although, then, Ripstein takes Kant's position on the issue of the institutional form of the right of nations to be unchanged between the two texts—both times, Kant is said to endorse a free league—Kant's reasons for endorsing a free league may differ between the two texts. For reasons of space, I note this possibility here without discussing it further.

So far as the *Doctrine of Right* is concerned, Ripstein believes Kant's endorsement of a free league to flow systematically from the three-fold distinction between the right of a state, the right of nations, and cosmopolitan right. Ripstein believes that distinction to be governed by the overarching general principle of freedom as independence. On Ripstein's account, freedom as independence—that is, the right not to be subject to another's power of choice—constitutes the master principle of Kant's entire philosophy of Right. If we apply this principle consistently across persons and states, we will find that there is no breakdown in the analogy which Kant draws between the state of nature among persons on the one hand and that between states on the other hand. What is at stake both times is the rightful claim to independence: just as individual persons in the state of nature pose a threat to each other's independence merely in virtue of their coexistence, so states threaten each other's independence merely in virtue of coexisting side by side. Both times the conflict is unavoidable in the sense that it is not ill will that puts persons and states in conflict with each other but the structural problem of mutual interference with one another's choices in the absence of law-governed interaction. Both times the relevant solution is entrance into a public condition governed by laws that apply to all equally.

The pertinent difference between the two scenarios lies in the fact that while in the case of individual persons the independence of each is *secured* through common submission under a coercive law, in the case of states the coercive imposition of law would *destroy* independence. States' independence consists in their juridical capacity to safeguard their citizens' rights.[9] Where a state is deprived of its constitutive lawmaking and—enforcing authorities, it loses the ability to safeguard its citizens' rights. But to compel a state into a supranational coercive federation is precisely to deny it its independent lawmaking capacity.[10] It follows that the solution to the state of nature between states cannot mirror that of the state of nature between private persons. At

[9] See also Flinkschuh (2010).

[10] In pursuing this line of analysis Ripstein in effect commits to a conception of sovereignty as non-divisible. On the divided sovereignty model, it would remain perfectly possible for states to continue to protect their citizens' rights even while decision-making powers relating to international relations are ceded to a supranational government. Proponents of the non-divisibility view, which I myself share, generally subscribe to the idea of a "supreme" authority; according to them, governmental decisions are separable but interlinked: the decision to go to war, for example, will invariably affect citizens' rights. Which sovereign competence then has supreme decision-making powers when areas of competence intersect or overlap?

an international level, the appropriate institution is a free league which places states under obligations of law without imposing a coercive enforcement authority above them: "The right of the state contrasts with the right of nations because the former concerns the rightful basis of a public authority in relation to those over whom its power is exercised, while the latter specifies the conditions of independence of a plurality of public authorities against each other" (Ripstein 2021: chap. 9, section IV).

As the earlier quote indicates, Ripstein works from both ends: on the one hand, states cannot be compelled without losing their rights-securing function vis-à-vis their citizens; on the other hand, the obligations of international law uphold these functions vis-à-vis other states. If the substantive content of international law is to protect each state's juridical independence from encroachments upon it by other states, then the institutional form of international law cannot be such as to deprive states of their juridical independence by compelling them into a coercive federation. To the contrary, "the structure [of international law] is simply one of upholding respective independence" (Ripstein 2021: chap. 3). In contrast to domestic law, international law thus relies on self-enforcement. Ripstein's point here is conceptual: a state, as an entity that claims juridical independence from other such entities, must in consistency extend recognition of *their* juridical independence to those other states and must comport itself in conformity with that recognition. It is part of the very logic of statehood that states recognize themselves and other states as bound by the principle of reciprocal respect for independence. This is why, for Ripstein, the non-coercive and customary nature of international law does not mark it out as a deficient form of law. To the contrary, the self-enforcing nature of international law is appropriate for the kind of entity that is subject to this type of law: "the law of nations is not voluntary. The mandatory rules are given by each nation's right to independence and territorial integrity." (Ripstein 2021: chap 9, section IV)[11]

For Ripstein, the right to defensive war similarly follows from the right to independence as master principle of Kant's legal philosophy. Any non-provoked act of war by one state against another constitutes an encroachment on the latter's independence. The right to independence includes a state's defensive right to resist such encroachments. By the same token, self-defense is the *only* legitimate ground of war. Any war waged by one state against another on some other ground—say, to compel the other to enter into a coercive league—would be inconsistent with each state's right to make up its own mind about these issues. No state can be the judge of another state's actions—this is why punitive wars are prohibited. While for Hruschka, the right to war includes the right to compel other states into a coercive federation, for Ripstein, such compulsion would constitute an act of aggression against others states' independence.

[11] Ripstein in fact strikes me as more ambivalent on this point than here suggested. Although he insists that the rules of international law are mandatory and that "no state can undertake to abandon the rightful condition and descend into barbarism" (ibid. chap. 9), he also concedes that "the lack of unitary enforcement mechanism and the perils of institutional corruption entail, unsurprisingly, that the international legal order will inevitably be defective in relation to its own ideal principles." (ibid.) Ripstein here suggests that there will inevitably be a gap between theory and practice of international law. Yet the absence of a unitary enforcement mechanism belongs to the level of "ideal principles" in Ripstein's account—so the deficiency of international law he mentions is constitutive of it, not an aspect of its imperfect implementation.

In sum, for Ripstein the principle of independence applies to relations between persons and states alike. The principle generates norms of interaction that are coercively enforceable at the individual level but not at the state level. At the state level, self-enforcement is mandatory because it is integral to the logic of sovereign statehood. While for Hruschka, Kant's endorsement of a coercive federation follows unambiguously from an appreciation of the implicit distinction between "mere" and "proper" league, for Ripstein, Kant's commitment to a free league of nations follows just as unambiguously from the idea of freedom as independence that ranges as master principle over all three forms of public right. Neither interpretation leaves room of ambiguity or ambivalence, yet both reach diametrically opposed conclusions with regard to Kant's final position in the *Doctrine of Right.*

5. Kant's Negative Endorsement of a Free League

As noted, I am not interested in the substantive differences between Hruschka and Ripstein as to the choice between free league and coercive federation—what interests me is each of their insistence, despite opposing conclusions, on the non-ambivalence of Kant's final position. Again, I take the two approaches to Kant's account of international right to be illustrative of a more general interpretive tendency that emphasizes the perspicuity of Kant's philosophy of right throughout. Ripstein repeatedly invokes a legal ideal in this regard: the law both demands and provides "closure." By this, he means that any actual legal system must be able to cover all cases falling within its domain and must authoritatively resolve all disputes brought before it. This is surely right so far as positive law goes. But can or should we expect closure in relation to a metaphysics of right? My impression is that both Hruschka and Ripstein take the view that we require closure at the metaphysical level so as to be confident about closure at the level of positive lawmaking: without certainty as to the a priori grounds of law, what assurance can we have that judgments in positive law correctly track the standards set by those grounds? My own view is that closure at the level of positive lawmaking does not require closure at the level of metaphysical grounding. To the contrary, the very fact that closure is required at the level of positive lawmaking may render lack of closure at the metaphysical level no bad thing: we precisely cannot be sure of positive law's juridical infallibility. To the contrary, faultless miscarriages of justice are not uncommon—faultless in the sense that substantive error can occur even despite due diligence and procedural correctness. One might say that the proposition that the law must determine what belongs to each with mathematical exactitude is defensible only against a background acknowledgment of law's permanent fallibility. And this acknowledgment of permanent fallibility—and hence revisability—may not be possible absent acknowledged lack of metaphysical closure: absent, that is, understanding of the ultimate inscrutability of the a priori grounds of right.

I shall return below to these more general considerations about the relation between closure at the positive level and openness at the metaphysical level. Before doing so I should clarify where I myself stand with regard to Kant on international right. Substantively, my interpretive position is close to that of Ripstein: I believe (cf.

Flinkschuh, 2010) that Kant endorses a free league over a coercive federation both in *Perpetual Peace* and in the *Doctrine of Right*. I take Kant's position to be more settled in the later text than the earlier one. In *Perpetual Peace*, Kant remains openly ambivalent, favoring either a free league or a coercive federation at different points in his argument. Even in *Perpetual Peace*, however, the argument regarding states' juridical function as rights-enforcers vis-à-vis their own citizens finally sways the decision in favor of a free league. In the *Doctrine of Right*, the juridical nature of state sovereignty is affirmed much more decisively, and this speaks once more in favor of a free league.[12] Kant's further acceptance of war as the only possible *juridical* means for the resolution of disputes between sovereign states further confirms his rejection of a coercive league.

I disagree with Ripstein over the reasons for Kant's endorsement of a free league. For Ripstein, recall, the principle of independence governs relations between states as much as it governs relations between persons. I do not myself believe the innate right to play a foundational role relative to state entrance among persons; nor, therefore, do I think that the analogy between persons and states appeals to independence as an overarching master principle.[13] To the contrary, my view is that Kant endorses a free league among states because the alternative—a coercive federation—is unavailable to him. More specifically, Kant cannot endorse a coercive federation because, as enforcers of citizens' rights, states cannot have a higher coercive authority above them: in relation to rights, states already are the highest authority. To say that Kant endorses a free league because he cannot endorse a coercive federation is not to say that the reasons for endorsing a free league are conclusive; it is at most to say that the reason for not endorsing a coercive federation are conclusive.

The problem with endorsing a free league is that it renders rights relations between states non-coercible. This is not a small problem: in the Introduction to the *Doctrine of Right*, Kant argues that the concept of coercion is "contained in" the general concept of right. Right is thus intrinsically coercive—that is, it is part of the very concept of right that rights are coercively enforceable. The reliance on self-enforcement at the level of international right is contrary to Kant's conceptual point about rights. Assuming that no one can compel himself, the demand that states comply with the demands of international right falls considerably short of coercible rights obligations. We saw that Ripstein emphasizes the mandatory nature of international law: indeed, on his account, the obligations of international law follow from reciprocal acknowledgment of each state's rightful claim to independence. However, mandatory law falls short of being coercible law. Moral laws are mandatory without being coercible, for example. Similarly, to say that international law is mandatory for states, such that they ought to follow it, falls short of saying that international law is coercible. Yet for Kant, "there is connected with right by the principle of contradiction an authorization to coerce someone who infringes upon it." (6:231).

In sum, on my reading of Kant's final position, his endorsement of a free league follows from his rejection of a coercive federation. Yet far from being systematically

[12] Again, I must reiterate that I am an adherent of the indivisibility of sovereignty view.

[13] It is in any case not clear to me that state sovereignty is easily equated with independence understood in terms of the capacity for self-determined choice. Sovereignty is surely more comprehensive than that.

unproblematic, the endorsement of a free league incurs the problem of the non-enforceability of intrinsically enforceable rights relations. Significantly, Kant ends up where he does not as a result of any faulty reasoning on his part. To the contrary, his reasoning is rigorous: the unenforceability of rights relations between states follows from the enforceability of rights relations between persons. It is, as it were, *because* rights relations between persons *are* coercible that rights relations between states *cannot* be coercible. And yet this yields the paradoxical conclusion that rights relations both are and are not coercively enforceable.

6. Law's Indemonstrable Origins

It turns out that international right, and the morality of right in general, is deeply problematic—problematic in the sense that a full understanding of the possibility of systematic rights relations across persons and states seems to elude us. This does not render systematic exploration futile. However, it does render the inquiry perplexing at many junctures. As noted earlier, the conflict between free league and coercive federation is merely one example within Kant's philosophy of right in which we seem to encounter, if not paradox, then deadlock or perplexity. Other well-known sources of perplexity include the relation between right and virtue, with the appeal to incentives in the one domain and their strict impermissibility in the other. They include the mentioned *lex permissiva* which, at least on one reading, authorizes a prohibited action so as to make rights relations practically possible. Finally, they include Kant's ambivalent position vis-à-vis nomadic peoples' whom he appears to exempt from a coercible duty of state entrance the a priori status of which he affirms elsewhere in the text.

In his detailed reconstruction of Kant's position on law and war, Ripstein himself repeatedly returns to the seeming aporia between Kant's rejection of war as "barbaric" and thus as unlawful, and his endorsement of rules of right conduct in war. This too is perplexing: how can something which is by its very nature unlawful nonetheless be subject to regulation by law—is this not akin to devising rules of right conduct for the mafia, thereby legitimizing the latter? Ripstein emphasizes that Kant leaves room for the possibility of lawful conduct even under conditions of general lawlessness. In one sense, the appeal may simply be to the separability of actions and hence of rights and wrongs: while it is wrong to wage non-defensive war, the commission of perfidy within war is a separate and additional wrong. At times, however, Ripstein points to something deeper. He suggests that even in war the possibility of peace remains: war without perfidy is closer to peace than war with perfidy. Here, there is the thought of the gradually civilizing function of law. While for now only perfidy and attack on non-combatants may be ruled out as unlawful, the very fact that they are makes an upward trend possible: it is possible that, in time, rights in war will give way to the cessation of war itself. According to this line of thought, the regulation by law of an unlawful, that is, "barbaric" activity such as war has a civilizing effect on that activity.

But there is a further thought; it concerns the law's arbitrary beginnings in time. This thought is distinct both from the separability of discrete rights and wrongs and from law's general "civilizing" role—it gets us back to the idea of law's self-referential authority. Consider Ripstein's frequent reminders that for law to be able to take authoritative hold, some antecedent *status quo* that is necessarily non-lawful must be accepted as the law's departing baseline. For example, a given distribution of land may obtain prior to its regulation by law that leaves some better off than others. Unless that distribution is nonetheless accepted as beyond dispute by all concerned, law cannot begin to take effect for the future.[14] Similarly, two warring states might be involved in a protracted territorial dispute. According to Ripstein (2021: chap. 8, section II), "the end of acquiring territory could never justify the use of war, but the fact that territory has been incorporated into a nation is sufficient to make it part of that territory." Unless the vanquished party accepts the war's outcome and agrees terms on that basis, peace will remain elusive. The general thought appears to be that for law to be able to take effect, past disputes must be accepted as beyond possible resolution by law.[15]

Note that this insight into law's necessary beginnings in lawlessness stands in some tension with Ripstein's commitment to independence as the relevant substantive criterion for law's rightfulness. If law's assigned function is to secure the innate right to independence of each, it is hard to see why parties should accept anything other than equal independence as the relevant baseline. In other words, no one who, in the *status quo ante*, lacks equal independence should have to accept the *status quo ante* as the relevant baseline for law's beginnings. I nonetheless think Ripstein right in suggesting that, necessarily, Kantian law-governedness emerges from lawlessness. However, given that Ripstein himself is committed to the right of each to independence as the decisive criterion of law's rightfulness, his remarks concerning law's beginnings take on a pragmatic flavor: ideally, the law departs from and secures the equal right to independence of each, but given that we have to start instituting the law at some point, we might as well start at any point—no matter how unequal the antecedent distribution of independence across persons. If, by contrast, we shed the idea of an independent criterion of rightfulness, the thought of law as originating in lawlessness takes on metaphysical significance. Kant then appears to be saying that, necessarily, justice—rightful law—emerges from lawlessness. I believe this latter to be Kant's position. My point is not that from injustice justice necessarily emerges; it is evidently not the case that justice cannot fail to come about (6:307–308). Rather, the thought is that to the extent to which justice (rightful law) does come about, it necessarily emerges out of lawlessness.

7. Law and Practical Political Judgment

What is the significance of saying that, necessarily, rightful law emerges from lawlessness? I take its significance to be twofold. Along the theoretical dimension, the claim echoes Kant's remarks regarding our final incomprehensibility of the a priori grounds of freedom and therefore of morality, including the morality of right; these grounds "lose themselves in the intelligible realm." From a theoretical perspective,

14 For a similar view, see Waldron (1992).

15 As Ripstein points out, Kant's argument against revolution has this general structure.

the possibility of law's emergence is ultimately unfathomable. As noted earlier, this does not show inquiry into the a priori grounds of law to be futile—it is after all only the inquiry itself that shows them to be ultimately unfathomable for us. This insight into the limits of human understanding is a substantial one, moreover: to understand one's limited understanding of a given domain of inquiry is to understand that domain a good deal better than one would absent the inquiry. Along the practical dimension, insight into theoretical indemonstrability secures agential freedom. This is a well-worn Kantian point—it is precisely because we cannot *know* that we are required to exercise *judgment*. By the same token, from a practical perspective, the future is essentially open: the possibility of law's emergence from lawlessness is never closed off; the first step from lawlessness toward law can always be taken and thus remains a permanent possibility for us.

The practical idea of law's emergence from lawlessness emerges most clearly in Kant's political teleology, which in effect charts a possible trajectory of law's gradual emergence from lawlessness. Throughout his teleological writings, Kant emphasizes the error-prone nature of human agency, its fickleness and ill-discipline as well as the human tendency to mistake short-term gains for lasting good. Kant is nonetheless committed to the emergence of lawfulness from lawlessness as a product of *human* agency, hence as necessarily unstable and unpredictable, with constant false starts, setbacks, and dead ends. The suggestion is not just of a long and arduous journey but of an essentially incompletable one. Yet for Kant, none of this is reason for fatalism or even only pessimism—it represents, to the contrary, the condition of human freedom.

Neither Hruschka nor Ripstein consider Kant's political teleology—indeed, Ripstein explicitly sets it aside, restricting himself to a systematic reconstruction of the laws of war as assessed from the vantage point of the mature *Doctrine of Right* which is itself remarkably free of any explicitly stated teleological commitments—it is, in that sense, an austerely *juridical* text. Yet the contrast between Kant's earlier and late political writings on this point should not be overstretched. As I have pointed out throughout this chapter, despite initial appearances to the contrary, the *Doctrine of Right* is hardly free of glaring inconsistencies, remarkable about-turns, and unsettling paradox. One way in which to account for these theoretically disturbing incongruities is to read the text, much like the teleological writings, as a sustained exercise in practical political judgment.[16] This is not a far-fetched proposal; Kant's philosophical thinking in general proceeds from an experiential, first-personal perspective, so never can ascend to the third-personal, all-seeing, God's eye standpoint. Kant's anthropocentric perspective is always, necessarily, a partial one such that there is, among other things, no looking round corners. This being so, not everything can be anticipated at once, and this means, in essence, that practical judgment is required at every juncture. Return once more to the dilemma between a necessarily coercive rights morality at the level of individual agency and a necessarily non-coercive rights morality, systematically related to the first, at the level of state agency. At the point at which Kant derives the moral necessity of coercible state entrance from the problematic act of unilateral acquisition, the problem of the right of nations lies as yet far in the distance.

[16] My thanks to Tom Bailey for opening up this possible line of interpretation. See also Bailey (manuscript).

By the time he arrives at the latter problem, he has incurred a set of commitments concerning the first problem which he cannot now simply disavow. Instead, he must tackle the second problem—the problem of a right of nations—while preserving those initial commitments intact. The end result is unlikely to be a sequence of linear conclusions.

Perhaps this practical way of reading the *Doctrine of Right* is naïve—surely, one might object, this is not just a handbook in political judgment. Well, but perhaps it is just that—not perhaps, a handbook in the usual sense of term (a sort of self-help book), but a sustained exercise in immanent political cum practical judgment. It seems to me that if one were to read the *Doctrine of Right* more in the spirit now proposed one might be less disturbed by Kant's at times breathtakingly audacious resolutions to arising bottlenecks as the text goes along—such as the very idea of a "lex permissiva," or of a non-coercible international law, or the nomadic exemption from an a priori duty of state entrance. The point would then be to learn how to preserve practical commitment to the idea of right and of rightful law without expecting that commitment itself to relieve one of the task of judging what, *in concreto*, such commitment requires of one in any given moment and political circumstance.

This leaves us with a final question: If the *Doctrine of Right* can in principle be read in the manner proposed, why are we so reluctant to read it in this way? Of course, there may be all sorts of reasons for resisting such a reading—including the possibility that it is in fact a mistaken reading. But the present point is more rhetorical: Why does the apparent certainty of law hold such appeal for us? By "us" I here largely mean fellow political theorists and philosophers—I doubt that members of the public generally display a similar degree of confidence in the law. My sense is that the current philosophical appeal of the law lies precisely in the promise of non-arbitrariness, impartiality, exactitude which we associate with the very idea. The law as idea promises to give us standards that are in an important sense free of the frailties of human judgment. And of course, so far as possible, actual law has to have these qualities if those subject to it are to be able to trust that law both secures them from arbitrary state power and defends the equal rights of each from incursion by others. These are good reasons for expecting positive law to be applied with precision and impartiality. However, they are ultimately instrumental reasons that do not really help to explain the current philosophical fascination with the very idea of law's precision and exactitude. It is of course true that the legitimate expectation that positive laws be impartial and exacting, etc., must stand in some relation to the very idea of law as exemplifying these qualities. Moreover, insofar as these qualities are associated with the very idea of law, the later can plausibly be understood as designed to compensate for human frailties: the impartiality and exactitude of law counteracts the partiality and infirmity of human judgment. Nonetheless, the law plainly cannot apply itself—to the contrary, each actual application of the law constitutes an act of finite human judgment. In that sense, the law can guide human judgment and agency, but it cannot ultimately replace it. But if the law cannot ultimately replace human judgment—if, to the contrary, law depends on human judgment in its application (and, indeed, in its design), then man-made law, too, ultimately is finite and fallible. In that case we should not, perhaps, place *too* much confidence in the law's precision and exactitude.

References

Bailey, T. (manuscript). "Separation and Unity in Kant's Doctrine of Right."

Bohman, J., & Lutz-Bachmann, M. (eds.) (1997). *Perpetual Peace. Essays on Kant's Cosmopolitan Ideal.* Cambridge, MA: MIT Press.

Brandt, R. (1882). "Das Erlaubnisgesetz, oder: Vernunft und Geschichte in Kants Rechslehre." In Brandt, R. (ed.), *Rechtsphilosophien der Aufklärung.* Berlin: de Gruyter Verlag, pp. 233–275.

Brandt, R. (1990). "Die Politische Institution bei Kant." In Göhler, G. (ed.), *Politische Institutionen im Gesellschaftlichen Umbruch.* Opladen: Westdeutscher Verlag, pp. 335–357.

Beck, L. (1965). "The Fact of Reason. An Essay on Justification in Kant." In Beck, L. (ed.), *Studies in the Philosophy of Kant.* New York: Bobbs Merrill, pp. 200–214.

Flikschuh, K. (2000). *Kant and Modern Political Philosophy.* Cambridge: Cambridge University Press.

Flikschuh, K. (2010). "Kant's Sovereignty Dilemma. A Contemporary Analysis." *Journal of Political Philosophy* Vol. 18(1): 469–493.

Hodgson, L. (2010). "Kant on the Right to Freedom. A Defense." *Ethics* Vol. 120(1): 791–819.

Hoeffe, O. (1995). Zum Ewigen Frieden. Berlin: Akademie Verlag.

Ludwig, B. (1988). *Kants Rechtslehre. Ein Analytischer Kommentar.* Hamburg: Felix Meiner Verlag.

Hruschka, J. (2004). "The Permissive Law of Practical Eeason in Kant's Metaphysics of Morals," *Law and Philosophy* Vol. 23(1): 45–72.

Hruschka, J. (2015). "Völkerstaat, Völkerbund under der permanente Staatenkingress in Kants Rechtslehre." In Hruschka, J (ed.), *Kant und der Rechtsstaat und andere Essays zu Kants Rechslehre und Ethik.* München: Verlag Karl Alber, pp. 129–150.

Kleingeld, P. (2004). "Approaching Perpetual Peace: Kant's Defence of a League of Nations and His Ideal of a World Federation." *European Journal of Philosophy* Vol. 12(1): 304–325.

O'Neill, O. (1992). "Vindicating Reason." In Guyer, G. (ed.), *The Cambridge Companion to Kant.* Cambridge: Cambridge University Press, pp. 280–308.

O'Neill, O. (1996). *Towards Justice and Virtue. A Constructive Account of Practical Reasoning.* Cambridge: Cambridge University Press.

Niesen, P., Eberle, O. (2011) *Immanuel Kant. Zum Ewigen Frieden.* Frankfurt: Suhrkamp Verlag.

Pallikkathayil, J. (2010). "Deriving Morality from Politics: Rethinking Kant's Humanity Formula." *Ethics* Vol. 121(1): 116–147.

Ripstein A. (2021, forthcoming) *Kant and the Law of War*, Oxford. Oxford University Press.

Waldron, J. (1992) "Superseding Historic Injustice," Ethics Vol. 103(1): 4–28.

11

The Right to Wage Private Wars of Subsistence

Its Nature, Grounds, and Place in Revisionist Just War Theories

Johan Olsthoorn

This chapter deals with the private use of coercion and armed force: more precisely, with wars fought by individuals or groups of individuals that lack political authorization.[1] Some philosophers have recently argued for the revisionist just war doctrine that individuals can have the right to initiate war in defense of their human rights when their government fails in its duty to protect them. It was a central tenet of several pre-Kantian just war theories too, that when judicial recourse is not available, individuals are entitled to enforce their basic rights by means of war. Almost inevitably, private wars fall foul of one of the standard requirements for just war: possession of legitimate authority. My aim in this chapter is not to assess revisionist arguments for dispensing with this requirement. Rather, I will take up two logically subsequent questions. How should we conceptualize private rights to secure basic rights by force of war? And where to fit these remedial rights within ethical theories of war?[2]

I will explore these questions by analyzing a controversial case of private warfare: wars undertaken in defense of human rights to subsistence. I critically contrast two ways to ground rights to wage so-called "subsistence wars": via "modern" duties of global justice and via "old" rights of necessity. Regarding the former, Cécile Fabre (2012: 97–129; 2014: 416–425) and other revisionists have argued that subjection to extreme deprivation can in principle give the severely deprived a just cause for waging defensive subsistence wars against those flouting their duties of global justice. The traditional right of necessity makes for a rival ground for legitimate subsistence wars. As conceived by Hugo Grotius (1583–1645), this right allows individuals to use and access privately owned resources and spaces to meet basic needs (without the owners' consent). Violation of the right of necessity generates a just cause for subsistence wars.

I propose to conceptualize "rights to wage private wars in defense of basic rights" in terms of "remedial rights." Remedial rights are, on my stipulative understanding,

[1] For helpful comments on earlier versions of this chapter, I would like to thank Anne Polkamp, Arthur Ripstein, Ronald Tinnevelt, the editors of this volume, and participants in the Yaoundé PhD seminar 2018 in Cameroon and the second *Public Uses of Coercion and Force* workshop at the University of Amsterdam. Special thanks to Alejandra Mancilla and Cécile Fabre for their generous feedback and ample corrections. The second half of the chapter contains material that has been published before, in Dutch (Olsthoorn 2019).

[2] On remedial war, see the chapter in this volume by Alon Harel.

Johan Olsthoorn, *The Right to Wage Private Wars of Subsistence* In: *The Public Uses of Coercion and Force*.
Edited by: Ester Herlin-Karnell and Enzo Rossi, Oxford University Press. © Oxford University Press 2021.
DOI: 10.1093/oso/9780197519103.003.0011

claims, privileges, powers, or immunities that persons acquire due to the violation or non-fulfillment of their non-remedial rights, or due to the violation or non-fulfillment of duties by others.[3] Remedial rights can entitle persons to engage in conduct that would have been impermissible were it not for the specific injustice they suffered. The right to wage war in defense of one's human rights is an extreme remedial right. Less extreme remedial rights entitle individuals to use coercion to resist injustices in ways that stop short of war: for example, destruction of property, smuggling, squatting, and piracy.[4] To bring out their emergent nature, and for ease of reference, I will occasionally refer to remedial rights as "second-order" rights; and to the underlying deontic notion as "first-order" rights or duties. It should be noted, however, that that terminology differs from standard rights-talk—in which privileges and claims are first-order rights, and powers and immunities second-order ones. My main contention is that which remedial rights persons acquire will depend on the nature and content of the first-order deontic notion whose violation or non-fulfillment generates the remedial right.

Distinguishing between the existence conditions of rights and the conditions of their lawful exercise, I argue that on the modern global-justice model it is nigh impossible for rights to wage private subsistence wars to arise in practice. This is ultimately because this model derives such rights from first-order enforceable *duties* (of global justice). The old right-of-necessity model, by contrast, grounds such resistance rights in first-order enforceable *rights* (of subsistence), allowing it to sidestep the challenges of indeterminate and underdetermined moral liability. While the permissible exercise of resistance rights is on both models equally strictly morally curtailed (by *ad bellum* norms of proportionality, necessity, and success), at least the existence conditions for such rights can be met in practice on Grotius's model.

I should stress that my analysis should not be taken for an endorsement of either the right of necessity or of the justifiability of (private) subsistence wars. My ambitions are exclusively theoretical: I canvas normative differences between old rights of necessity and modern human rights to subsistence in order to probe the idea of remedial rights theoretically. By exploring what it means for rights and duties to be privately enforceable, my analysis elucidates the structure of revisionist just war theories that jettison the legitimate authority condition.

The chapter unfolds as follows. Section 1 discusses Fabre's and Grotius's take on the legitimate authority condition and shows how its suspension figures in the debate on subsistence wars. Section 2 explains why just subsistence wars are destined to remain a mere theoretical possibility on the modern global-justice model. Section 3 reveals that the right of necessity is, for better or worse, a more capacious ground for remedial rights to uphold subsistence rights through armed conflict—even though lethal force will on this account too be seldom all-things-considered morally permissible in

[3] On the distinction between the violation and non-fulfillment of rights, see Pogge (2008: 53). Remedial *rights* should not be confused with remedial *responsibilities*, a notion developed by David Miller (2001) (cf. Tomalty 2017). Remedial responsibilities are special obligations to put a morally bad situation right, where "special" indicates that these obligations are not shared equally among all agents.

[4] For explorations of non-bellicose rights to resist global economic injustice, see Gargarella (2007) and Caney (2015).

practice. Section 4 distinguishes between the existence conditions of rights and the conditions of their lawful exercise to bring this point home.

1. Revisionism about the Legitimate Authority Condition

It has long been a staple of orthodox just war theory that a necessary condition for *jus ad bellum* is that the war must be launched by a legitimate authority. Whereas the authority to initiate wars was traditionally attributed to either states or sovereign rulers, modern orthodox ethicists of war have argued that non-state entities can in principle also legitimately start a war on behalf of some community—think of political liberation movements. This has induced reinterpretations of the legitimate authority condition.[5] Some revisionists have gone further and discarded the legitimate authority requirement entirely. Fabre has argued that from a cosmopolitan perspective, "there are very good reasons for dropping the requirement altogether. . . a war need not be waged by a legitimate authority in order to count as a just war" (2008: 965). In her view, "[g]roups of actors which act in unstructured, disorganized ways, as well as individuals themselves, can have the right to go to war" (2012: 144).[6]

Fabre's rationale for discarding the legitimate authority requirement lies in her cosmopolitan conviction that the sole justification for resorting to war is the protection of human rights against violations by others. She argues that individuals generally have "meta-rights" to protect their rights.

> For it would be incoherent, on the one hand, to claim that individuals' fundamental interests in life-saving resources and basic freedoms are important enough to be protected by rights, and on the other hand, to deny that individuals' interests in the protection of those rights is important enough to be protected by a right to defend those rights themselves (Fabre 2012: 115).

This meta-right to protect one's human rights, including *in extremis* by lethal force, is a precondition for a minimally decent life and hence itself a human right:

> the right to protect oneself from violations of one's human rights by others is a *human* right, in the sense that it is a right to a freedom (to wit, the freedom to defend oneself, without interference, against others) which we need in order to lead a minimally flourishing life. By extension, the right to wage a war in defence of one's human rights should also be conceived of as a human right. (Fabre 2008: 969; also 2012: 145, 286)

[5] On the need and meaning of the legitimate authority condition, see, e.g., Parry (2017) and Benbaji (2018).

[6] This argument is compatible with the view that legitimate authority is morally required for wars waged *in name of some community*—as being either authorized by or having political authority over a person seems a general precondition for acting in their name (Benbaji 2018: 309–311; cf. Fabre 2008: 973, 2012: 151–156). Fabre's view could then be that just wars can be waged by belligerents lacking legitimate authority provided these wars are not undertaken on behalf of a community.

Legitimate states exercise this meta-right on behalf of their citizens (Fabre 2014: 408). Yet if a state fails in its duty to protect its members, then "the meta-right to protect fundamental human rights reverts back to individuals" (Fabre 2012: 115, also 148). It would be inconsistent with cosmopolitan principles to deny individuals due protection of their fundamental rights because of their lack of group memberships. Yet we would be doing so if, on account of the legitimate authority requirement, we would only accord to political communities the right to protect human rights by force of war (Fabre 2012: 147, 286). Proponents of the legitimate authority condition thus gainsay individuals a human right.

The revisionist doctrine of just private warfare relies on a controversial semantic premise: Is it linguistically justified to subsume under "war" some forms of private use of lethal violence? In common parlance, any non-public use of armed force is routinely labeled "terrorism" and thereby morally condemned. To refrain from settling a substantive philosophical question by definitional fiat, I will assume that the idea of a just private war is not conceptually foreclosed. By private war, I will mean any temporally prolonged resort to lethal violence by groups of individuals that lack legitimate authority. Private defensive wars differ from simple cases of individual self-defense in at least two ways. The former, but not the latter, consist in temporally and spatially protracted conditions of hostility, in which belligerents kill in defense not only of their own rights but also of those of their comrades (cf. Fabre 2012: 55).

That individuals can be entitled to enforce their basic rights by force of war was a tenet of the just war theory of Grotius, too. According to the Dutch thinker, morality "does not prohibit all Manner of Violence, but only that which is repugnant to Society, that is, which invades another's Right" (1625: I.2.1.3). Proportionate defensive force used against unjust aggressors, for example, neither undermines society nor violates anyone's rights. It is hence morally permitted. Distinctive of Grotius's just war theory is that this fundamental moral principle governing the use of force applies equally to all agents. By nature, every individual can by right (i.e., without committing an injustice) engage in that violence which violates no one's rights. Grotius explicitly accepted the possibility of just private wars (1625: I.3.1).[7] Indeed, he subsumed under "war" *any* condition of hostility, "not even excluding single Combats" (1625: I.1.2.1). The unilateral right of individuals to repel injuries by force is curtailed by civil law, which prohibits vigilantism and offers legal remedies against wrongdoing. However, where judicial recourse is not available, such as on the high seas or due to the imminence of the threat, individuals retain the right to repel injuries by armed force (1625: I.3.2).

Grotius was not the first thinker to conditionally allow individuals to engage in private warfare. According to Francisco de Vitoria (1485?–1546), "any person, even a private citizen, may declare and wage defensive war." This principle renders the legitimate authority requirement partially obsolete: "From this we may gather that any person may wage war without any other person's authority, not only for self-defence but also for the defence of their property and goods" (Vitoria 1991: 299). Only the

[7] Grotius distinguished between public wars (authorized by the civil sovereign) and (unauthorized) private wars (1625: I.3.1–5). Both kinds of war can be just (i.e., permitted by the law of nature) but only public wars can be solemn (i.e., have legal implications deriving from the voluntary law of nations). For an overview of Grotius's just war theory, see Olsthoorn (2018a).

state, however, has the right to wage *offensive* wars: that is, punitive wars and wars waged to recover what is due (Vitoria, 1991: 299; cf. Grotius 1625: II.1.2.1). Grotius broke new ground by denying that the right of punishment is essentially political in origin (1625: II.20.40.4). In the state of nature, it is morally permitted for any private individual to punish wrongdoers for the public good. After all, such private punishment upholds society and does not violate the (forfeited) rights of wrongdoers (1625: II.20.8–9). Civil law curtails these individual rights of punishment (for good reason). Yet private parties retain the right to initiate both defensive and offensive wars when judicial recourse is unavailable. As Tuck (1999: 85) concludes, for Grotius, "there is no significant moral difference between individuals [outside of the state] and states . . . both may use violence in the same way and for the same ends."

Both Fabre and Grotius thus countenance the theoretical possibility of just private wars, in part because both deny that having legitimate authority is a necessary condition for just war (whether civil or international). Indeed, for Fabre, all rights can be reduced to rights held by individuals (*qua* individuals or *qua* group members). "The right to wage war, therefore, is a right as held by individuals to kill in defence of their rights" (Fabre 2012: 55). Grotius was not a reductivist about rights of war. His endorsement of private warfare rather followed from the premise that the moral principles determining the permissibility of use of force apply equally to individuals and to states. Both Fabre and Grotius hold that private individuals have forgone the right to initiate war within legitimate states. But, both concur, in certain conditions individuals regain the right to defend their basic rights by waging private wars.[8]

As Ripstein's *Kant and the Law of War* amply reveals, for Kant, the very idea of enforcing rights by means of war is morally fraught: "reason, from the throne of the highest morally legislative power, delivers an absolute condemnation of war as a procedure for determining rights" (8: 356). In the state of nature, individuals may rightly use armed force against hostile others—not however in order to assert their rights, but to coerce enemies to mutually enter a rightful condition where adjudication is available (8: 439n). This chapter will bracket Kantian worries about private rights enforcement.[9] Instead, I will explore diverging ways of conceptualizing individual rights to enforce one's human rights by force of war. Revisionist defenses of the possibility of just private wars—presupposing the rejection of the legitimate authority condition—render such remedial rights salient. Not so much because enforcing rights by means of war is the most extreme enforcement method that exists. But rather because any right to wage war that (groups of) individuals lacking legitimate authority will have are likely to be reducible to rights to protect their basic rights (unless one endorses the Grotian and Lockean doctrine of individual natural rights of punishment).[10] The

[8] Fabre outdoes Grotius on this point. Grotius denied that individuals can permissibly take up arms in order to avenge injuries "which the Prince neglects to revenge" (1625: I.3.5). Only necessity can justify private warfare within a state. Fabre, by contrast, maintains that "If the state is unable *or unwilling* to wage war on behalf of its individual members in those cases where there is an all-things-considered justification for it, the right to do so reverts to the latter" (2012: 148, emphasis added).

[9] For a Kantian critique of Fabre's cosmopolitan defense of rights to unilaterally resort to armed force, see Stilz (2014: 323–335).

[10] Fabre does not discuss punishment as a just cause for war. Her reductivism about rights of war commits her to either denying the permissibility of punitive wars, or to attributing rights of punishment to individuals.

public use of armed force, by contrast, can justly serve more purposes. The international community, for example, has the right to authorize armed interventions in third countries for peacekeeping purposes.

This chapter thus turns to theories of just private warfare because these theories offer an opportunity to examine a relatively ignored philosophical question: What does it mean for rights and duties to be enforceable? My contention is that whether individuals can in practice have a right to wage war in defense of their subsistence rights depends on how we conceptualize the underlying deontic notion (whose violation or non-fulfillment triggers the remedial right).[11] A comparison of Fabre's and Grotius's justifications for private wars waged in defense of subsistence rights brings this out nicely.

2. Remedial Rights and Modern Duties of Global Justice

Subsistence wars are wars initiated in order to realize subsistence rights.[12] Such wars can be waged by actors endowed with legitimate authority: as when a state plagued by famine attacks a neighboring country that refuses to share its large food surplus, in order to procure resources needed for its people's survival (Luban 1980: 177). The current debate over the permissibility of subsistence wars focuses, however, on its application to agents that lack legitimate authority. Protagonists of the in-principle justifiability of subsistence wars have argued that dereliction of duties of global distributive justice can make the affluent morally liable to suffer defensive wars by the global poor thus wronged (Fabre 2012: 97–129; Fabre 2014: 416–425; Lippert-Rasmussen 2017: 447). Subjection to extreme poverty can thus give the poor a just cause for waging defensive war against the foreign affluent to force them to comply with their duties of global justice. Orthodox just war theories do not acknowledge this just cause and therefore stand in need of revision. Indeed, according to Fabre, we must "broaden the range of just causes for waging war to include violation of individual basic rights" of all kinds (2008: 969). For reasons outlined in this section, this revisionist just war doctrine has limited, if not nil, practical significance.

Existing defenses of the in-principle justifiability of subsistence wars assume that the well-off have stringent duties of justice to do their share to alleviate dire poverty worldwide. These duties are the counterparts of human rights to subsistence. For Fabre, that right is held against the whole of humanity and entitles each "to the material resources without which they could not hope to live a minimally decent life" (2012: 97).[13] Compare Henry Shue's notion of "basic rights": "everyone's minimum

[11] On privatization and warfare, see also the chapter by Malcolm Thorburn in this volume.

[12] This chapter will ignore potential differences between "redistributive wars" (Lippert-Rasmussen 2013, 2017; Pogge 2013), "resource wars" (Tadros 2014), and "subsistence wars." All will be taken to refer to armed hostilities started in response to grave economic injustices, including subjection to severe material deprivation.

[13] Compare the ostensibly stricter definition offered in Mancilla (2019: 2): "the right to subsistence can be defined as the right to those material provisions needed for one's self-preservation, i.e., those material provisions required for enjoying a minimal physical and physiological well-being." The prevalence of just subsistence wars will depend, in part, on the amount of material resources we take individuals to be entitled to by human right to subsistence alone.

reasonable demands upon the rest of humanity" (1980: 19). Basic here means that individuals need to have these rights secured (i.e., rendered non-precarious) before they can exercise any other rights. Rights to subsistence (i.e., to minimal economic security) are basic in this sense, as are rights to physical security. Subsistence rights, on Shue's account, entitle humans to "unpolluted air, unpolluted water, adequate food, adequate clothing, adequate shelter, and minimal preventive public health care" (1980: 23; also Mancilla 2016: 66–67). Subjection to severe deprivation through the foreseeable effects of unjust global economic institutions and trading regimes violates subsistence rights.

What kind of remedial rights do the global poor acquire by dint of being subjected to severe deprivation? Do they or their allies ever obtain a right to resort to armed force to secure their basic rights? If so, against whom? The philosophical literature explores these questions by analyzing various kinds of duties of global justice (whose systematic dereliction violates human rights).[14] Such duties can be assistance-based: the mere ability to help the destitute creates obligations of justice to do so at some personal cost. Or contribution-based: we ought to alleviate poverty created or sustained by our actions as a matter of justice.[15]

Can dereliction of assistance-based duties of global justice be a just cause for defensive war? Luban has answered in the affirmative, accepting the permissibility of "a war for subsistence" (1980: 177). If country "*A* is morally obligated to give food to [country] *B*, and assuming that *B*'s sole purpose in fighting is to procure food, a defence by *A* would be an unjust war." More generally, Luban regards socially basic rights—both to security and subsistence—as "worth fighting for" (1980: 175). Christian Barry concurs: "In some extreme cases, we claim even that those in need might be permitted to kill those who are failing to assist them as a means of compelling them to act" (Barry 2019: 79; also Barry & Øverland 2016: 42–43). Fabre explicitly maintains that "breaches of their right to receive help give the very deprived a just cause for waging a war against the affluent" (2012: 110). Given how grievous the rights-violations in question are—thousands of children die of poverty-related diseases globally each day—she deems it ad hoc to hold that violations of assistance-based duties of justice never warrant resort to armed force.

Contribution-based duties of justice are generally considered more stringent and demanding than assistance-based ones. The stringency of duties is determined by the extent to which they constrain the obligor through disallowing courses of conduct. Duties are demanding to the extent that, upon their dereliction, the obligor is required to accept costs to remedy the situation of any obligee thereby disadvantaged (Barry &

[14] It can be argued that the severely deprived have remedial rights to secure their basic rights by force of war even against innocent threats. Remedial rights, after all, do not only arise from the culpable violation of first-order rights but also from their protection. Grotius, for one, denied that killing in self-defense is lawful only in response to wrongdoing: "this *Right* of *Self-Defence*, arises directly and immediately from the Care of our Preservation, which *Nature* recommends to every one, and not from the Injustice or Crime of the *Aggressor*" (1625: II.1.3, also III.1.2.1). Whether it can ever be morally permissible to intentionally kill innocent threats and innocent aggressors is a highly controversial question (e.g., Otsuka 1994; Barry & Øverland 2016: 177, 185–187).

[15] These are not the only possible bases capable of grounding responsibilities of justice to alleviate global poverty. Barry & Øverland (2016: 3) further mention association- and benefiting-based responsibilities. Since neither have so far been taken up in the debate on subsistence wars, I bracket them.

Øverland 2016: 6–8). Enforceable duties are typically both stringent and demanding. Enabling harm is morally worse than merely allowing harm to happen. Contribution-based duties of global justice, which command the affluent to alleviate the global poverty created or sustained by their own actions, are therefore more stringent and demanding than assistance-based ones. Systematic dereliction of contribution-based duties hence generates a wider set of remedial rights (i.e., they are "more enforceable") than assistance-based ones. Observe that the assistance-based and contribution-based duties of global justice under discussion are both counterparts of the human right to subsistence. Their differing stringency and demandingness reflect that the affluent can be to blame for non-realization of human rights to subsistence in two morally distinct ways: by allowing or by enabling severe global poverty to persist.

To what extent individuals in rich countries contribute to the endurance of global poverty is a contested empirical question. Thomas Pogge has argued that "we are harming the global poor by foreseeably subjecting them to avoidable severe poverty" (2005: 55). High import tariffs effectively shut off Western markets to agricultural and manufacturing products from Africa, while agricultural subsidies allow Western producers to dump their crops in the Global South far below cost price, outcompeting local farmers. The affluent have a duty to compensate the global poor, Pogge contends, in virtue of having imposed on them "an institutional order that foreseeably produces avoidable human rights deficits" (2005: 61).

Bracketing disputed empirics, Gerhard Øverland (2011)[16] and Kasper Lippert-Rasmussen (2013, 2017) have defended the conditional thesis that *if* the well-off have stringent contribution-based duties to alleviate severe global poverty, *then* their violation makes them liable to suffer some defensive harm. Øverland discusses several challenges to the conditional, from appeals to the insignificance of individual contributions to the opacity of causal connections between individual contributions and human rights abuses. Each challenge seeks to disentangle liability to defensive force from contribution-based duties of global justice. Øverland contends that all such attempts to explain away liability simultaneously render duties of global justice non-stringent, thus begging the question. Lippert-Rasmussen (2013: 67) has similarly argued that, if Pogge's analysis of the causes of global poverty is correct, then redistributive wars are in principle morally permissible.

These arguments support the revisionist thesis that unjust subjection to severe deprivation can in principle be a just cause for war. The thesis does not imply that the global poor can in practice lawfully embark on armed resistance to secure their subsistence rights. Just subsistence wars, though possible in principle, are highly unlikely for two reasons. First, having a just cause is a necessary but not a sufficient condition for permissible defensive war. Not "any infringement of socially basic human rights is a *casus belli*," Luban writes: "here as elsewhere in the theory of just war the doctrine of proportionality applies" (1980: 175). To be all-things-considered permitted, resort to armed force must be proportionate, a last resort, and have a reasonable prospect of success (bracketing the legitimate authority requirement). These conditions are unlikely to be met by the severely deprived. For a start, their lack of resources forestalls

[16] A revised version of Øverland (2011) is published in Barry & Øverland (2016: 173–194).

reasonable hope to achieve their war objectives by fair means (Pogge 2013: 106). Fabre admits that subsistence wars will most probably be futile. What reasonable hope do the global poor have to succeed in an armed fight against the imposition of an unjust economic system, for example, through guerrilla or "terror-like" wars (Øverland 2011: 284)? The answer is straightforward: "None whatsoever" (Fabre 2012: 117). Even if this moral obstacle can be circumvented by engaging powerful allies to fight on behalf of the destitute (Fabre 2012: 117), due to the institutional rather than kinetic nature of subjection to extreme destitution, taking up arms in support of human rights of subsistence will almost always fall short of the last resort condition (Pogge 2013: 103–105).[17] Nevertheless, Fabre insists, "subsistence wars can on principle be just"—and this is a theoretically significant outcome, mandating revision of standard just war theory (Fabre 2012: 129).

There is a second set of reasons why just subsistence wars are, for better or worse, destined to remain a mere theoretical possibility. Intentionally or foreseeably using armed force against non-liable (i.e., innocent) persons is morally worse, and hence harder to justify, than using the same force against liable persons. Moral liability is a technical term. For Anne to be liable to suffer some harm by Belle means that Anne has forfeited the right against Belle to not have that degree of harm imposed on her. Persons are not wronged by having that harm inflicted on them which they are liable to suffer. It is impermissible to resist the imposition of harm one is liable to suffer (McMahan 2009: 8–10). These are all widely accepted conceptual claims.

Let us assume that Øverland is right: we cannot coherently claim that the affluent have stringent contribution-based duties of justice to alleviate global poverty *and* hold that victims cannot have a right to defend themselves with force against corresponding economic injustices. Regardless of whether we think the affluent are culpable or merely morally responsible for the imposition of unjust economic structures upon the global poor,[18] two concerns arise regarding their ensuing moral liability. First, what renders a person a legitimate target of defensive force is contributing to avoidable death and hardship from poverty-related conditions, not being affluent as such (Øverland 2011: 288). The well-off will be responsible for the plight of the global poor in differing degrees, reflecting their individual contribution to sustaining extreme poverty and the degree to which they are derelict in their global justice duties. For many, individual responsibility will be slim, generating liability only to a small degree of defensive harm. The infliction of any harm in excess of this wrongs the affluent and, for it to be all-things-considered justified, these wrongs must be outweighed by the moral good armed force can be reasonably expected to produce. The foreign affluent may well at most be liable to such a small degree of defensive force that initiating a subsistence war will inevitably be disproportionate (Lippert-Rasmussen 2017: 451–452). If this holds true generally, then dereliction of duties of global justice is not grave enough a kind of wrongdoing to constitute a just cause for war.

[17] Lippert-Rasmussen (2017: 461–462) acknowledges this but argues that the last resort and success conditions as commonly understood are too restrictive and must be revised.

[18] As a determinant of moral liability, moral responsibility is a broader notion than culpability: we can be morally responsible for some rights-infringement without having done anything wrong. On this distinction, see McMahan (2005).

Compounding this issue of underdetermined moral liability, we cannot normally "know of most individual members of the affluent communities whether they are in fact derelict in their duties of justice towards the distant poor" (Fabre 2012: 123). The institutional rather than kinetic nature of subjection to extreme poverty makes it hard to determine who is morally responsible for the plight of the global poor and to what degree. And in case of reasonable doubt, we ought to regard people innocent. Hence even if we believe that economic injustice in principle warrants defensive war, since it is exceedingly hard to determine who is a lawful target, any such war will in practice be impermissible. For all these reasons, just subsistence wars remain "a theoretical exercise [rather] than a practical possibility" (Mancilla 2016: 10).

As the next sections reveal, indeterminate and underdetermined moral liability render the private use of coercion to secure basic rights impermissible only because of the *kind* of remedial rights subsistence war-theorists accord to the global poor. The argument Fabre et al. develop does not invoke any remedial right to use force to obtain resources to meet basic needs. Rather, their claim is that the severely deprived and their allies may *enforce* duties of global justice of the affluent. Theirs is a right, not to obtain the resources to realize their basic rights, but to use force to get the foreign affluent to fulfill their duties (Øverland 2011: 283; Fabre 2012: 121). Lawful exercise of these remedial rights hence requires specification of due liability—for armed force can be used only to ensure that those currently derelict in their global justice duties henceforth comply. The next section offers an alternative model for envisioning the remedial rights of the severely deprived.

3. Remedial Rights and the Old Right of Necessity

The main proponent of the in-principle justifiability of subsistence wars, Cécile Fabre, has argued that dereliction of either assistance—or contribution-based obligations of justice can make one a legitimate target for defensive war. In addition to these two kinds of wrongdoing, Fabre endorses in passing a third justification for subsistence wars: the traditional right of necessity (2012: 103–105; 2014: 416). The existence of rights of necessity was widely accepted by philosophers and theologians in the late medieval and early modern periods, although disagreement persisted over their ground, nature, and conditions of lawful exercise. On Grotius's account, on which I focus here, the right of necessity entitles the destitute to use resources and spaces privately owned by others without their consent.[19] Violation of this right constitutes a *casus belli*. This section shows that Grotius's right-of-necessity-model sidesteps the problems of indeterminate and underdetermined moral liability that render just subsistence wars a merely theoretical possibility for modern revisionists. For better or

[19] Kant vehemently opposed rights of necessity: "a right of necessity, which, as a supposed *right* to do *wrong* when in extreme (physical) need, is in any case an absurdity" (8: 300). However, he seems to have understood this right not as a right to use another's goods in times of need (which can hardly be called absurd), but as "an authorization to take the life of another who is doing nothing to harm me, when I am in danger of losing my own life" (6: 235).

worse, the old right of necessity is a more capacious ground for rights of private warfare than modern duties of global justice are.

Grotius grounded the right of necessity in a theory of private property and territorial rights.[20] The earth, he argued, initially belongs to all of humanity in common. All exclusionary property and territorial regimes are established by convention. Given that the earth was originally collectively owned, Grotius wondered, how extensive are the private property and territorial rights that people have agreed to set up? It would have been irrational and unreasonable, he claimed, to institute exclusionary regimes so strict that the destitute are morally necessitated to starve in case the affluent fail to provide due assistance. As Pufendorf (1729: II.6.5) later eloquently put it: "Can any Human Institution bind me with such a force, that in case another Man neglects his Duty towards me, I must rather die, than recede a little from the ordinary and the regular way of acting?" The institution of property was therefore set up with the equitable restriction that "in a Case of absolute Necessity, that antient Right of using Things, as if they still remained common, must revive, and be in full Force" (Grotius 1625: II.2.6.2). As the "as if" [*tanquam si*] clause attests, necessity does not annihilate private property rights. Rather, the right of necessity grants the destitute claim-rights to *use* privately owned resources and spaces (without the owner's consent) to relieve pressing needs. It generates "a Right to enjoy in common those Things that are already become the Properties of other Persons" (1625: II.2.6.1). Resources thus taken do not come to be owned by the needy. "This then is a Sort of Right to take, without a Right of acquiring" (1625: III.13.1.1). Grotius could therefore coherently impose a duty of restitution upon anyone compelled to invoke rights of necessity—the owner has a right to be compensated for the use of what remain *her* goods (1625: II.2.9, III.12.1.1, III.17.1). The right can only be exercised as a last resort and resources may not be taken from someone equally badly off (1625: II.2.7–8).[21]

Grotius conceptualized the enforceable use-rights persons have in dire need as themselves a type of property rights: rights to use resources owned by someone else. Property rights in external resources do not presuppose ownership, as any tenant knows. Early modern jurists had a distinct legal category for property rights in resources owned by others: so-called "servitudes."[22] In his influential introduction to Dutch jurisprudence, Grotius classified servitudes and other use-rights as forms of *dominium* (1926: II.33–44). The right of necessity thus generates property rights in resources owned by others which restrict the range of property and territorial

[20] Grotius did not derive the right of necessity from a natural right of self-preservation, as Hobbes would later do: "it is permitted by the right of nature for people living in extreme need to take other people's things, even by force" (Hobbes 1668: 30.18). Nor from a right to the means necessary to discharge natural obligations (to oneself or others), as per Vattel (1758: II.9.117–120).

[21] On the structure and ground of Grotius's right of necessity, see Olsthoorn (2018b: 196–212). That paper examines the contrasting ways in which Mathias Risse and Hugo Grotius have used the idea of collective ownership of the earth to ground moral restrictions on exclusionary property regimes. It bears noting that the right of necessity, as traditionally understood, is of limited application (Mancilla 2017: 608). Since it merely establishes enforceable use-rights in resources legally owned by others, it cannot express or replace distributive rights of the global poor. A principle of global distributive justice remains necessary to give the global poor moral ownership of their fair share of the earth's resources and spaces.

[22] E.g., Pufendorf (1729): I.1.19: "A *Power* over other Mens Things, is what the *Civilians* term *Servitudes*" [potestas in rem alienam *servitus*].

rights: private ownership comes "with this favourable Exception" of use-rights by others in acute need (1625: II.2.6.4). The property rights persons can claim in virtue of owning a resource are never so expansive as to include a right to withhold the resource from the severely deprived.[23] In turn, the right of necessity grants the destitute limited property titles (use-rights) in resources or spaces privately owned by others. The right of necessity is therefore not a matter of charity; rather, it generates enforceable rights (1625: II.2.6.4). Indeed, unjustly withholding necessities from the indigent is a just cause of war: for whenever judicial recourse is unavailable, we may by war defend what is ours—including our use-rights in things owned by others (1625: II.1.2.1–2, II.2.10).

The early modern proprietary conception of rights is conceptually alien to us. Recent attempts to revive and globalize the age-old notion of a right of necessity have overlooked some of these foreign features, distorting the ideas of Grotius to straight-jacket them into a more familiar framework (e.g., Risse 2012; Mancilla 2016, 2017). It may seem more intuitive to portray the right of necessity as a moral prerogative to "encroach" upon the property rights of others (Mancilla 2016: 76). Yet this suggests a rights conflict—between owners and the needy—that wasn't there on Grotius's account. Mancilla treats the right of necessity as a remedial right—a right to take, use, or occupy resources belonging to others triggered by the unjust non-realization of subsistence rights (2016: 3). Perhaps for this reason, she focuses on rights deficiencies brought about by "human-made, structural causes" (Mancilla 2016: 65).

On my reading, the old right of necessity was not a remedial right but rather a first-order right that all persons have in virtue of the social contract of property. Wrongdoing is hence not needed to trigger the right. Thus, the landlord on whose land the destitute are squatting cannot justify their eviction by saying that *he* has fulfilled his duties of justice. For the squatters' rights do not arise in response to wrongdoing of him or anyone else: theirs is a first-order, not a remedial right. However, if the penurious are barred from using privately-owned resources in dire need, then *that* injustice (a violation of their right of necessity) does generate remedial rights, including a right to use proportionate armed force against resisting owners as a last resort.

Mancilla has invoked the right of necessity to conditionally justify "practices like shoplifting, pick-pocketing, urban squatting, rural illegal encampments and occupations" (Mancilla 2017: 608). Can Grotius's right of necessity be extended to justify subsistence wars? I think so. Consider: it is not unimaginable that climate change will render parts of the earth uninhabitable. The right of necessity entitles climate change refugees to migrate to habitable parts of the planet (Risse 2009). Territorial rights, established by the social contract of property, are not so extensive as to permit states to exclude such refugees from their territories. States that forcibly keep out climate change refugees violate the rights of necessity of the latter. On Grotius's account, wrongful exclusion gives climate change refugees a just cause to wage subsistence wars against these states (who are liable to suffer proportionate defensive force). Even as orthodox a just war theorist as Michael Walzer (1983: 46) agrees that necessity can create a just cause for war: "If, driven by famine ... thousands of people were to fight

[23] As Pufendorf would later put it, "the Power and Privilege of *Propriety* doth not reach so far as that the things may not be taken away without the Owner's Consent" when in dire need (1729: II.6.5).

their way into a [country] otherwise closed to them, I doubt that we would want to charge the invaders with aggression."

Can any war fought in defense of rights of necessity by the global poor be all-things-considered morally permissible in practice? Not likely. The right of necessity does not render less strict the *ad bellum* proportionality, last resort, and success conditions. As these conditions render the private use of armed force by the global poor in defense of their human rights of subsistence impermissible, so they will render unjust any defensive war waged to protect their right of necessity. Just subsistence wars will be rare on both accounts. Nonetheless, as the next section reveals, the two models portray the normative conditions of those unjustly subjected to severe deprivation rather differently.

4. Comparing the Two Models

Let us compare the remedial rights triggered by violations of human rights of subsistence and of old rights of necessity, respectively. My contention is that structural differences between the last two rights entail that the remedial rights they generate differ as well. As noted, the old right of necessity generates use-rights to *specific* resources/spaces owned by others. The global poor can lay claim to each and any resource which they desperately need to realize their basic rights. This is because the right of necessity grants the severely deprived limited property rights in resources and spaces exclusively owned by others. It is thus, paradoxically, *their* property right (in goods owned by others) that the global poor are defending against affluent owners. By denying the global poor access to resources and territories, the affluent violate the property rights of the poor. The crime they thereby commit is akin to theft. This has important implications for the remedial rights of enforcement that arise from violations of the right of necessity. These violations directly and unmistakenly establish moral liability to suffer proportionate defensive harm: anyone unwarrantedly withholding their resources from the needy becomes *ipso facto* liable for that harm. The problem of indeterminate liability hence does not occur. Furthermore, since the violation of the right of necessity will always involve kinetic wrongdoing by some identifiable agent (who is unjustly withholding resources), underdetermined liability is less likely to figure than on the global-justice model (where human rights deficits are due to myriad agents flouting stringent duties of global justice to varying degrees).

By contrast, modern remedial rights derive, in case of assistance-based duties, from the first-order moral right to receive some *unmarked* resources from others to alleviate critical need. In case of contribution-based duties, the remedial right trades on the first-order right not to be subjected to severe deprivation. In both cases, any remedial right to use armed force faces the dual problem of indeterminate and underdetermined liability. The global poor have, according to Fabre, a just cause to use force against the affluent for "getting them to comply" (Fabre 2012: 121). But this requires determining who has failed to comply with their duties of justice, and to what extent. Individual wrongdoing is moreover likely to have been far too slight to justify armed force in response.

Modern subsistence-war theorists thus still focus on the duties of the affluent, and relatedly, on the remedial rights of the global poor to enforce those duties. Indeed, these remedial rights primarily express the enforceability of duties of global justice, rather than the enforceability of individual rights to subsistence. The modern global-justice model by implication recognizes no remedial rights of the global poor to safeguard their subsistence rights in cases when no one is morally responsible for their severe deprivation. By contrast, the old right-of-necessity model takes as primary the right to subsistence—more precisely, the right to unilaterally take resources and access territories to satisfy dire need. Consequent remedial rights are grounded in this right, rather than in corresponding duties of justice (even if the last resort condition implies that in the present world, the right of necessity cannot justify coercive measures unless global justice duties are systematically flouted).

This discrepancy between the two ways of modeling remedial rights to forcibly defend subsistence rights has significant implications for how the normative condition of the severely deprived is portrayed. To grasp this, we must distinguish between the existence conditions of rights and the conditions of their lawful exercise (i.e., their moral practicability). This is a distinction with a difference insofar as we hold that some moral norms constrain the exercise of rights without being constitutive of these rights. This distinction arguably applies to the right to wage war—an equivocal notion. This right can either be understood broadly as denoting an all-things-considered moral permission of A to wage war against B. Or narrowly, as expressing that A has a just cause to initiate war against B (where B is morally liable to suffer proportionate harm inflicted by A and has no right to resist that harm). It is meaningful and morally informative to say that A has a right (narrowly) vis-à-vis B to wage war against B, while simultaneously holding that A cannot permissibly exercise this right at present (i.e., A lacks an all-things-considered moral permission to wage war against B). Circumstances might be such that A cannot at present reasonably expect to achieve their war objectives by fair means (rendering resort to war futile). Alternatively, any resort to armed force may currently fall foul of the *ad bellum* proportionality requirement. However, these conditions do not annul the liability of B to suffer some defensive lethal harm by A—harm which B cannot rightly resist. That liability, incurred by wrongdoing, is the correlative of A's (narrow) right vis-à-vis B to wage against B. Insofar as A cannot permissibly exercise that right vis-à-vis B, third parties (but not B) can have a duty to thwart A's (all-things-considered impermissible) war against B. The point in recognizing A's right vis-à-vis B to wage war against B is to express that B is morally responsible for a wrong against A grave enough to constitute a *casus belli*. The right of war, narrowly understood, thus consists in having a just cause against an identifiable and morally liable enemy.

Fabre introduces a facially similar distinction between "the right to wage war in general" and "the right to wage *that particular war*" (2012: 148, also 2008: 971–972). Suppose some state is derelict in its duties to protect the human rights of its citizens against violations by a third party. The right to protect one's human rights by force of arms reverts, according to Fabre, in that case to the citizens themselves. While the violation of their human rights gives these citizens a just cause to protect their human rights by means of private war against the aggressor third party, as a disparate group of individuals, chances are they will fail to meet the success condition. Indeed, for this reason, it may seem that only entities capable of having legitimate authority (e.g., states or politically organized liberation movements) can have the right to wage war since only such collectives are generally large enough to be able to comply with all just war requirements (including the success requirement). However, to draw this conclusion would be to conflate, Fabre contends, "the right to wage war in general" with "the right to wage some particular war." That a group of individuals cannot permissibly wage some particular war does not mean that the group lacks the right to wage war in general.

My distinction is compatible but different. The thought I seek to capture is that A can have a right vis-à-vis B to wage war against B even if other just war norms (*ad bellum* proportionality, necessity, success) render it impermissible to exercise that right. A just cause arises in response to some grave wrong and always pertains to some particular war. Some readers may frown on my proposal to express the condition of having a just cause to wage war against another party in terms of having a right vis-à-vis that party to wage war against it (the legitimate exercise of which right being curtailed by the other just war requirements). Rather than convincing skeptics of the independent soundness of this conceptual proposal, I will illustrate its theoretical usefulness by letting it bring out a significant difference between Fabre's and Grotius's accounts.

Fabre argues that the severely deprived have a *general* right to wage war to protect themselves from violations of their human rights by others. However, for reasons outlined in section 2, they are unlikely to have a right to wage any *particular* war against a specifiable and liable enemy party. Subjection to severe deprivation violates human rights to subsistence. It is grave enough an injustice, according to Fabre, to justify in principle the use of armed force in protection of these rights. Yet as Caney (2015: 59) points out, no "clearly defined unified agent" is responsible for the plight of the global poor. In the absence of a legitimate target, responsible for the injustice, the severely deprived have no specifiable remedial right to wage a specific war in defense of their subsistence rights. Insofar as it cannot practically be determined who is derelict in their global justice duties, rights to enforce these duties by means of war are nonexistent.

Indeterminate liability thus bears on the existence conditions of a right to wage some specific war. Underdetermined liability does so too. To be sure, underdetermined liability makes it harder for the use of armed force to be proportionate (it being morally worse to intentionally kill non-liable than liable persons). And the *ad bellum* proportionality requirement constrains the legitimate exercise of rights to wage war: it is impermissible to initiate war if it foreseeably creates more evil than moral good. But below a certain threshold, underdetermined liability also affects the existence conditions of a right to wage war: only grave wrongs constitute a *casus belli*. The non-realization of subsistence rights of the global poor is morally shocking. But when responsibility for this injustice is structurally diffuse, creating individual liability to only a small degree of defensive harm, then dereliction of duties of global justice may not be grave enough an injustice to form a just cause of war.

By contrast, remedial rights generated by the right of necessity *can* be fully specified, even if these rights cannot be exercised because of additional moral restrictions governing resort to armed force (restrictions that constrain, not constitute, the remedial right). This is ultimately because the right of necessity endows the global poor with enforceable rights to use necessary resources and spaces. These use-rights can be exercised in non-ideal conditions (including amidst widespread dereliction of duties of global justice), even if they cannot always permissibly be defended by force of war. Since violating the right of necessity inevitably involves forcibly and kinetically denying the destitute access to particular resources or spaces, indeterminate liability cannot occur while underdetermined liability is much less likely than on the global-justice model.

5. Conclusion

Global justice theorists have long focused on the nature, extent, and grounds of duties of the affluent to alleviate the plight of the global poor and to realize justice worldwide. The last few years have seen a flurry of works that shift perspective to the agency and remedial rights of the global poor. Some have taken issue with the disempowering way in which the severely deprived are commonly portrayed in the global justice literature: as passive right-holders awaiting fulfillment of their human rights by others (Deveaux 2015). Others have argued that the prevailing focus on the duties of the affluent leaves an important moral question unaddressed: What are the poor morally permitted to do to hasten realization of their basic human rights?

This chapter has explored the revisionist just war doctrine that the severely deprived can have a just cause to protect their subsistence rights by armed force. I have analyzed two remedial rights to engage in private war that have recently received scholarly attention: one deriving from the old right of necessity, the other from the modern human right not to be subjected to severe deprivation. I have compared these two remedial rights with respect to their existence conditions and conditions of legitimate exercise. I have pointed out that the problems of indeterminate and underdetermined moral liability pertain only to the modern global-justice model. This is so because that model derives remedial rights from enforceable duties of justice; and not—as the right-of-necessity model does—from enforceable individual rights.

The distinction I made between the existence conditions of a right vs. its moral practicability helps dispel a doubt regarding the permissibility of private warfare. In practice, it might generally be impermissible for individuals and disorganized groups of individuals to protect their human rights against violations by means of war insofar as its private nature forestalls reasonable hope of achieving their war objectives by fair means. (Dubbed terrorists by those endorsing the legitimate authority requirement, they might find few allies.) I have suggested that the proportionality and success requirements should be understood as limiting the permissible exercise of rights vis-à-vis liable enemies to wage war against them (rather than as being constitutive of such rights). If we accept this suggestion, then the difficulties private parties face in meeting the success condition for just war is no reason to deny that they can have a right to wage war. It would stop us from reintroducing the legitimate authority condition by the backdoor: by interpreting the other just war requirements in such a way that only belligerent parties endowed with legitimate authority can meet them.

Let me conclude with a final theoretical observation on the nature of remedial rights. These second-order rights (arising from the violation or non-fulfillment of first-order rights or duties) have been somewhat neglected by philosophers. My analysis has suggested that what form remedial rights take (to what and against whom they give claims, privileges, powers, and immunities) depends on the nature and content of the underlying substantive deontic notion. In particular, I have suggested that remedial rights arising from the non-fulfillment of enforceable obligations differ structurally from remedial rights arising from the violation of first-order rights—even if these obligations entail correlating rights. Clarifying what it means for rights and obligations to be enforceable will require thinking through the different structures remedial rights take.

References

Barry, C., & Øverland, G. (2016). *Responding to Global Poverty: Harm, Responsibility, and Agency.* Cambridge: Cambridge University Press.

Barry, C. (2019). "Harm, Responsibility, and Enforceability." *Ethics & Global Politics* Vol. 12(1): 76–97.

Benbaji, Y. (2018). "Legitimate Authority in War." In Lazar, S., & Frowe, H. (eds.), *The Oxford Handbook of Ethics of War.* Oxford: Oxford University Press, pp. 294–314.

Caney, S. (2015). "Responding to Global Injustice: On the Right of Resistance." *Social Philosophy and Policy* Vol. 32(1): 51–73.

Deveaux, M. (2015). "The Global Poor as Agents of Justice." *Journal of Moral Philosophy* Vol. 12(2): 125–150.

Fabre, C. (2008). "Cosmopolitanism, Just War Theory and Legitimate Authority." *International Affairs* Vol. 84(5): 963–976.

Fabre, C. (2012). *Cosmopolitan War.* Oxford: Oxford University Press.

Fabre, C. (2014). "Rights, Justice and War: A Reply." *Law and Philosophy* Vol. 33(3): 391–425.

Gargarella, R. (2007). "The Right of Resistance in Situations of Severe Deprivation." In Pogge, T. (ed.), *Freedom from Poverty as a Human Right: Who Owes What to the Very Poor?* Oxford: Oxford University Press, pp. 359–374.

Grotius, H. (1926) [1631]. *The Jurisprudence of Holland.* Lee, R.(ed.). Oxford: Clarendon Press.

Grotius, H. (2005) [1625]. *The Rights of War and Peace* (3 vols.). Tuck, R. (ed.). Indianapolis: Liberty Fund.

Hobbes, Th. (2012) [1668]. "The Latin *Leviathan*." In Noel Malcolm (ed.), *Leviathan* (3 vols.) Oxford: Clarendon Press.

Kant, I. (1996). *Practical Philosophy.* Gregor, M. (ed.). Cambridge: Cambridge University Press.

Lippert-Rasmussen, K. (2013). "Global Justice and Redistributive Wars." *Law, Economics and Philosophy* Vol. 1(1): 65–86.

Lippert-Rasmussen, K. (2017). "Pogge, Poverty and War." *Politics, Philosophy & Economics* Vol. 16(4): 446–469.

Luban, D. (1980). "Just War and Human Rights." *Philosophy & Public Affairs* Vol. 9(2): 160–181.

Mancilla, A. (2016). *The Right of Necessity: Moral Cosmopolitanism and Global Poverty.* London: Rowman & Littlefield.

Mancilla, A. (2017). "What the Old Right of Necessity Can Do for the Contemporary Global Poor." *Journal of Applied Philosophy* Vol. 34(5): 607–620.

Mancilla, A. (2019). "The Human Right to Subsistence." *Philosophy Compass* Vol. 14(9): 1–10.

McMahan, J. (2005). "The Basis of Moral Liability to Defensive Killing." *Philosophical Issues* Vol. 15: 386–405.

McMahan, J. (2009). *Killing in War.* Oxford: Clarendon Press.

Miller, D. (2001). "Distributing Responsibilities." *The Journal of Political Philosophy* Vol. 9(4): 453–471.

Olsthoorn, J. (2018a). "Grotius and the Early Modern Tradition." In May, L. (ed.), *Cambridge Handbook of the Just War*. Cambridge: Cambridge University Press, pp. 33–56.

Olsthoorn, J. (2018b). "Two Ways of Theorizing Collective Ownership of the Earth." In Penner, J., & Otsuka, M. (eds.), *Property Theory: Legal and Political Perspectives.* Cambridge: Cambridge University Press, pp. 187–213.

Olsthoorn, J. (2019). "Mondiale Rechtvaardigheid Afdwingen: Oorlog, Noodzaak, en Verzetsrechten van's Werelds Misdeelden," *Algemeen Nederlands Tijdschrift voor Wijsbegeerte* Vol. 111(1): 45–62.

Otsuka, M. (1994). "Killing the Innocent in Self-Defence." *Philosophy & Public Affairs* Vol. 23(1): 74–94.

Øverland, G. (2011). "602 and One Dead: On Contribution to Global Poverty and Liability to Defensive Force." *European Journal of Philosophy* Vol. 21(2): 279–299.

Parry, J. (2017). "Legitimate Authority and the Ethics of War: A Map of the Terrain." *Ethics & International Affairs* Vol. 31(2): 169–189.

Pogge, T. (2005). "Severe Poverty as a Violation of Negative Duties." *Ethics & International Affairs* Vol. 19(1): 55–83.

Pogge, T. (2008). *World Poverty and Human Rights: Cosmopolitan Responsibilities and Reforms. Second Edition*. Malden, MA: Polity Press.

Pogge, T. (2013). "Poverty and Violence." *Law, Economics and Philosophy* Vol. 1(1): 87–110.

Pufendorf, S. (1729) [1672]. *Of the Law of Nature and Nations.* Kennett, B. (trans.). London: printed for J. Walthoe et al.

Ripstein, A. [forthcoming]. *Kant and the Law of War.* Oxford: Oxford University Press.

Risse, M. (2009). "The Right to Relocation: Disappearing Island Nations and Common Ownership of the Earth." *Ethics & International Affairs* Vol. 23(3): 281–300.

Risse, M. (2012). *On Global Justice.* Princeton, NJ: Princeton University Press.

Shue, H. (1980). *Basic Rights: Subsistence, Affluence, and US Foreign Policy.* Princeton, NJ: Princeton University Press.

Stilz, A. (2014). "Authority, Self-Determination, and Community in *Cosmopolitan War.*" *Law and Philosophy* Vol. 33(3): 309–335.

Tadros, V. (2014). "Resource Wars." *Law and Philosophy* Vol. 33(3): 361–389.

Tomalty, J. (2017). "Remedial Responsibility for Severe Poverty: Justice or Humanity?" *Journal of Applied Philosophy* Vol. 34(1): 89–98.

Tuck, R. (1999). *The Rights of War and Peace: Political Thought and the International Order from Grotius to Kant.* Oxford: Oxford University Press.

Vattel, E. (2008) [1758]. *The Law of Nations.* Kapossy, B., & Whatmore, R. (eds.). Indianapolis: Liberty Fund.

Vitoria, F. (1991). *Political Writings.* Pagden, A., & Lawrance, J. (eds.). Cambridge: Cambridge University Press.

Walzer, M. (1983). *Spheres of Justice: A Defense of Pluralism and Equality.* New York: Basic Books.

12

Between Wormholes and Blackholes

A Kantian (Ripsteinian) Account of Human Rights in War

Aravind Ganesh

1. Introduction

Despite conceiving war in more terrifying terms than even the most zealous "divergence" theorists, Kant's theory of the law of war nevertheless has the resources to allow for "human rights" claims to be made in the barbaric conditions of armed conflict. This chapter has four parts. Part I recounts certain prominent accounts of human rights in war, which generally conceive of law as instrumental means for attaining moral ends. On this basis, these approaches criticize as wrongly decided a line of cases finding human rights obligations to obtain despite the obligor state having little to no capacity to enforce its writ, due either to the territory being embroiled in active civil war or under the control of another state. Part II then introduces Kant's conception of rights and war, the hallmark of which lies in law being conceptually prior to morality. This results in a conception of "human rights" as possible only where there are political institutions capable of enforcing law, and a conception of war as a condition where there are no rights at all.

In a sudden twist, however, part III then argues that individuals may nevertheless make claims of human rights in time of war. Where an applicant can make out even a risible claim that a defendant state has held itself out as an authority over it, that state may not be heard to deny concomitant obligations of human rights. Nor may it deny that it was holding itself out as an authority if doing so would amount to admitting a war crime. The limit is reached, however, when the facts alleged by the applicant are *utterly* barbaric, disclosing not even a scintilla of the possibility of a rightful condition. In such situations, the applicant becomes a bare life: her inability to state a cause of action mean that she can, in effect, never have any causes of action.

In place of a conclusion, part IV demonstrates how the structure of the argument in parts II and III mirrors the architectonic of the *Doctrine of Right*.

2. "Standard" Approaches: Law as Efficient Means to Morality

Academic legal discussions of the relationship between humanitarian and human rights law standardly begin with the following passage from the *Nuclear Weapons* advisory opinion:

Aravind Ganesh, *Between Wormholes and Blackholes* In: *The Public Uses of Coercion and Force.* Edited by: Ester Herlin-Karnell and Enzo Rossi, Oxford University Press. © Oxford University Press 2021. DOI: 10.1093/oso/9780197519103.003.0012

> ... the protection of the International Covenant of Civil and Political Rights does not cease in times of war ... In principle, the right not arbitrarily to be deprived of one's life applies also in hostilities. The test of what is an arbitrary deprivation of life, however, then falls to be determined by the applicable *lex specialis*, namely, the law applicable in armed conflict which is designed to regulate the conduct of hostilities. (1996:para. 25)

The underlying assumption, evidently, is that the baseline rules on how persons should treat one another are contained in the body of legal doctrine known as human rights, and that humanitarian law "elucidates" the meaning of these rights—in particular the right to life—in the exceptional situation of war. (Dill, 2020:204)[1] But what exactly makes war so exceptional? The prevailing "reductive individualist" school holds that war is "special" only in that it throws up an impenetrable fog preventing us from *discerning* what is morally required, which otherwise remains the same in all situations. As Jeff McMahan puts it:

> There is really only one morality. There is no special morality that supplants ordinary morality whenever conditions of war obtain ... [War] differs from other conflicts, including individual self-defence and other-defence, only in scale and complexity. (McMahan, 2010:505)

McMahan naturally has his challengers, the most prominent being (1) "realist" theories denying the existence of any morality at all in war (e.g., Shue, 2010), and (2) "collectivist" theories holding that a transformation from everyday morality to that which governs in wartime is brought about by the aggregation of individuals into collective units called states (e.g., Walzer, 1977). For reasons that cannot be set out here, the reductivists' answer to the second set of theories appear—to this author at least—largely successful: basic morality is no more altered by aggregation than the substantive elements of the tort of negligence are changed in class actions (McMahan, 2006).[2] Of more interest are "realist" theories, particularly as expressed by Henry Shue, who agrees with McMahan that "morality is all of a piece," but differs from him in arguing that "what one ought to do is complicated by, among other considerations, the functional requirements for good laws" (Shue, 2008:91). Because the functional requirements for morality are simply not present in war, "there is no function to be performed by a competing 'morality' of war"; so instead all we can hope for is a "morally justified set of laws of war" (Shue, 2008:89). The laws of war have no distinctive morality behind them other than damage control. Or, as Shue puts it graphically elsewhere, war is a place where "shit happens," such that the only thing we can hope for is a set of laws of war that, again in Shue's words, helps "constrain the 'shit' ..." (Shue, 2010:516).

Nevertheless, it may be said that McMahan's basic tenets of reductive individualism have come to define the terrain of current theoretical discourse about the laws of war.

[1] Note that Dill is critical of this interpretation.

[2] But see Frowe, 2015:187–188, arguing that aggregation is not an "illegitimate move" for a reductive individualist because it "is a familiar feature of ordinary life, and so to think it relevant to war is not inconsistent with reductivism."

If the difference between war and self-defense or other-defense is one of degree rather than kind, the project of theorizing the morality of war essentially becomes a matter of working out what our everyday ethical intuitions would require under conditions of extremity or limited knowledge. Accordingly, recent works about the law and morality of war feature innumerable hypotheticals each more imaginative than the last, involving runaway trolleys, hidden tunnels, vaporizing laser guns, and angry rapists. As for reductive individualist conceptions of the role of law and political institutions, McMahan's original strand condemns central doctrines of contemporary humanitarian law—chiefly, the symmetrical treatment of aggressors and defenders and the distinction between combatants and civilians—as morally indefensible (McMahan, 2009:chaps. 2, 5; Frowe, 2014: chaps.5–6). A newer strand, however, constructs "prudential" defenses of the same *jus in bello* doctrines out of various empirical suppositions about combatant incentives (Haque, 2017:chaps. 3–6; Dill, 2019). In a recent piece, Janina Dill argues that even though human rights law is the "better" law in that it provides more accurate guidance in complying with our moral obligations to one another, it is "displaced" in time of war by the far less demanding tenets of humanitarian law for two reasons. Firstly, war is beset by "epistemic gaps" preventing us from gathering enough information to make the right decision. Secondly, war presents "incurable volitional defects" that make our incentives to disobey our moral duties so irresistible that law would be disobeyed if it were to require compliance with them, thus defeating the point of the institution. Accordingly, she conceives the two bodies of law as carrying out a "moral division of labour" of essentially the same project, with human rights law applying "symmetrically" only in non-protracted international armed conflicts, and humanitarian law applying in protracted international conflicts and non-international armed conflicts (Dill, 2020:200–201, 229–230).

Roughly the same approach is reflected in much human rights scholarship, particularly on the concept of "human rights jurisdiction." Most contemporary human rights treaties include as one of their very first provisions a clause on "state jurisdiction" or "human rights jurisdiction," specifying that the state parties to the treaty bear obligations under the treaty only toward persons "under their jurisdiction," or some similar formula (*European Convention on Human Rights (ECHR)*: Art 1; *International Covenant on Civil and Political Rights (ICCPR)*: Art 2(1)). As such, human rights jurisdiction is a "threshold criterion" that must be satisfied before a claim of a human right can even be made (Besson, 2012:861; *Ilaşcu*, 2004:para. 311).

For instance, in a 2011 monograph that remains the most widely cited text on the subject, Marko Milanovic begins with the idea that human rights obligations are by default owed by states universally to all persons regardless of borders. This being practically unattainable, provisions on "state jurisdiction" are included in international human rights treaties to "limit the obligations of contracting states to those situations where these obligations could realistically be met and human rights effectively protected" (Milanovic, 2011:56). Under Milanovic's model, therefore, valid human rights claims might be made in wartime situations, because states always retain "negative" duties not to cause harm, and because international humanitarian law only has a "weak" *lex specialis* effect (Milanovic, 2011:229–261; 2016:84).

This may be contrasted with Samantha Besson's theory of extraterritorial human rights jurisdiction, which identifies a state's human rights obligations not with its crude ability to affect material welfare but with its provision of "authority." On Besson's explanation, obligations under the ECHR arise when a state acts in a "normative fashion to give reasons for action" rather than when it engages in "mere coercion" (Besson, 2012:872). Although she agrees with Milanovic on little else, she does, however, conceive of human rights norms as presumptive baselines. In her view, the moral reasons human rights law supposedly represents or advances are valid regardless of our institutional context, such that private individuals retain "residual or subsidiary" human rights duties "when their political institutions have failed or before they have set up their institutions" (Besson, 2013:418–21). Even though these moral reasons are binding upon us without more, reasons of "allocative" efficiency nevertheless require the setting up of political institutions to "perfect" them through relations of right and obligation (Besson, 2013:418). To wit, the law's supposed task of providing normative guidance cannot be accomplished if the putative duty-bearer cannot enforce its guidance, and so, there is no point in imposing obligations upon it. As such, despite identifying a state's human rights obligations with its claims of authority rather than its capacity to exert power, Besson nevertheless argues that human rights duties of political institutions become *legal* obligations only where those institutions have some meaningful capacity for the latter: the claim of authority by the putative duty-bearer must be accompanied by "effective power" and "overall control [which] should be effective and exercised, not merely claimed" (Besson, 2012:873). Thus, the feasibility of an obligation is not just relevant to the question of whether the putative obligor has breached it, but to *whether it even exists at all.* At times she clothes this thought in language borrowed from Kant: "the feasibility of duties is [both] part of their concrete requirements (following the principle 'ought implies can') and a condition for their existence" (Besson, 2012:868).

In short, despite their seemingly vast differences, the above-mentioned authors all share the same structure of argument. The same principles of morality apply across all scenarios. Whether these give rise to "rights" and "obligations," however, depends upon secondary, "allocative" questions of information and resource distribution. Because a right is a social guarantee to the actual enjoyment of the "substance" of a right (Shue, 1996:13), it follows that there is no point in making demands to those substances in the exceedingly difficult conditions of war, as a result of which those guarantees necessarily lapse.

3. Some Difficulties with the Practice

This picture is at some variance with the actual legal practice concerning human rights in war. There is much in the Strasbourg jurisprudence—the most developed body of case law on the subject—to indicate that epistemic gaps or volitional difficulties are wholly irrelevant. According to the leading *Al-Skeini* decision, "state jurisdiction" under Article 1 ECHR arises if the state party wields "authority and control" over the persons concerned, or "effective control over territory" (*Al-Skeini*, 2011:paras. 133–140). Applying this test, a drone strike upon a civilian wedding party in a distant land would not give rise to Article 1 ECHR state jurisdiction (and therefore obligations under the convention) even if the state party ordering the strike *knew full well* that the targets were civilians (Milanovic, 2012:130).[3] Contrariwise, several decisions have found human rights jurisdiction and obligations to obtain despite the state concerned facing extreme practical difficulties, epistemic and worse. The prime example is a case from 2004, *Ilaşcu*, where the Strasbourg Court held that Moldova had a positive obligation to use all diplomatic, economic, judicial, and other measures within its ability to secure the ECHR rights of individuals held in custody in a breakaway region under the control of the Russian-sponsored Transdniestrian puppet government (*Ilaşcu*, 2004:paras. 332–52). A similar rationale was adopted by the Inter-American Court in *Pueblo Bello Massacre*, where Colombia was held responsible for human rights violations committed by paramilitaries in a region where the Colombian government not only exercised no control, but where conditions were so violent they arguably constituted a Non-International Armed Conflict (NIAC) (*Pueblo Bello Massacre*, 2006:paras. 119–127, 139–140).

Most enigmatic of all, however, is the *obiter* discussion toward the end of *Al-Skeini*, where the Court considered a gloss on the alleged facts of an earlier case involving an alleged incursion by Turkish soldiers into Iraq where they shot several shepherds. The court opined that Article 1 ECHR state jurisdiction would not have arisen on those facts, but that things would have been different if the Turkish forces had instead taken the shepherds to a nearby cave before shooting them (*Al-Skeini*, 2011:para. 136). What possible epistemic or volitional gaps could there be that might remove an individual shepherd's "right" not to be shot where he stood? If such gaps did exist, how could they be so easily cured by a short jeep ride?

One may of course deride these results as incoherent products of rotten compromises, or as unprincipled judicial activism. Milanovic is a most thoroughgoing critic in this regard: his research agenda begins with a polemical attack against the *Banković* decision, where the Strasbourg Court held inadmissible a claim on behalf of civilians killed as a result of the aerial bombardment of a radio tower during the Kosovo War, and ends by criticizing *Ilaşcu* for "undermin(ing) the broader principle that it is only the state's capacity to act, and not its territorial title, that should matter" (*Banković*, 2001; Milanovic, 2011:264, 107). While Besson accommodates the *Banković* result (but not its reasoning), her stipulation that normative guidance be accompanied by effective and actually exercised power requires the reversal of the *Ilaşcu*. The same also appears true of Dill, who cites *Ilaşcu* in the course of observing that the presence of "organised armed violence on a state's territory" is one indication that it lacks the element of "internal control" necessary for performing law's first moral task (Dill, 2020:244 n. 151).

The most well-developed critique of *Ilaşcu*, however, is in a recent article by Lea Raible, which advances from the premise that the word "jurisdiction" has not one, but many meanings (Raible, 2018). As used by human rights lawyers, Raible claims that "jurisdiction" refers to a threshold or "trigger" activating states' obligations (Raible, 2018:320). As used by general international lawyers, however, that same word refers instead to the bundle of entitlements states ordinarily have in their territory, which Raible

[3] Applying the *Al-Skeini* test, Milanovic remarks that "While the ability to kill is 'authority and control' over the individual if the state has public powers, killing is not authority and control if the state is merely firing missiles from an aircraft."

describes as "not entirely unlike property" (Raible, 2018:322). Importantly, Raible argues that the two meanings are independent and severable: "title to territory and jurisdiction in international human rights law serve different purposes" (Raible, 2018:323). The law of territory, in Raible's view, is primarily concerned with maintaining international stability (Raible, 2018:322–323). Human rights law, in contrast, concerns the "allocation" of rights and obligations between individuals and states, which makes effective control over territory the crucial determinant (Raible, 2018:325, 327–328). As such, occupation of a state's territory by another cannot in and of itself extinguish title, so the state retains "jurisdiction" within the meaning of general international law of territory. However, where a state's territory has been occupied by another, it no longer has any ability to advance the welfare of, or provide normative guidance to the people there, so the question of "jurisdiction" within the meaning of human rights law does not even arise. According to Raible, the problem with *Ilasçu* then, is that it treats human rights obligations as "allocated" automatically along with title to territory as determined in international law, such that "title always results in [human rights] jurisdiction, regardless of whether the latter's criteria are met or not . . ." (Raible, 2018:329).

Despite such universal condemnation, however, the Strasbourg Court shows no indication of overruling *Ilaşcu*, either implicitly or explicitly. Presented with a broadly similar set of facts eight years later in *Catan*, the Court reaffirmed its position, holding that

> Although Moldova has *no effective control* over the acts of the [breakaway entity] in Transdniestria, the fact that the region is recognised under public international law as part of Moldova's territory gives rise to an obligation, under Article 1 of the Convention . . . (emphasis added). (*Catan*, 2012:para. 110)

This, as argued below, is for good reason.

4. The Kantian (Ripsteinian) Alternative: The Separate Rationality of War

In a nutshell, the hallmarks of the "standard" approach shared by Dill, McMahan, Haque, Milanovic, and Besson were (1) that the same moral rights that apply in everyday life also apply in war; and (2), that law, rights, and obligations are "efficient means" for specifying, directing, or perfecting these moral rights in particular factual contexts (Zylberman, 2016a:321). We also briefly discussed Shue's realist conception of war as a moral vacuum. Kant, as developed by Ripstein, articulates a path between McMahan and Shue by essentially inverting the structure of their arguments.[4] Against McMahan, Kant's objection is that the everyday morality of the reductive individualists is anything but reductive; it is the *artificial* product of a certain configuration of political institutions. Moral reasons are not just difficult to discern or inconvenient to comply with outside this context; without institutions, they are internally contradictory, and hence

[4] On Kant's political philosophy as neither "ethics-first" nor realist, see Rostbøll (2019:22).

radically incomplete (Ripstein, 2018:29). Against Shue, the Kantian objection is that war is not like mud, dirt, manure, or some wholly "empirical" thing lacking criteria against which it may be compared and evaluated (Ripstein, 2018:9, 34), such that all that can be done is to ameliorate or "humanize" it. Instead, the *ideal* rather than functional requirement of war is simply *peace*; that is to say, "the possibility of leaving the state of nature among states . . . and entering a rightful condition" (RL 6:347, 485).

4.A. The Difference between Legal and Moral Labor

The source of the difference between Kant on the one hand, and the reductivists and realists on the other, lies in their conceptions of the fundamental legal status of a human being. McMahan and Shue broadly conceive of this as lying in the satisfaction of a range of basic interests, which implies an entitlement to their protection from harms or threats. For Kant, however, interests and harms or threats thereto cannot be analytically basic. To use a common example, there would be something objectionable in the condition of a slave even if her interests were not just met but extravagantly catered for by a kindly master. As such, the single organizing principle for Kant's philosophy of right is freedom, understood as *independence* from the determining choice of another (RL 6:237, 393). Crucially, Kantian freedom-as-independence must be distinguished from more common "liberal" notions of freedom as "freedom from constraint" (Pettit, 1997; Skinner, 1998:chaps. 1–3).[5] The slave's condition would be defective even if her master went away on permanent holiday: should someone attack her outside the house, she would neither be able to seek recourse before the police or the courts, but would have to rely either on her strength or on the kindness of passersby. Whatever "freedom" she enjoys is no different from that of an unleashed dog. The free*dom* of a free person is different: it is a positive status entitling her to marshal her community's capacity for force to prevent her from domination and instrumentalization by others. As such, Kantian freedom-as-independence implies that unless you are a subject of an *authority*—a common powerful will entitled to use force—you have no rights.

So far, the argument is reasonably familiar, as are the objections to it. A less well-explored implication of this theory of freedom is the way law conceptually prior to morality, rather than the other way round.[6] Rights and obligations cannot be "allocated" by reference to individuals' moral duties, because the same *force* that is crucial for the maintenance of rights is also inimical to the fulfillment of moral duties (Rostbøll, 2019:6; Willaschek, 2009:52). A shopkeeper's legal obligation to deal fairly with his customers is in no way a function of his moral duty to be honest, because compelled honesty is morally worthless. Instead, the law's purpose is to enable persons to channel coercion into forms of interaction where they are "externally" free to make choices with the means at their disposal in a manner compatible with a like

[5] Kantian freedom has many points of similarity with the concept elaborated in Neo-Roman and Republican literature. There are differences, but these need not be explored here.

[6] I am taking a liberty. In the Introduction to the *Metaphysics of Morals*, Kant first distinguishes laws of morality from laws of nature, and only then specifies juridical laws and ethical laws as the two components of the former. RL 6:214, 375–376. "Morality" as used in this chapter corresponds to what Kant calls "ethics."

ability on the part of all others to do the same (RL 6:230, 387). Whether or not the "internal" *substance* of those choices is ethically commendable is a separate, and subsequent question. This explains the structure of the *Metaphysics of Morals*: the *Doctrine of Right* comes before the *Doctrine of Virtue*. We must first elaborate the conditions under which freedom is assured for all persons before can we begin to talk about what moral reasons apply to us. Right precedes Virtue because unless you are free, your choices are not yours to be praised or blamed for.

Rather than refer to ultimate moral reasons to provide order and coordination to his theory of interpersonal legal relations, Kant turns instead to Newtonian physics (Ripstein, 2018:39, 42 n 26). Under the caption "Right is connected with an authorisation to use coercion," Kant explains in distinctly mechanical terms that

> Resistance that counteracts the hindering of an effect promotes this effect and is consistent with it. Now whatever is wrong is a hindrance to freedom in accordance with universal laws. But coercion is a hindrance or resistance to freedom. Therefore, if a certain use of freedom is itself a hindrance to freedom in accordance with universal laws (ie wrong), coercion that is opposed to this (as a hindering of a hindrance to freedom) is consistent with freedom in accordance with universal laws, that is, it is right. (RL 6:231, 388)

Persons may be likened to billiard balls rolling around a table, with the salient difference that they get to choose whether to roll around or stop. The first iteration of force—someone else pushing you to move, or to come to a stop—is incompatible with freedom, and therefore "wrong." The second iteration of force—as deployed to reverse the first wrongful variety—is consistent with reciprocal independence, and therefore "right." In this way, the operation of force in interpersonal legal relations constructed "by analogy with presenting the possibility of bodies moving freely under the law of the *equality of action and reaction*" (RL 6:233, 389). Peter Birks expresses the same thought—albeit in more optical terms—when he describes a remedy as "the same thing as the right, looked at from the other end" (Birks, 1997:24). If I wrongfully take your thing, you may rightfully deploy force against me to make me give it back to you—*compensation*. If I wrongfully use your thing, you may rightfully deploy force against me to make me surrender my gains to you—*restitution*.

As such, law does not perform any "moral labour" at all; the task of a theory of law is simply, to set out "with mathematical exactitude" (RL 6:233, 390) how embodied rational beings the limited confines of the world may, in an entirely unmoralized fashion, roll around or come to a stop compatible with a like ability of all other rational embodied beings. Once we work this out, we can finally begin the difficult work of examining which choices are good, and which are bad. Kant illustrates this insight in relation to war when he describes the category of Cosmopolitan Right as simply the "rational idea of a *peaceful*, if not friendly, thoroughgoing community of all the nations on the earth ..." (RL 6:352, 489). Peace is made in the first instance between enemies, not friends. Afterwards, they may decide to become friends, or they may not. Only after peace is conceived is it worthwhile to speak of companionship, civility, or any of the other moral reasons that do bind us. The catastrophe of war, then, is that it stops this process before it can even begin.

4.B. Wormholes and Blackholes: Law without Institutions

The next step is to explain how this system of reciprocal coercion malfunctions without political institutions, as is the case in war. For Kant, war is distinctive not because epistemic, volitional, or logistical difficulties are present, but because institutions with legitimacy to settle disputes on the merits are absent (RL 6:347–348, 485–486). To illustrate what this leads to, consider the following hypothetical set in a "state of nature" lacking political institutions, entitled *Toothbrush*:

> In a fit of panic, you 'aggress' upon another by taking their toothbrush. There are no epistemic gaps. You knew perfectly well that the toothbrush was being habitually possessed and used by that other, not you.

The "obvious" moral thing to do is to return the toothbrush, because taking things habitually possessed and used by others is incompatible with your own habitual use and possession of things to the exclusion of others. However, a problem arises at this point. Suppose you decide to return the toothbrush "unilaterally," out of the kindness of your heart, or even out of a suddenly remembered sense of duty to the moral law. Because it is *ex gratia*, this attempt to do the right thing still leaves the other's use and possession of the toothbrush at your mercy, and so, goes nowhere in reversing the initial violation of reciprocal independence. At the same time, however, your independence also prevents you from allowing her to reclaim the toothbrush from you forcefully, for this would be to recognize a unilateral power on her part to designate you as a wrongdoer and to punish you accordingly. Even if she is "correct" in this determination, as is the case here, this remains an act of subjection to another's determining choice. You would therefore be entitled—perhaps even required—to resist her by force of your own.

In this way, morality outside political institutions issues utterly contradictory demands, and so, no demands at all. Without the presence of a common powerful will to protect individuals in their possession of things, no one can use and possess them exclusively, which is to say they have no rights in them. At best, such an entitlement lasts only so long as it can be defended physically. Thus, the state of nature is one where right is determined by might, which is to say that rights do not exist at all.

Such are the defects of a system of pure "private" right. The solution to them lies in magical thinking—"postulation." To close the circle in *Toothbrush*, we must magic into existence (1) a legislature that "omnilaterally" transforms each person's unilaterally acquired external objects into peremptory property rights, (2) a judiciary to adjudicate disputes of rights according to a procedure built around the presumption that no one has done wrong, and (3) an executive to maintain these rights by force (RL 6:307, 452; Ripstein, 2009:169–174).[7] In this way, morally defensible interpersonal relations require—that is, can only be contemplated under—what Kant calls the

[7] Note that for reasons that cannot be explored here, the institutional configuration for the domestic rightful condition between natural persons is different from that of the rightful condition between states as natural persons.

"Postulate of Public Right"; that is, the minimal institutional framework where private rights are publicly and systematically legislated, adjudicated, and enforced (RL 6:307, 451–452).

The legitimacy of public authorities arises not from their instrumental effectiveness in securing interests or advancing values, but because they are "partly constitutive of our independence" (Zylberman, 2016b:296). Likewise, human rights are not timeless, pre-institutional entitlements inhering automatically in human beings in the state of nature, but can obtain as *legal* rights only in a rightful condition. Recall Besson's contention that private persons retain residual or subsidiary human rights duties to fulfill welfare rights after political institutions have failed or before those institutions come into being. This cannot be right: even on a loose understanding of a "duty" as simply a reason to criticize a duty-bearer for failure to do something, persons caught in the middle of a civil war cannot have any duties to build a bit of a hospital, because they cannot be faulted for fleeing altogether. Without public institutions invested with political authority, such rights cannot even be thought. In this way, the idea of human rights as juridical means of ensuring independence presupposes a context of authority and obedience: "Jurisdiction"—law-speaking—constitutes not a functional, but a *conceptual* threshold that must be crossed before a claim of human rights can even begin to be made.

For the same reasons, it "is difficult even to form a concept of [right] or to think of law in this lawless state [of war] without contradicting oneself (*inter armas silent leges*)." (RL 6:347, 485) War, then, is not a fog rendering rights indiscernible or inconvenient, but a tear in the fabric of the *Doctrine of Right*, through which all the contradictions of the state of nature re-emerge. Or, to update Kant's Newtonian imagery to that of contemporary astrophysics, war is like a juridical "wormhole": an anomaly where "ordinary" rationally available principles of right simply do not apply. The insights gained from innumerable trolley problem variations are therefore wholly inapplicable, because war is not a peacetime activity (Walzer, 2006:43). Likewise, it is inappropriate to transpose standards from the human right to life into wartime, discounting only for epistemic gaps and volitional difficulties (cf Dill, 2020:232–235). Just as in *Toothbrush*, aggressors and defenders in war are symmetrically situated: they do *one another* no wrong, because nobody has any rights. Instead, they "do wrong in the highest degree"—that is, wrong, *period*—simply by the fact of being a condition in which might determines right (RL 6:307–308, 452) In this way, Kant provides a distinctive rationale for Haque's insight that the laws of war contain no permissions, only prohibitions (Haque, 2017:chap. 2).

A wormhole, however, always admits of the possibility of an exit. War may be wrongful in the highest degree, but if it is waged at all, it must be done in a manner that leaves open the possibility of making peace. Thus, the core of the concept of a wrong in war lies in closing this exit, rendering a future peace impossible (ZeF 8:347, 320; RL 6:308, 452, 6:347, 485). In the Sixth Preliminary Article of *Toward Perpetual Peace*, as well as in a footnote to the *Doctrine of Right*, Kant illustrates how such a wrong might obtain with the example of a false surrender (ZeF 8:347, 320; RL 6:308, 452). Attacking the enemy after staging a false surrender explodes the distinction between fighting and surrendering, and therefore also the distinction between war and peace (Ripstein, 2018: chap. 4). The problem is not just that perfidy tends to drive the parties into wars of extermination that terminate in the "vast graveyard of the human race" (ZeF 8:347, 320). Rather, it is that the state of war persists even if one side successfully exterminates the other, because the very concept of peace has been destroyed. The same conflation of war and peace is effected by attacking civilians in violation of the (aptly named) principle of distinction. Although Kant describes war as wrongful in the "highest" degree, war crimes appear wrongful in a degree yet higher still, because they hand everything over to "savage violence" (RL 6:308, 452). Now the wormhole turns into a black hole, from which nothing can escape.

As such, the Kantian account is exactly the opposite of that of the *Nuclear Weapons* opinion. Human rights law and humanitarian law are not continuous locations on a single spectrum, with the former body of law shading into the other. Neither law nor morality is all of a piece. Instead, Kant would agree with the description given in the first known work suggesting the theory of humanitarian law as *lex specialis* to human rights law, but ultimately rejecting it in favor of the view that the two belonged to distinct domains (Calogeropoulos-Stratis, 1980:95–96, cited at Milanovic, 2016:98). It might therefore seem that Kant must reject the "convergence" theory of human rights law as applying alongside humanitarian law, and recommend instead that they obtain only in "law enforcement" scenarios where there is some semblance of the Postulate of Public Right. And so, despite agreeing very little with the standard approach described above in part I, it would also appear that Kant would nevertheless fully agree with their rejection of the *Ilașcu* line of cases. After all, where there is no sign of the defendant state's political institutions being capable of using force to enforce private rights in these cases, it would appear that no claim of human rights can be made, and hence no human rights obligations.

This would be a bit too quick.

5. Human Rights Obtain Even in War After All

Regardless of the "empirical" facts of the situation, a claim of human rights in war may be constructed out of implied restrictions upon what states "may be heard to say," once the individual claimant has successfully "characterized" the relation between herself and that state as one of authority and obedience. This argument has two limbs, both of which have already been introduced. The first is that the legal rights and obligations going under the name of "human rights" in contemporary legal practice require an underlying relationship of authority and obedience. Jurisdiction is the threshold that must be crossed before a human rights claim can be made. The second limb arises from the thought that the final end of war is peace. As such, where the applicant can successfully "characterize" her complaint as a violation of right occurring in a condition of peace, it is not permissible for the defendant state to defend itself by admitting the deeper wrong of waging war, or worse, a crime in war.

5.A. The Fiduciary Conception of Human Rights and the *Nemo Auditur* Limitation

To explain how these two limbs come together, we require some further examination into what sort of rights "human rights" are.[8] Recall that a human being's fundamental juridical status lies in independence from the arbitrary choice of another. Insofar as legal rights envisage the use of force by one person against another compatibly with the freedom of each, the only way in which one person may rightfully assert *authority* over another—that is to say, deal with another unilaterally, non-consensually, and hierarchically—is within the framework of a "fiduciary" relationship. In the *Doctrine of Right*, Kant introduces a third category of private right entitled "rights to persons akin to rights to things." The reasons for this curious terminology will interest us later. The prime example of this category is parental right, which arises from that fact that procreation is an act by which two parents "have brought a person into the world without his consent and on [their] own initiative..." (RL 6:280–281, 429–430). Kant's account of the legal obligations of parents has nothing to do with the interests or needs of the child. Rather, the argument is formal, and modeled upon the *negotiorum gestio* in Roman law: one person unilaterally disposing of the affairs of another is consistent with reciprocal independence only if the former administers those affairs competently and faithfully on behalf of the latter. In this way, parents

> cannot destroy their child as if he were something they had *made* (since a being endowed with freedom cannot be a product of this kind) or as if he were their property, nor can they even just abandon him to chance, since they have brought not merely a worldly being but a citizen of the world into a condition which cannot now be indifferent to them even just according to concepts of right.

Conversely, the obligation toward the child then gives rise to a right "of parents to *manage* and develop the child... not only to feed and care for him but to educate him," and also to bring the child "back even against his will into his parents" possession from another's possession (RL 6:281–82, 430). As such, a fiduciary's obligations toward the beneficiary are inextricably linked to their rights over the beneficiary. In this way, a fiduciary's rights and obligations are "inseverably compound." (Birks, 2000:33).[9] They are simply the two sides of the same coin.

At the beginning of the section on Public Right, Kant uses precisely this image of parenthood to describe "natives of a country" as "born of the same mother (the republic)..." (RL 6:343, 482). Most people no more choose their country than they do their parents. Moreover, like parents, states claim the entitlement to determine their subjects' legal rights and obligations as well as to bind them by their pronouncements.

[8] The description of human rights set out here is not intended to be valid across all historical and social contexts. Instead, it is merely an "immanent" description of the role, logic, and rationale animating the category of legal claims going under the name of human rights in the practice of contemporary human rights lawyers. Put differently, it is an attempt to identify what these legal rights *do* in legal practice, and on this basis see where they "fit" within the scheme of Kant's general theory of law.

[9] Note that Birks means something different by his use of that phrase. He intends it to describe the relationship between the fiduciary duties of competence and good faith.

Accordingly, just as private law contains a set of principles to ensure that private fiduciary relationships do not degenerate into abuse and exploitation, public law contains a set of principles to ensure that relations of authority and obedience do not degenerate into domination.[10] These are human rights (Criddle & Fox-Decent, 2016:106–113). Despite its deep theoretical differences with Dill and Besson, the Kantian account thus arrives at a superficially similar description of human rights as intimately linked with political authority.

We may now explain how human rights may nevertheless obtain in the inhospitable contexts of active or near-active conflict by analogy to *negotiorum gestio*. Suppose your neighbor has left to go traveling, leaving his lands untended. You exploit this opportunity to farm his land. When your neighbor returns, he has two choices. First, he can claim a civil wrong against you (in trespass, *iniuria*). Alternatively, he may characterize what you did not as a wrong, but as a unilateral choice on your part to administer his affairs, such that you must now account to him for all profits, as well as compensate for any damage to the land. If your neighbor elects for the second option, you may not be heard to say, "No my lord, there is no fiduciary relationship here; I was stone cold robbing him!" Remedial force is rightful only if it "hinders" a prior "wrongful" use of force incompatible with reciprocal freedom. If persons could plead further wrongs as a defense to liability for wrongs, force would end up exacerbating original hindrances. As such, the *nemo auditur* principle as described here works a hindrance of a hindrance of a hindrance of a hindrance to freedom.

Something of this can be discerned in the operative passages of *Al-Skeini*, where the Strasbourg Court held the death of an Iraqi individual in the custody of British forces operating in Basra to give rise to the UK's Article 1 ECHR state jurisdiction, simply because

> the United Kingdom ... assumed in Iraq the exercise of some of the public powers normally to be exercised by a sovereign government ... In these exceptional circumstances, the Court considers that the United Kingdom, through its soldiers engaged in security operations in Basra during the period in question, exercised authority and control over individuals killed in the course of such security operations, so as to establish a jurisdictional link between the deceased and the United Kingdom for the purposes of Article 1 of the Convention. (*Al-Skeini*, 2011: para. 149)

These passages obviously cannot be read to mean that the UK's liability under the convention arose because British forces had *actually* established even some minimally stable system of governance; the English Court of Appeal had earlier ridiculed that proposition as "utterly unreal" (*Al-Skeini v Secretary of State for Defence*, 2005:para. 124). Rather, a preferable reading is that the UK's obligations of accountability arose because it had *pretended* to those public powers; that is to say, "held itself out" as an authority over persons in the region, including the individuals killed. The question of whether the United Kingdom had the wherewithal to exercise these public powers is irrelevant at the "jurisdictional" stage, because having assumed the office of an

[10] See RL 6:306, 451, stating that public law contains "no further or other duties" than can be conceived at private law.

authority, a defendant state cannot be heard to deny the obligations that come with it.[11] What, however, if the United Kingdom denied assuming any "public powers," and argued instead that the civilians concerned had been killed in the course of wartime activity? Now, this might very well be true as a matter of fact, but it cannot be heard to say this before a court of law because to do so is to admit to a war crime.[12]

5.B. "Jurisdiction" Means Just One Thing

We may now construct a defence of *Ilasçu* and *Catan* along the same lines. The way forward may be made clearer by drawing a contrast with Raible's critique, which is that those decisions conflate the two distinct meanings of "jurisdiction"—as the source of human rights obligations, and as the set of property-like rights over territory. On the conception of states as public fiduciaries, it follows that their fiduciary office is an inalienable aspect of their "status"; that is to say, of their very persons. As such, they may neither unilaterally disclaim their office nor farm out their duties to others, as if these things were *property* they could dispose of at will. (*Lake Lanoux Arbitration*, 1957:128) This rationale governs the three fundamental characteristics of sovereigns: personhood under international law, jurisdiction over territory, and rulership over their subjects.

Pace Raible, territory is entirely unlike property, but instead more akin to the "body" of the state (cf Ripstein, 2021: chap. 2, section III).[13] Sovereign states are not real estate conglomerates. Even though prominent textbooks continue to depict territory as transmissible property, often by reference to colonial and imperial precedents (Brierly & Clapham, 2012:168–169; Shaw, 2017:363), more recent developments emphasize both the intimate connection of territory with rulership, as well as its default inalienability (e.g., *President of the Republic et al. v. Ali Ayyoub et al.*, 2017:38–39, reported in Le Moli, 2019:793).[14] As this chapter was being written, the Danish prime minister rebuffed an offer by the president of the United States to purchase Greenland with the words "Thankfully, the time where you buy and sell other countries and populations is over" (Helmore, 2019). Prime Minister Fredriksen's response was entirely fitting, because she captured fully the sense in which treating territory as property to be bought or sold is the same as treating the people living on it as chattels to be bought or sold. Accordingly, wantonly alienating independence, territory, or subjects would be tantamount to a person voluntarily enslaving or prostituting himself.[15]

[11] Speaking extrajudicially, Lord Dyson has expressed the similar thought that if a "Contracting State has taken over control of the civil administration of the foreign territory then its inability to control the situation is not a ticket out of the Convention" (Dyson, 2014:20).

[12] It is difficult to imagine how, on the facts of the case, the United Kingdom could have argued that such deaths might have met the necessity and proportionality limitations.

[13] For reasons that cannot be explored here, Ripstein resists the full analogy of territory with body.

[14] In a fascinating decision, the Supreme Administrative Court of Egypt recently struck down as unconstitutional a boundary agreement that had the effect of transferring certain islands in the Red Sea to Saudi Arabia because "the territory of the homeland . . . is not the property of any authority of the State."

[15] Both are illegal for *juridical* rather than ethical reasons. On the illegality of contracts for slavery, see TP 8:293, 294; On sexual intercourse outside of marriage as illegal self-alienation, see RL 6:359–360, 494–495.

A claim of sovereignty over territory just is the same thing as claiming authority over persons there, which in turn just is the same thing as having obligations to account to them. Once again *pace* Raible, "jurisdiction" means exactly the same thing at general international law and at human rights law—"law-speaking." The rights and obligations are inseverably compound—two sides of the same fiduciary coin. Thus expressed, the fundamental problem in *Ilaşçu*, *Catan*, and *Pueblo Bello Massacre* is that it is inconsistent with right for a state to lay claim to a territory without bearing any obligations of accountability to the people on it. Just as parents may not abandon their children to chance, states may not unilaterally disclaim their subjects (Verdross, 1937:576). It makes no difference if the territory was not abandoned but wrongfully seized by another state or rebel group. As long as a state exists, its subjects have a right to its "status" as their public fiduciary, which is neither in the power of the state to give away unilaterally, nor defeasible by the wrongs of third parties such as other states or rebel groups. The right to authority survives its violation, as do the obligations of authorities.

As such, civilians caught in NIACs retain the ability to make claims in human rights against their sovereign, regardless of the difficulties that sovereign may have in fulfilling them. Accordingly, Moldova owed obligations under the ECHR to its subjects in Transdniestria notwithstanding its loss of control over the territory, just as Colombia owed obligations under the Inter-American Convention to its citizens in Pueblo Bello despite the existence of an NIAC. Contrary to the standard approaches described, practical difficulties short of the extinction of the state or the population cannot negate the *existence* of obligations, because obligations are not contingently "allocated" with the obligor's convenience in mind. The very concept of "duty" lies in its being morally required regardless of one's inclinations. This is not to say that volitional and epistemic difficulties are irrelevant; rather, they come into the analysis during the specification of the concrete demands of the obligation in the particular circumstances of the case; that is, at the *merits*.[16] Given its total lack of physical control over Transdniestria, Moldova could not do very much for its purported subjects there, but it still owed them *something*. Hence, it owed an obligation, incumbent upon no other state, to take whatever "diplomatic, economic, judicial or other measures" feasible within its limited power (*Ilaşcu*, 2004:para. 331).

5.C. The Absolute Limit

This brings us to the absolute limit. Recall the *obiter* discussion in *Al-Skeini* of two hypothetical scenarios, the first involving shepherds simply being shot by enemy soldiers, and the second involving those shepherds first being driven to a cave by those soldiers before being shot. Recall also that the court opined that human rights jurisdiction would not have arisen on the first set of facts, but that it might have done on the second. An explanation, elusive on any of the mainstream approaches, can be provided on the Kantian account. In the second situation where the shepherds are ordered into a van and driven to a cave for a brief period, it can be said, however risibly, that they were *governed*.

[16] See Ripstein (2009:170), on the need to "state a cause of action" before one can even raise the question of "how to apply concepts of right to particulars."

Obviously, the facts of the second hypothetical do not even begin to resemble the minimal institutional framework of the Postulate of Public Right. This, however, is immaterial. In the manuscript, Ripstein remarks that Kant "sets a very low bar" for the Postulate of Public Right, and that it remains a "nice question" as to how defective institutions can be before it can no longer be characterized as a rightful condition (Ripstein, 2021: chap. 2). One possible explanation of the first *Al-Skeini* hypothetical is to say that if the bar to proving the existence of a rightful condition is very low, then the bar for disproving it must be correspondingly high. Alternatively, perhaps part of this nicety arises because what needs to be characterized is not reality as it *actually* stands, but the "constructed" reality of how things stand *as between the applicant and the defendant*. In a recent article, Haim Abraham draws upon Hobbes to illustrate the constructed nature of war: the civil condition between individual subjects of a single state coexists alongside a state of nature between states (Abraham, 2019:819). By extension, whereas a state of war may exist between two belligerents and their combatants, no such thing obtains between a belligerent and its own subjects, or between it and the other's civilians. The relevant *juridical* reality is constructed upon the parties' pleadings, which is in turn determined by what they can be heard to plead. As such, the question in the second hypothetical scenario is whether the state concerned may be heard to deny the existence of the Postulate once the applicant has fully pleaded its elements, no matter how implausibly. It cannot, because doing so would be tantamount to admitting the very worst wrong of all, a war crime.

In the first scenario, however, where the shepherds are shot where they stand, there is no wiggle room left even to *allege* the existence of a rightful condition. The shepherds have merely been "affected," in exactly the same way as the victims in *Banković* or of a drone strike. They have not even the beginnings of a claim in human rights, or indeed, any right. Force has entirely supplanted law, handing everything over to savage violence. Accordingly, these are not human rights violations, but something much, much worse.

6. Conclusion

In the appendix to the *Doctrine of Right*, Kant makes clear the three-step pattern he has repeated throughout the main text. First, a binary distinction is drawn between two juridical categories (A and B). Second, he applies this distinction against itself, creating two more categories (AB and BA). Third, of the four possibilities, one is degenerate (BA), leaving just three valid ones (A, B, and AB) (RL 6:357–358, 493; Ripstein, 2019).

The first instance of this pattern occurs in the section on Private Right. Kant starts with the binary distinction between the familiar categories of a "right to a thing (*ius reale*)" and a "right against a person (*ius personale*)"—roughly tort/property and contract. Apply this distinction against itself, and the first result is the category we scripted earlier as fiduciary relations, which Kant calls a "right to a person akin to a 'right to a thing, (*ius personaliter reale*), that is, possession (though not use) of a person as a thing . . ." (RL 6:260, 412). In this way, Kant elaborates fiduciary relations by triangulating contract with property: the fiduciary "possesses" the beneficiary in the sense of binding them by their pronouncements, but cannot "use" them self-interestedly. Overlaid across these three categories are three "bases of acquisition": unilateral, bilateral, or omnilateral, or *facto*, *pacto*, and *lege* (RL 6:260, 412). Tort/Property rights are acquired *facto*, contractual rights *pacto*, whereas fiduciary obligations are imposed *lege*—by operation of law.

A fourth possibility arises by triangulating from the opposite direction: a "right to a thing akin to a right to a person" (RL 6:358, 493). However, because it is nonsensical for you to owe obligations to a thing (your wallet), or for your wallet to be obligated to make some performance (roll over on command),[17] the possibility of a property right conceived as a contractual right "drops out without further ado" (RL 6:358, 493). This is even though it is perfectly feasible to keep a certain class of things as property, and to force them to perform tasks: slaves. Despite its "empirical" viability, the *juridical* incoherence of slavery makes it illegal (Ripstein, 2019:6). We are left with just three valid categories of private right. The fourth is empty, or *vacat*. (RL 6:241, 396).

The second instance of this pattern is in Public Right. Here, Kant begins with a binary distinction between the "right of a state" (municipal law) and the "*right of nations* (*ius gentium*)." He applies this distinction against itself to derive a third category of a "*right for a state of nations* (*ius gentium*)," also described as "cosmopolitan right (*ius cosmopoliticum*)" (RL 6:311, 455). Thus expressed, cosmopolitan right covers *private* international law; that is, the rules governing recognition of property, contract and status rights created domestically by foreign states, but which must be recognized by the forum state when individual visitors bring them into its jurisdiction (Ripstein, 2019:7). Municipal law arises *facto*—from the consolidation of power over a territory, international law arises *pacto*—from the voluntary consent of nations, and the rules of private international law arise *lege*. The fourth possibility—a right of nations conceived as the right of a single state—is empirically possible in a world state, or in an imperial order where one state determines the external relations of another as a colony. Both, however, are juridically incoherent, and so this category is *vacat*.[18]

The argument of this chapter also developed in this manner. We started with the binary distinction between peace (rights exist) and war (only prohibitions exist). The laws of war arise *facto*—one belligerent aggresses upon another. Peace is achieved *pacto*—belligerents conclude a peace treaty. Apply this distinction against itself, and you first derive a conception of a condition of war seen as a

[17] Your dog might roll over on command, but it probably does so volitionally rather than out of a sense of legal obligation.

[18] Kant fulminates against both colonialism and "universal monarchy" in *Toward Perpetual Peace*, but famously observes that the latter can only be a "soulless despotism, [which] after it has destroyed the seed of good, finally deteriorates into anarchy" (ZeF 8:367, 336). "Anarchy" and "despotism" are technical terms: in the fourfold categorization of forms of government Kant sets out in the *Anthropology*, Kant defines *anarchy* as a defective form where law and freedom obtain without force and *despotism* as where law and force obtain without freedom. The other two are *republic*—where force, freedom and law obtain harmoniously; and *barbarism*—where there is force without law or freedom (Kant, 2006d:235 (AA 7:330–331)). Although he states that only a republic "deserves to be called a true civil constitution . . ." (AA 7:331, 236), it does appear that despotism and anarchy also qualify as defective rightful conditions deserving of freedom from external interference (Ripstein, 2021: chap. 9). The form of government Moldova and Colombia exercised over the claimants in *Ilaşcu* and *Puerto Bello Massacre* would qualify as Kantian anarchy, meaning that claims of right and obligation can validly obtain. A *universal* despotism or anarchy, however, would be intolerable in a way their regional correlates aren't, because there is no escaping the former.

condition of peace. Notwithstanding actual conditions of conflict, if an applicant can make a colorable claim of a rightful condition between herself and the defendant state, the latter may not be heard to deny it. This *nemo auditur* restriction applies *lege*, as a hindrance of a hindrance of a hindrance of a hindrance to freedom. The fourth possibility is where a condition of peace is treated as a condition of war. A white flag formally signifies peace but soldiers can raise it perfidiously as a tool of war. Unarmed shepherds may not be part of the war effort but may be shot as if they were soldiers. These are all empirically possible, but they hand everything over to savage violence; that is, to a juridical emptiness from which nothing can escape.

Acknowledgment

My deepest thanks to my fellow participants at the *Workshop—The Public Uses of Force: Europe and Beyond*, organized by Ester Herlin Karnell and Enzo Rossi at University of Amsterdam (June 7, 2019), as well as to my colleagues at Oxford Brookes University School of Law. References to Kant's works list the sigla used by *Kant-Studien* first, the page numbers of the *Akademie* compilation of Kant's collected works second, and those of Gregor's translation third.

References

Abraham, H. (2019). "Tort Liability for Belligerent Wrongs." *Oxford Journal of Legal Studies* 39(4): 808–833.

Besson, S. (2012). "The Extraterritoriality of the European Convention on Human Rights: Why Human Rights Depend on Jurisdiction and What Jurisdiction Amounts to." *Leiden Journal of International Law* 25(4): 857–884.

Besson, S. (2013). "The Allocation of Anti-poverty Rights Duties: Our Rights, but Whose Duties?" In Nadakavukaren-Schefer, K. (ed.), *Poverty and the International Economic Legal System* Cambridge: Cambridge University Press, pp. 408–431.

Birks, P. (1997). "Definition and Division: A Meditation on Institutes 3.13." In Birks, P. (ed.), *The Classification of Obligations*. Oxford: Clarendon Press, pp. 1–35.

Birks, P. (2000). "The Content of Fiduciary Obligation." *Israel Law Review* 34(1): 3–38.

Brierly, J. (2012). *Brierly's Law of Nations: An Introduction to the Role of International Law in International Relations*. Oxford: Oxford University Press.

Calogeropoulos-Stratis, A. (1980). *Droit humanitaire et droits de l'homme: La protection de la personne en période de conflit armé*. Geneva/Leiden: IUHEI Press.

Criddle, E., & Fox-Decent, E. (2016). *Fiduciaries of Humanity: How International Law Constitutes Authority*. New York: Oxford University Press.

Dill, J. (2019). "Distinction, Necessity, and Proportionality: Afghan Civilians' Attitudes toward Wartime Harm." *Ethics & International Affairs* 33(3): 315–342.

Dill, J. (2020). "Towards a Moral Division of Labour between IHL and IHRL during the Conduct of Hostilities." In Bohrer, Z., Dill, J., & Duffy, H. (eds.), *The Law Applicable to Armed Conflict, Max Planck Trialogues on the Law of Peace and War*. Cambridge: Cambridge University Press, pp. 197–265.

Dyson, J. (2014). "Essex Lecture 2014: The Extraterritorial Application of the European Convention on Human Rights." Essex: University of Essex Press.

Frowe, H. (2014). *Defensive Killing*. Oxford University Press.

Frowe, H. (2015). "Can Reductive Individualists Allow Defense Against Political Aggression?" In Soble, D., & Valentine, S. (eds.) *Oxford Studies in Political Philosophy*. Oxford: Oxford University Press, pp. 173–193.

Haque, A. (2017). *Law and Morality at War*. Oxford: Oxford University Press.

Helmore, E. (2019). "Trump cancels Denmark trip after PM says Greenland is not for sale." *The Guardian* [online]. August 21, 2019.

Kant, I. (2006a) (1797). "The Doctrine of Right." In Gregor, M. (ed.), *Practical Philosophy: The Cambridge Edition of the Works of Immanuel Kant*. Cambridge: Cambridge University Press, pp. 386–506.

Kant, I. (2006b) (1793). "On the Common Saying: That May Be Correct in Theory, But It Is of No Use in Practice." In Gregor, M. (ed.), *Practical Philosophy: The Cambridge Edition of the Works of Immanuel Kant*. Cambridge: Cambridge University Press, pp. 273–309.

Kant, I. (2006c) (1795). "Toward Perpetual Peace." In Gregor, M. (ed.), *Practical Philosophy: The Cambridge Edition of the Works of Immanuel Kant*. Cambridge: Cambridge University Press, pp. 311–351.

Kant, I. (2006d) (1798). In Louden, R. (ed.), *Kant: Anthropology from a Pragmatic Point of View*. Cambridge: Cambridge University Press.

Le Moli, G. (2019). "*President of the Republic et al., v. Ali Ayyoub et al.*" *American Journal of International Law* 113(1): 791–798.

McMahan, J. (2006). "Killing in War: A Reply to Walzer." *Philosophia* 34(1): 47–51.

McMahan, J. (2009). *Killing in War*. Oxford: Oxford University Press.

McMahan, J. (2010). "Laws of War." In Besson, S., & Tasioulas, J. (eds.), *The Philosophy of International Law*. Oxford: Oxford University Press, pp. 493–510.

Milanovic, M. (2016). "The Lost Origins of *Lex Specialis*: Rethinking the Relationship between Human Rights and International Humanitarian Law." In Ohlin, J. (ed.), *Theoretical Boundaries of Armed Conflict and Human Rights*. New York: Cambridge University Press, pp. 78–117.

Milanovic, M. (2011). *Extraterritorial Application of Human Rights Treaties: Law, Principles, and Policy*. Oxford: Oxford University Press.

Milanovic, M. (2012). "*Al-Skeini* and *Al-Jedda* in Strasbourg." *European Journal of International Law* 23(1): 121–139.

Pettit, P. (1997). *Republicanism: A Theory of Freedom and Government*. Oxford: Oxford University Press.

Raible, L. (2018). "Title to Territory and Jurisdiction in International Human Rights Law: Three Models for a Fraught Relationship." *Leiden Journal of International Law* 31(2): 315–334.

Ripstein, A. (2009). *Force and Freedom: Kant's Legal and Political Philosophy*. Cambridge, MA: Harvard University Press.

Ripstein, A. (2019). "Bringing Rights and Citizenship under Law on a Globus Terraquaeus." Presented at the 2019 World Kant Congress, Oslo.

Ripstein, A. (2021, forthcoming). *Kant and the Law of War*.

Rostbøll, C. (2019). "Kant and the Critique of the Ethics-First Approach to Politics." *Critical Review of International Social and Political Philosophy* 22(1): 55–70.

Shaw, M. (2017). *International Law* (8th ed.). Cambridge: Cambridge University Press.

Shue, H. (2010). "Laws of War." In Besson, S., & Tasioulas, J. (eds.), *The Philosophy of International Law*. Oxford: Oxford University Press, pp. 511–564.

Shue, H. (1996). *Basic Rights: Subsistence, Affluence, and U.S. Foreign Policy* (2nd ed.). Princeton, NJ: Princeton University Press.

Shue, H. (2008). "Do We Need a 'Morality of War'?" In Rodin, D., & Shue, H. (eds.), *Just and Unjust Warriors: The Moral and Legal Status of Soldiers*. Oxford: Oxford University Press, pp. 87–111.

Skinner, Q. (1998). *Liberty Before Liberalism*. Cambridge: Cambridge University Press.

Verdross, A. von (1937). "Forbidden Treaties in International Law." *American Journal of International Law* 31(4): 571–577.

Walzer, M. (1977). *Just and Unjust Wars: A Moral Argument with Historical Illustrations* (4th ed). New York: Basic Books.

Walzer, M. (2006). "Response to McMahan's Paper." *Philosophia* 34(1): 43–45.

Willaschek, M. (2009). "Right and Coercion: Can Kant's Conception of Right Be Derived from His Moral Theory?" *International Journal of Philosophical Studies* 17(1): 49–70.

Zylberman, A. (2016a). "Why Human Rights? Because of You." *Journal of Political Philosophy* 24(3): 321–343.

Zylberman, A. (2016b). "Human Rights and the Rights of States: A Relational Account." *Canadian Journal of Philosophy* 46(3): 291–317.

Other References

Al-Skeini and ors, R (on the application of) v. Secretary of State for Defence [2005] EWCA (Civ) 1609.

Al-Skeini and ors v. United Kingdom, Merits and just satisfaction [2011] ECHR 1093; (2011) 53 EHRR 18.

Banković and ors v. Belgium and ors, Admissibility ECHR 2001-XII; [2001] ECHR 890.

Case of the Pueblo Bello Massacre v. Colombia, del Carmen Álvarez Blanco and ors v Colombia, Merits, reparations and costs [2006] IHRL 1524.

Catan v. Moldova and Russian Federation, Merits and just satisfaction [2012] ECHR 1827; (2013) 57 EHRR 4.

European Convention for the Protection of Human Rights and Fundamental Freedoms, Nov. 4, 1950, 213 UNTS 221, n.d.

Ilaşcu and ors and Romania (intervening) v. Moldova and Russian Federation, Admissibility ECHR 2004-VII; [2004] ECHR 318.

International Covenant on Civil and Political Rights (ICCPR), adopted Dec. 16, 1966, 999 U.N.T.S. 171, n.d.

Issa and ors v. Turkey, Merits [2004] ECHR 629; (2004) 41 EHRR 567.

Lake Lanoux Arbitration, France v. Spain (1957) XII (1963) RIAA 281.

Legality of the Threat or Use of Nuclear Weapons, Advisory Opinion [1996] ICJ Rep 226.

President of the Republic et al. v. Ali Ayyoub et al, (2017) Judgment, on appeal No 74236/62 J S.

13

Kant and the Criminal Law of War

Malcolm Thorburn

Introduction

On the surface, Arthur Ripstein's *Kant and the Law of War* looks very much like a number of other recent philosophical writings trying to make moral sense of the law of war. Like them, Ripstein's account explains why certain doctrines are morally justified, suggests that some others are morally dubious, and interweaves tidbits of the political and intellectual history of the law of war in with his moral account. But Ripstein's treatment of the law of war is radically discontinuous from almost all of the contemporary philosophical literature in the field because it works with a very different understanding of what it is to provide a moral justification for legal ordering. The standard approach, whether it is in the law of war or in more traditional areas of domestic law, is to compare the substantive norms of a particular area of law with what the author takes to be the demands of morality and to use that comparison as the grounds for moral evaluation of the field. On Ripstein's Kantian approach, however, that is to put the cart before the horse. Before we ask what norms a given area of law ought to include, we need first to recognize that these are norms put in place as between some authority and its subjects. So the first normative questions we must confront concern that relationship of authority that obtains between them: On what grounds does the lawmaker have legitimate authority over its subjects? And what are the limits of that legitimate authority?

This Kantian shift in focus toward legal relationships of authority has massive implications for our understanding of all areas of law. In Ripstein's Kantian account of private law, *Private Wrongs*, he argued that most of the norms of private law property, contracts, tort, and unjust enrichment are best understood as facets of how persons can relate to one another while remaining independent from one another, no one taking on a position of authority over anyone else (Ripstein 2016). In *Force and Freedom*, his groundbreaking work on Kant's general theory of law, Ripstein argued that it is both possible and necessary for a state to exercise justified authority over its subjects, and he outlined what sorts of matters are within its jurisdiction in doing so (Ripstein 2009). In *Kant and the Law of War*, Ripstein extends the Kantian analysis to relationships among independent states in war and in peace, and to the implications of that analysis for the individuals caught up in these relationships. As we follow Ripstein's Kantian analysis, we find that not only can some of the doctrines that have most troubled other moral theorists be explained by the Kantian account, but they turn out to be central implications of it.

Malcolm Thorburn, *Kant and the Criminal Law of War* In: *The Public Uses of Coercion and Force*. Edited by: Ester Herlin-Karnell and Enzo Rossi, Oxford University Press. © Oxford University Press 2021. DOI: 10.1093/oso/9780197519103.003.0013

In this chapter, I focus on Ripstein's account of a doctrine that has caused a great deal of trouble to moral philosophers of law over the years: the equal criminal immunity of combatants (which he treats primarily in chapter 6). It is the doctrine according to which soldiers are held immune from criminal prosecution for their war-making activities irrespective of whether they fought for a just cause (defending their country) or as part of their country's unjustified aggression.[1] It is probably the best example of a doctrine that appears to most moral philosophers to be either totally repugnant or else explicable only as a sort of depressing concession to reality. To Ripstein's Kantian eye, however, it comes through as a clear and necessary aspect of what international legal ordering is all about.

* * *

This chapter proceeds in three parts. In part 1, I set out the contours of the Kantian approach to the morality of law, which begins with the relationship of public authority and only later proceeds to the evaluation of how that authority has been exercised. In part 2, I consider Ripstein's application of that approach to the legal equality of combatants in war. Although his account suggests a comprehensive justification of the doctrine, his chapter does not spell this out fully. In this part, I suggest a way in which he can do so. Finally, in part 3, I conclude with some comments on the vision of international relations and of individual responsibility reflected in Ripstein's Kantian account and I draw some lessons for my own field of study, criminal law theory.

1. The Priority of Authority Relations

Ripstein's Kantian account of law challenges the most basic understanding of what normative legal philosophy is all about. Like many moral philosophers, Ripstein is interested in showing how legal ordering and legal institutions can be morally justified. But the approach he takes to that question is radically different from the one that most of his contemporaries have taken. For most of them, the way to justify a legal order is to show how it can best serve as a tool for bringing about greater compliance with morality's demands. The usual way to do this is in two steps. At the first step, we determine what morality demands of each one of us, and we use this as a guide to what the law's first-order norms ought to be.[2] *Ceteris paribus*, the closer the law cleaves to the demands of morality, the closer it is to being fully morally justified. At the second step, we ask what means it would be morally acceptable for legal officials to use to induce people to comply with the law's demands. In short, we try to reduce the law to a

[1] The *Geneva Convention relative to the Treatment of Prisoners of War* (the Third Geneva Convention) was adopted in 1949, replacing the Geneva Convention of 1929; it governs the treatment of prisoners of war. Under the Convention, enemy combatants may only be captured as prisoners of war, to be released upon the end of hostilities. They may not be punished for any killing (so long as it took place according to the norms of IHL requiring discrimination between combatants and noncombatants and proportionality between harm and military advantage).

[2] The locus classicus of this view in the modern literature is Raz (1988:53ff), but it is at the core of many other influential writers' accounts, such as Antony Duff, John Gardner, and others.

set of questions about what morality requires of each individual (whether subject or enforcer) and we try to ensure compliance with those demands as much as possible.

According to this sort of approach, law is simply a tool by which to transmit moral norms—and, like most tools, it can be taken up by anyone who is good at handling it. If the norms are good ones, there should be no principled reason (on the standard account) why it must be our national government that tells us to follow them, rather than a foreign government, a private corporation or the busybody next door. The object of moral concern (on the standard account) is the set of norms themselves as they apply to subjects and to officials, rather than any special relationship between the putative authority and the putative subject. On Ripstein's Kantian account, by contrast, this way of thinking gets things backward. For Ripstein, we cannot ask any interesting questions about the moral merits of specific legal norms until after we have settled whether the party issuing those norms stands in a relationship of legitimate authority over its putative subjects. There are simply no genuine legal norms to evaluate unless they issue from within a relationship of legitimate authority. And for Kantians, the legitimacy of a putative authority is a distinct and prior question to the merits of the norms it has put in place over its putative subjects.

Now, of course, one might deny the distinctness of these two questions in the manner of Joseph Raz. According to his influential "Normal Justification Thesis," the legitimacy of a putative authority just is a function of the moral merits of the norms it puts in place. Where the putative authority provides good guidance to its putative subjects on what morality requires of them, it thereby has practical authority. But it is precisely this elision of legitimate authority and the moral merits of the norms a putative authority puts in place that Ripstein's Kant denies. In moral life, each person should ask herself "how ought I to act?," and it is up to each of us to conform to morality's demands as best we can. In law, however, we ask the irreducibly *relational* question "what is this authority entitled to obligate me to do?" Of course, moral obligations are often relational in one sense: they often owed *to someone* in particular.[3] Nevertheless, should we fail to discharge our moral obligations, there is no one with standing to enforce those demands on us. To be sure, the person we have wronged may have good grounds to call out our moral wrongdoing and to criticize us for it, but she does not have the standing to force us to live up to those demands. In law, however, there is always an authority that has a special moral standing not merely to call us to account, but to use coercive force to require us to live up to our legal obligations.

The Kantian account responds very differently to the anarchist's challenge that has given shape to so many contemporary accounts of authority.[4] For most other contemporary scholars, the anarchist's challenge has been insurmountable: they have not been able to explain how anyone could ever legitimately be entitled to have the normative power to decide what another person must do and to be entitled to coerce her to do so, where necessary. Without an account of public authority of this kind, they have tried to work around its absence by proffering a sort of *ersatz* authority in its place. They insist that no one is entitled to stand in such a position over others; but

[3] The two most influential recent accounts of this phenomenon are: Scanlon (2000) and Darwall (2009).

[4] Joseph Raz structures his account of political authority in *The Morality of Freedom* as a response to Wolff's (1970) anarchist challenge.

(they insist), it is enough to say that we are all at liberty to use some coercion to induce others to comply with freestanding moral norms that apply to them. If we sidestep the deeply illiberal implications of this view of coercion in the service of morality, we might think we have found a way around the problem of authority. But even so, the enterprise we have thereby justified is quite different in structure from our ordinary understanding of legal relations. On this picture, there is no entity with a special standing to set norms over us; there is no direct connection between legal norms and their coercive enforcement; and the only legitimate form of political debate is to try to find those laws that most closely approximate the truth about our moral obligations. These and many other features of the standard account of law's claim of authority make it deeply unsatisfying.

For Ripstein's Kant, the question of public authority is prior to and distinct from the merits of the norms the authority might put in place. But if, as Ripstein's Kantian account insists, the proper starting place for the moral evaluation of a legal system is the existence of a relationship of legitimate public authority, how should we determine if we are dealing with such a legitimate public authority? As we have already seen, Ripstein's Kant rests legitimate public authority on what might seem like the flimsiest of grounds: effective control. This is because the moral question of justified public authority is answered by effective control: we need a single, effective authority that can provide *some* answer to the question of how we may relate to one another around here. To have some answer to that question is *almost* all that is required in order to solve the basic moral question of public authority. Once there is a single, effective authority in place, of course, then we can ask questions about the moral merits of the legal norms around here. Until then, however, there just is no system of legal norms to evaluate.

Effectiveness is not the only requirement for a valid claim of public authority, however. If public authority were just a solution to the problem of public order, then perhaps *any* effective solution might be enough to meet its demands.[5] But that is not the problem to which Ripstein's Kant sees public authority as providing the answer. According to Ripstein's Kantian picture, legal authority is required for the possibility of individual freedom because the alternative of living without any public authority is not independence for all but rather dependence of each person on everyone else (or, as John Locke once put the point, "wherever law ends, tyranny begins . . ."[6]) That is, in the absence of any public authority, each person is at the mercy of all others; it is a stable public authority that gives rise to a stable set of rights that realize our independence.

A claim of public authority, for Ripstein's Kant, is distinctive, not reducible to brute power over others. The effective solution to the question of "who is in charge of what around here?" must be one that does not reduce independent persons to mere things at the mercy of another's brute power. In order for the entity with effective power to constitute valid authority, rather than mere tyrannical power, it must present itself as a solution to the problem of how persons may live among others while preserving their independence. Indeed, a valid public authority must present itself as *nothing*

[5] By contrast, Bernard Williams (2005:3) insists that the "first political question" is simply that of securing "order, protection, safety, trust, and the conditions of cooperation. . . ."

[6] John Locke, "Second Treatise of Government" in *Two Treatises of Government*, ed. Peter Laslett (Cambridge: Cambridge University Press, 1988), ch. 18, s. 202.

more than a solution to that problem. Anytime that a public official decides to exercise public power for his own purposes, rather than as the part he has been assigned in the solution to the problem of independence of each, he exceeds the bounds of his lawful authority. This idea is captured in the idea of a legal office.[7] Public officials must never simply be given powers or rights as such; instead, we must conceive of them as occupying an office that is made up of a set of rights, powers, duties, privileges, etc., that exist simply to serve the purpose that defines their office. So long as this is the case, we are dealing with a genuine authority—and as soon as the officeholder exceeds those purposes, it is acting outside its jurisdiction. Of course, this means that there is very considerable room for rather awful but nonetheless legitimate authorities. They might address the questions that define their authority in stupid, cruel, wasteful, and small-minded ways—indeed, a study of human history would suggest that that is usually the case—but so long as they are exercising their powers for the purposes that define their office, rather than at their own whim, they remain genuine legal authorities and not mere tyrants.

One signal advantage of the Kantian approach is that it makes considerable room for vigorous political disagreement within an otherwise legitimate legal order.[8] We can insist that a given institution has genuine authority over us even though we think that many of its laws are misguided. The legitimacy of a particular institution, on this account, is a formal question about whether it stands in the right sort of relation to us. Although the moral merits of certain laws are very important, this is a distinct question from whether the authority who put them in place is legitimate.[9] And that means not only that institutions that put in place bad laws might still be genuine authorities, but also that institutions that purport to put in place morally impeccable "laws" but who lack the proper standing to do so are still not genuine authorities. The first question of public authority, for Ripstein's Kant, is whether the authority stands in the right relation to those over whom it claims authority—a question that is answered by reference to its effectiveness and by whether its efforts are set out as answers to the questions that define its authority. This is not only a prior question, but also a conceptually distinct question from whether the laws the authority puts in place are good ones.

2. Justifying the Equality of Combatants

Arthur Ripstein has done more than almost anyone in recent times to develop the general Kantian account of law and to set out its implications across a variety of different

[7] Arthur Ripstein, "Property and Sovereignty: How to Tell the Difference," 18 *Theoretical Inquiries in Law* 243 (2017). See also my "Policing and Public Office" 70 *University of Toronto Law Journal* 248–266 (2000).

[8] Jeremy Waldron's *Law and Disagreement* (Oxford: Oxford University Press, 1999) gave rise to a large literature on the importance of political disagreement. His *Political Political Theory: Essays on Institutions* (Cambridge, MA: Harvard University Press, 2016) points out the importance for political theory to leave room for politics (rather than dictating substantive answers to matters of political dispute).

[9] This stands in stark contrast to the currently fashionable moralist position, summarized succinctly (*The Morality of Freedom* at 104) by Joseph Raz as follows: "The puzzling aspect of our conclusions is in the refusal to give a yes or no answer to the question: is the authority of the government legitimate? We concluded that it is legitimate to various degrees regarding different people."

doctrinal areas. In *Kant and the Law of War*, he applies that framework to a number of important questions about the relations among nations in war and in peace, often to brilliant effect. In this chapter, I focus on one area of the law of war that has deeply troubled many contemporary just war theorists and that is of particular interest to criminal law theorists: the equal criminal immunity of combatants. Under modern International Humanitarian Law (IHL) as set out in the Third Geneva Convention, soldiers are immune from criminal prosecution for their otherwise criminal acts of killing, maiming, property damage, and the like, so long as they abide by IHL norms requiring discrimination between combatants and noncombatants and proportionality between harm and military advantage.[10] What is more, this immunity from criminal prosecution and punishment applies equally to combatants on both sides of a war. That is, it is a matter of indifference to modern IHL whether a soldier's killing, maiming, etc. was part of a morally justified effort to defend his country from attack or whether it was part of a totally unjustified act of aggression by his country. To take a particularly pointed example, Wehrmacht soldiers invading Poland in 1939 and Polish soldiers defending their homeland from German aggression both enjoyed the same immunity from prosecution under the legal equality of combatants doctrine.

This doctrine has been deeply troubling to a great many contemporary moral philosophers for rather obvious reasons. For we must keep in mind that the strategy most of them use when making moral sense of the law is to compare its content to their best understanding of interpersonal morality. In this case, the conduct in question includes the worst crimes imaginable that are normally subject to very severe punishment: murder, attempted murder, aggravated assault, the large-scale destruction of property, and much more. So what good moral reason might we have to hold soldiers immune from prosecution for doing these terrible things? If it were a situation of genuine national self-defense (as with Polish soldiers defending their country), we might perhaps think that there was at least something close to a moral justification available. But the equal immunity of combatants doctrine applies even to soldiers who lack any reasonable grounds to believe that they were defending their country from attack. German Wehrmacht soldiers benefit from combatant immunity irrespective of whether they personally believed (much less whether they *reasonably* believed) that they were genuinely defending their country from Polish aggression. It seems that this doctrine gives individuals *carte blanche* to engage in killing just because they are in an army that is at war. It seems hard to imagine how that fact could have the requisite moral significance to justify murder and the other horrors of war.

How, then, can we justify this doctrine of combatant immunity? Solutions to this puzzle have usually come in one of two flavors. Traditional just war theorists have provided ever more elaborate arguments to show that, when all is said and done, individual soldiers on both sides of a conflict really are morally justified in doing all that they do. Michael Walzer has argued that a soldier on either side of a conflict "can be personally attacked only because he is already a fighter. He ... has allowed himself to be made into a dangerous man" (Walzer 1992:145). But that seems like a preposterous argument. A person who is ready to defend his homeland is, indeed, "a dangerous

[10] See n. 1 supra.

man" toward those attacking his homeland but why should that give the aggressor justification in killing him? Should we also let aggressive criminals attack and kill armed police officers executing arrest because they, too, are "fighters" and "dangerous men"? Of course not. Tom Hurka has suggested a slightly different strategy. "[Soldiers'] situation," he writes, "is like that of boxers who, in agreeing to a bout, permit each other to do in the ring what would be forbidden as an assault outside it. And just as the boxers' interaction is governed by formalized rules, so is the soldiers' . . ." (Hurka 2007:210). Here, too, the suggestion seems preposterous. We do not generally allow individuals to consent to being targeted for killing in any setting. What is more, the doctrine applies whether or not the soldiers on both sides were conscripted into service, rather than actually consenting to fight. So Hurka's argument from consent fails, too.

In recent years, most just war theorists have given up on these and other efforts to show that killing by both sides in war really is morally justified. Uniting under the banner of "revisionist just war theory," they have insisted that the laws of war are wildly out of step with the demands of morality and cry out for radical revision. Jeff McMahan, one of the leading revisionist just war theorists, explains their position as follows: "It is imperative that we work to bring the law of war into closer conformity with the morality of war, as we have succeeded, to a considerable degree, in bringing domestic criminal law into conformity with the morality of interpersonal relations in domestic society" (McMahan 2011:107–108). Despite this critical attitude toward the law of war, however, McMahan and many other revisionist just war theorists insist that there is a good explanation for the legal equality of combatants. Their explanation focuses on the rules of enforcement, rather than on the morality of the underlying conduct. Although aggressors are wrongdoers in ways that defenders usually are not, they argue, it is best not to punish either side because prosecuting and punishing them would do more harm than good. As McMahan puts it, "the law of war is designed not to protect rights but to prevent harm" (McMahan 2009:107). Adil Haque has extended this argument by analogy to the doctrine of diplomatic immunity. Countries agree not to prosecute one another's diplomats for crimes they commit abroad, Haque suggests, "on prosaic considerations of treaty and custom, reciprocity and impartiality, marginal incentives and aggregate consequences" (Haque 2017:28). That is, on the revisionist account, there is no deep moral justification for combatant immunity or for diplomatic immunity. It is just that granting immunity in this way is a useful convention, for it tends to reduce the total amount of harm we cause in the world.

Ripstein rejects both the traditional just war attempts to show that soldiers are justified in killing in war, and the revisionist attempt to reduce the doctrine of combatant immunity to nothing more than a useful convention. Nevertheless, his account is much closer to the revisionist account than it is to the traditional just war view. For Ripstein, like the revisionists, sees combatant immunity as something quite distinct from the moral merits of the underlying conduct. It is, instead, concerned with the decision to prohibit, to prosecute, and to punish. Where he differs from the revisionists, however, is in their understanding of the doctrine as a mere convention, based on "prosaic considerations" of aggregate harm reduction. If we think of combatant immunity (or diplomatic immunity) as just a convention based on harm reduction, Ripstein argues, "[t]he difficulty is not just that [its] empirical justifications are only as good as the hypotheses on which they rest, and that the generality of those hypotheses is uncertain. A more significant difficulty lies in the thought that this really is a compromise, rather than a matter of principle.... [W]ill diplomacy in general be undermined if we retaliate against their diplomats or will it instead perhaps be strengthened providing tit-for-tat incentives? These questions are not easy to answer, but they also seem like the wrong questions to ask" (Ripstein 2021: chap. 6, section II).

Ripstein argues that the broader Kantian approach to law that he has set out in his earlier works provides a neater, more elegant, and altogether more convincing account of the equal legal immunity of combatants from criminal prosecution. His argument does not depend on an implausible account of the morality of killing in war (in the manner of traditional just war theorists); neither does it turn on the unpredictable effects of one legal rule or another on the conduct of others (in the style of McMahan and other revisionist just war theorists). Instead, Ripstein's argument, once more, arises from the relationship of authority between ruler and ruled as prior to any discussion of the moral merits of the norms the ruler might put in place. The question for Ripstein is neither "was the killing morally justified?" nor "is there good moral reason not to punish these wrongdoers?" Instead, it is the prior question of authority: "Does anyone have the moral standing to prohibit that conduct in the first place?"

To see how Ripstein's Kantian argument proceeds, it is best to begin with a simpler case of criminal liability outside the context of war. As we have seen earlier, the Kantian account of justified public authority is based on the morally necessary function that only states can perform. That is, in order for individuals to live together while still retaining their status as independent, they must live under public laws that define the scope of each person's jurisdiction. That means that it is not only permissible but necessary for states to put in place laws governing the relations among persons and ensuring through coercion that those laws are effective. Those laws are of various sorts, but they normally include criminal laws prohibiting all of the various horrors of war: murder, attempted murder, large-scale intentional destruction of property, and so on. Outside the context of war, then, it is normal to find states around the world prohibiting individuals within their jurisdiction from engaging in all these sorts of conduct.

If the normal course of affairs is to prohibit all such conduct, what could possibly give us reason to grant a general immunity to soldiers for committing such atrocities as part of their country's war efforts? In true Kantian fashion, Ripstein insists that both diplomatic immunity and combatant immunity are necessary features of an international order made up of independent sovereign states. "[D]iplomatic immunity," Ripstein argues, "is not an exception to the principles of political independence and territorial integrity (and so territorial jurisdiction) but rather an instance of its underlying rationale" (Ripstein 2021:chap. 6, section II). Ripstein puts the same point perhaps even more starkly as follows: "the general prohibition on aggressive war is a direct implication of each legal order's entitlement to [...] independence, understood as non-subordination by another legal order. *This same claim against subordination of one legal order to another* stands in the way of punishing a prisoner of war or defeated soldier for fighting without just cause" (ibid.). That is, Ripstein wants to show that combatant immunity (like diplomatic immunity) is conceptually required by the very idea of independence that animates the structure of international relations and, in particular, the general Kantian prohibition against aggressive war.

The basic Kantian story here is actually fairly straightforward. Soldiers, when they are simply doing the normal work of fighting a war—attacking enemy soldiers and other military targets, without imposing intentional or disproportionate harm on civilians in the process—should be thought of not as individuals doing individual criminal acts, but as the physical embodiment of their state, engaged in war against another state. Although states are fully entitled to prohibit this sort of killing and destruction of property when committed by individuals in their private capacity, they do not stand in the right sort of authority relation vis-à-vis other states to prohibit them from doing so. States are one another's equals on the world stage, none in a position of authority over any other. For that reason, their actions are simply outside the scope of any other state's authority to prohibit or punish. That categorization of soldiers' war-making as acts of states, rather than as acts of individual wrongdoers is the central move in the Kantian argument for combatant immunity. To articulate the position, however, is not (yet) to provide an argument for it. In order to justify that position, Ripstein needs to show a number of things. First, he must show why we should ever think of soldiers as just the embodiment of their state, rather than as individuals doing criminal acts. And second, he must explain why states must defer to wildly implausible determinations by others states that they are under attack and need to take defensive action. Although Ripstein sets out all of the basic building blocks for the Kantian account of combatant immunity, he does not draw them together tightly in an explicit argument in chapter five of *Kant and the Law of War*. This is understandable, since he is busy there with a plethora of implications of the Kantian account of war and international relations more generally. Nevertheless, it leaves Ripstein's argument for this crucial and controversial doctrine rather sketchy and unsatisfying to the many just war theorists who are already skeptical of his Kantian starting points. In what follows, I draw together the strands of the argument and make them more explicit.[11]

Looming in the background of Ripstein's Kantian argument about combatant immunity is his defense of the larger Kantian view about the legitimate grounds for going to war in the first place. A key Kantian innovation that Ripstein makes much of in the earlier chapters of his book (primarily but not exclusively chapter 3—national defense) is that the grounds of war championed by earlier writers on just war theory—to adjudicate a legal dispute among nations, or to punish wrongdoing by nations[12]—can never justify a nation going to war. The only justification for going to war, according to Kant, is national defense.[13] As one might have come to expect from Ripstein, this flows from Kant's understanding of the relations of authority and independence between states. States, unlike natural persons, do not have purposes of their own. As we saw before, each state, on the Kantian picture, exists only to solve the moral problem of public authority: to put in place a rightful condition among the persons

[11] In what follows, I borrow from Thorburn (2019).

[12] Hugo Grotius, Samuel Pufendorf and Emer de Vattel argued that war was justified as a procedure for adjudicating legal disputes (either punishment or for resolving a matter of private rights) among nations. Ripstein discusses Kant's rejection of this view in chapter 3 of his manuscript.

[13] Kant's understanding of jus ad bellum, which was heterodox at the time, is now enshrined in the United Nations Charter s. 51: "Nothing in the present Charter shall impair the inherent right of individual or collective self-defence if an armed attack occurs against a Member of the United Nations . . ." The permission for defensive war occurs within the context of a general prohibition on war-making. For a detailed treatment of the outlawing of war and its historical context, see Hathaway & Shapiro (2017).

in its jurisdiction. Still, however limited the purposes of a state on the Kantian picture might be, each people is entitled to govern itself in its own way through its public authority. That is, just as each individual is entitled to be independent of other individuals, so, too, each state is entitled to govern itself free of the meddling of other nations. States, like individuals, have a right to independence.

If all states are entitled to independence of this sort, then this has two important implications for the Kantian account. First, it means that no state is entitled to claim authority over any other—each is sovereign within its own jurisdictions (or, as Ripstein (2021: chap. 6, section II) puts it, "one legal order cannot impose its legal order on another . . ."). Second, it means that every state is entitled to use force to resist attempts by other nations to interfere with it. For a nation to defend itself is not to claim authority over the invader; it is simply to act to ensure that one's own independence continues uninterrupted. Nations cannot set and enforce rules of conduct on other nations, for this would be to assume a position of authority over them that we do not legitimately have. But they can use force to repel attacks on their territory, for that is simply upholding their own independence in the face of attack, rather than assuming a position of authority over others.

When we put these two claims together, however, they lead to a rather curious result. In many cases, it is fairly easy to make sense of the fact that no state stands in a position of authority over any other: it is not the business of the government of Poland how the government of Germany chooses to organize its internal affairs. Where things become difficult, however, is when a state answers for itself the question whether it is under attack and needs to undertake preemptive force against the "attacking" army. For here, the norm of deference—that no state is in a position of authority over any other—requires that even if that answer is totally preposterous (as it was in the case of Germany's claim of self-defense against Poland in 1939), other countries (including the countries it thereby attacked, such as Poland) have no standing to question its determination of the matter.

This is already a remarkable result: *no nation ever* has the standing to adjudicate whether another country's claim to be defending itself is correct—even where (as in the case of Germany in 1939) it is patently obvious that the whole thing is a charade. So far, Ripstein's Kantian account concerns only the relations among nations. But in the combatant immunity doctrine, we are concerned with the criminal liability (or immunity) *of individual soldiers* for the acts they commit in wartime. How do the Kantian principles of international relations generate this peculiar doctrine concerning the liability of individual soldiers? It is here that Ripstein's argument leaves out the two crucial steps outlined earlier, which I will try to supply in what follows.

2.A. The Public Office of Soldier

The first missing step in Ripstein's account is why it is that states should consider members of one another's armed forces simply as parts of their state apparatus, rather than as individual criminal wrongdoers. The first feature of diplomats and soldiers is, Ripstein (2021: chap. 6, section III) suggests, that they are "official representatives of the sending country." As such, they are not mere subjects of their own country's laws (who can potentially be subject to another country's laws) but are flesh-and-blood parts of the public authority governing their country. It is beyond the scope of this short chapter to delve into all the intricacies of this matter, but it is important to recognize that the relationship that obtains between soldiers or diplomats and their government is not just the ordinary relationship of state and subject. Indeed, it is not even exhausted by the relationship of employer and employee. Soldiers are, in an important sense, embodiments of their state itself: whereas diplomats are the mouthpieces (and, to some limited extent, the brains) of their country as it engages in diplomatic relations with other countries, soldiers are the arms (and, to a very limited extent, the brains) of their country as it engages in warfare with other countries.

The idea of public office plays a crucial role in Kant's understanding of the state, and it is a Kantian idea that Ripstein himself has explored elsewhere (Ripstein 2017). It is regrettable, then, that he does not make full use of this idea in his treatment of combatant immunity, for without it, the account seems radically incomplete. The idea, roughly, is this. To be a soldier is not just to be a person to whom a state has granted the permission to use otherwise impermissible coercive force. Indeed, we shouldn't really think of the state granting permissions to soldiers at all, for that seems to suggest that these permissions somehow *belong* to individual soldiers. Instead, the Kantian conception of the state is that it *just is* a set of offices linked together through certain relationships of authority. The state as a whole is a way of organizing human conduct around a guiding purpose (on the Kantian account, this purpose is putting in place a rightful condition of independence for all),[14] and it is made up by a network of offices, each with its own defining purpose. Put another way, the state is not an entity distinct from natural persons; it is, rather, a way of relating to one another. When we enter into a public office of state, we thereby take up the rights, duties, powers, and liabilities associated with that office. But we take them up just insofar as these allow us to carry out the purposes that define our office. To say that "the state" carried out some action in the world must always be a way of conceiving of the actions of particular human beings. One can carry out acts *as* the occupant of a public office (and therefore, one's acts can be attributed to the office, and therefore to the state of which it is a part) or one can undertake them in one's private capacity. State acts are just human acts carried out from within a public office.

To be a soldier, then, is not just to be an individual who is permitted to do things that other individuals are not. Instead, one should say that the office of soldier is part of the network of offices that make up a state. When one enters such an office, it becomes possible to exercise the rights, powers, and privileges of that office (and to be subject to any special duties that pertain to it) for the purposes that define that office. Now, the reason why we should characterize things in this way, rather than just as personally taking on these new legal relations, is that there is a permanence and

[14] Since Thomas Hobbes, many philosophers have endorsed this idea of sovereignty as an office, directed at a defining purpose. In the *Leviathan* Chapter 30, Hobbes himself defined the purpose of the sovereign as securing the "the procuration of the safety of the people."

a self-standing structure to offices. The rights, powers, liberties, and duties of office are not organized by relation to a particular human being who bears them; instead, they are structured around the defining *purposes* of the office, which exist prior to any particular occupant and will endure after any given occupant has left the office. The office of soldier is organized around the purpose of defending a state from aggression by other states, as this can be integrated into a larger network of offices connected to its centers of decision-making power. For individual soldiers, however, it is not a matter of deciding for themselves whether their country is under attack and in need of defending. Their job, instead, is to defer to the decisions of their superiors on this question and even on the question of how the purpose of defending their country should be carried out. The bounds of an individual soldier's discretion to decide how she may act are quite limited, and they are more limited the lower down the ranks the soldier finds himself.

All of this means that when we are dealing with a soldier who is acting within the proper bounds of his office, his acts should be attributable to the state rather than to him as a private individual. Accordingly, in the domestic setting, he may be subject to a variety of public law constraints on the proper exercise of his power that private actors are not, but he may also do a great many things that private persons are prohibited from doing. (It is for this reason that we brand private individuals as vigilantes and criminals when they try to undertake tasks that are reserved to occupants of a certain public office, such as policing or the military.) In the international context, this means that other countries should treat individuals acting within their public office as being the arms of their state, rather than as acting in their own private capacity.

2.B. Deference to Implausible Claims of National Self-Defense

The idea that public officials are subject to different norms from private actors is familiar from domestic law. What is less familiar in the domestic context, however, is the demand that countries should defer even to wildly implausible and self-serving claims by other states (e.g., that the acts they have ordered their soldiers to carry out are really part of a larger effort of national self-defense). Since combatant immunity applies to soldiers insofar as they can properly be thought of as public officeholders, carrying out their state's business, we must also make sure that the orders the soldier is carrying out are ones that are within any state's jurisdiction to give in the first place. As we saw above, the only legitimate ground for going to war, on Ripstein's Kantian account, is national self-defense. Any other ground for war is either a direct assault on the independence of other nations (a pure expression of domination) or the illegitimate assumption of a position of authority over other states (as when a state claims to be punishing another state for wrongdoing or extracting some obligation from the other state). So if a state were to announce publicly that it was going to war for any reason other than self-defense, this would be to announce that it was going beyond the legitimate bounds of its jurisdiction as an independent state in a community of equal independent states. Accordingly, the acts of any individual carrying out his country's aggressive war could not be characterized as the acts of a public authority.[15] Further, if an individual soldier carried out acts that could not be characterized as part of an act of national self-defense against the aggression of another state, this could not be subject to combatant immunity, either. This would be the case if a soldier targeted civilians who are not part of the interstate conflict, for targeting civilians is not part of a soldier's task of defending his country from attack from another country. These acts could only be characterized as those of a private individual who is otherwise liable to the criminal courts of whatever country otherwise has jurisdiction over him. That much is clear. But why should soldiers who are fighting what is, in fact, an aggressive war be entitled to immunity from prosecution simply because their superiors declared the war to be one of national self-defense?

The point here is that it is up to each state to address the questions that are within its jurisdiction as it sees fit. Once again, the Kantian focus on relationships of authority means that the normative question at the heart of the law of war is not "was this or that decision the morally right one to make?" but rather, "was this or that question one that it was up to this public authority to decide?" to be followed up with the question "and was the person carrying out the act a duly authorized member of the state apparatus, carrying out the decision through permissible means?" In short, on the Kant/Ripstein account of the equality of combatants, the actual moral rights or wrongs of soldiers' actions is neither here nor there. Once we have determined that the *question* of whether or not their country is under an attack requiring the use of defensive force to repel it, is within their state's jurisdiction, then its decisions about how to answer that question—whatever they might be—constitute valid (even if morally unjustifiable) state action. That is, the immunity of soldiers from prosecution for their war-making acts does not turn on whether they personally were morally justified in their actions, nor does it turn on whether their state leaders were correct (or even in good faith) in deciding to launch a putatively defensive war. All that matters is that their actions are state actions, properly understood. And so long as they were properly ordered to carry out a war, and the war was properly launched on grounds that bring their decision within their state's jurisdiction, then they cannot be held personally liable for their actions. Their actions, however reprehensible and morally unjustifiable, belong (as a matter of legal responsibility) to the state and not to the individual soldier.

The Kantian argument turns almost entirely on the question of jurisdiction: whose call it is to decide whether a country is under attack and needing to use force to repel it. What is so unusual is that this question of jurisdiction is treated as a crucial moral matter, and not just as a technical issue. This stands in stark contrast to almost all other contemporary moral philosophers of law. For most of them, jurisdiction is "regarded as a secondary and essentially administrative question of bureaucratic competence, subject to adjustment as necessary or at least prudent, without consideration of the underlying power to generate and apply the norms to a particular person in the first place..." (Dubber, 2013:277).

[15] Even Nazi Germany recognized the need to characterize their wars as defensive. This is why they fabricated "Operation Himmler"—what appeared to be attacks from Polish soldiers on August 31, the day before their invasion of Poland. These attacks were designed to give some semblance of plausibility to their claim to be acting in national self-defense.

3. Conclusions and Implications

Kant and the Law of War is a big, ambitious book. Although its title suggests that it is only a contribution to just war theory, it is in fact much more than that. Although Ripstein treats the usual topics in just war theory—*jus ad bellum* (the ground for going to war), *jus in bello* (the proper conduct of war), and *jus post bellum* (making and maintaining peace at the end of war)—he sets his treatment of these specific issues within a larger statement of the law of nations more generally. It treats issues such as colonialism, civil wars, diplomatic relations, and an account of *jus cogens* in the nature of public authority and the ways in which such entities must relate to one another. In this way, the book completes Ripstein's three-part exploration of Kant's theory of law. He explored private law through *Private Wrongs*; domestic public law through *Force and Freedom*; and he has provided a general statement of Kant's account of public international law in this book. That is, Ripstein is not just setting forth a theory of law, but of the relations among states more generally.

As Ripstein's treatment of combatant immunity makes clear, the Kantian account is remarkable in its ability to make sense of certain doctrines that appear to be morally indefensible. This might sometimes seem like a strength of the account, but in cases where the doctrine is so morally repugnant, it might seem to be more of a liability. If the Kantian account can make moral sense of the criminal immunity of Wehrmacht soldiers invading Poland, then we might be tempted to conclude, "so much the worse for Kantian just war theory." But if we have that reaction, it is a sign that we are still in the thrall of a moralistic view of law. The Kantian picture of public authority does not tell us *how* to exercise our authority; it tells us only what powers are within our legitimate authority in the first place.

References

Darwall, S. (2009). *The Second Person Standpoint: Morality, Respect and Accountability.* Cambridge, MA: Harvard University Press.

Dubber, M. (2013). "Criminal Jurisdiction and Conceptions of Penality in Comparative Perspective." *University of Toronto Law Journal* Vol. 63(1): 277.

Haque, A. (2017). *Law and Morality at War.* Oxford: Oxford University Press.

Hathaway, O., & Shapiro, S. (2017). *The Internationalists: How a Radical Plan to Outlaw Remade the World.* New York: Simon & Shuster.

Hobbes, T. (1994). *Leviathan* (Curley, E., ed.). Indianapolis: Hackett.

Hurka, T. (2007). "Liability and Just Cause." *Ethics & International Affairs* Vol. 21(2): 199–218.

McMahan, J. (2009) *Killing in War.* Oxford: Oxford University Press.

Raz, J. (1988). *The Morality of Freedom.* Oxford: Oxford University Press.

Ripstein, A. (2009). *Force and Freedom: Kant's Legal and Political Philosophy.* Cambridge, MA: Harvard University Press.

Ripstein, A. (2016). *Private Wrongs.* Cambridge, MA: Harvard University Press.

Ripstein, A. (2017). "Sovereignty and Property: How to Tell the Difference." *Theoretical Inquiries in Law* Vol. 18: 243.

Ripstein, A. (2021, forthcoming). *Kant and the Law of War.*

Scanlon, T.M. (2000). *What We Owe to Each Other.* Cambridge, MA: Harvard University Press.

Thorburn, M. (2019). "Soldiers as Public Officials: A Moral Justification for Combatant Immunity." *Ratio Juris* Vol. 32: 395–414.

Walzer, M. (1992). *Just and Unjust Wars: A Moral Argument with Historical Illustrations* (2nd ed.). New York, NY: Basic Books.

Williams, B. (2005). "In the Beginning Was the Deed." In Geoffrey Hawthorn, ed., *In the Beginning Was the Deed: Realism and Moralism in Political Argument.* Princeton, NJ: Princeton University Press.

Wolff, R. (1970). *In Defence of Anarchism.* New York: HarperCollins.

14
EU Solidarity as Collective Self-Defense?
Constitutionalism and the Public Uses of Force

Ester Herlin-Karnell

1. Introduction

In Arthur Ripstein's new manuscript on *Kant and the Law of War* (Ripstein 2021), the idea of self-defense plays an essential role as the only legitimate ground for waging war.[1] As Ripstein explains, for Kant there is a special morality that governs force and the state constitutes a public rightful condition in this context (Ripstein 2021: chap. 1, section 1). Therefore, as Ripstein elucidates, for Kant war is public in nature and only lawful war is permissible exactly because it is simply protecting each person's right to independence from the determining the choice of another.[2] The only ground for using force is therefore the prevention of force (Ripstein 2021: chap. 3). Likewise in "just war" theory, the use of force plays an important law-enforcement or self-help function, as means to prevent and punish certain violations of international law (Meisels, 2018; Shany, 2009).

In contemporary international law, the premises for using force are strongly grounded in the Kantian tradition. For example, force is only allowed by states to remedy a wrong, i.e. to repel an attack or uphold an important right held by a state (Ripstein 2021: chap. 3, section I). The UN establishes a nearly complete ban of the use of force and proclaims in Article 2 UN that all members shall settle their international disputes by peaceful means and that states shall refrain in their international relations from the threat or use of force against the territorial integrity or political independence of any state. Yet the UN Charter authorize familiar departures from this prohibition in two circumstances: Security Council authorization to restore international peace (Article 42 UN) and self-defense (Article 51 UN). All use of force needs furthermore to be justified in terms of proportionality and necessity (Vidmar, 2017; Benbaji & Statman, 2019:1–5). There are two further exceptions of the ban of the use of force, albeit somewhat controversial. First, there is the rescuing mission (emergency rescue) of foreigners abroad. Secondly, there is the closely connected question of humanitarian intervention (Forteau, 2015), concerning the conditions under which intervention could be considered "just" (Tan, 2017: chap. 10). The issue

[1] Manuscript June 2018. References are updated, as far as possible, to reflect the final version of the manuscript 2021.
[2] See also the chapters by Massimo Renzo and Malcolm Thorburn in this volume.

Ester Herlin-Karnell, *EU Solidarity as Collective Self-Defense?* In: *The Public Uses of Coercion and Force.*
Edited by: Ester Herlin-Karnell and Enzo Rossi, Oxford University Press. © Oxford University Press 2021.
DOI: 10.1093/oso/9780197519103.003.0014

of emergency rescue and humanitarian intervention highlight, however, the question as to how to balance the right or duty, to protect individuals against the right and independence of states (Ripstein 2021: chap. 7). As Yuval Shany explains, the entrusting of the Security Council with the primary responsibility for maintaining international peace and has important implications on the scope of the right to self-defense and it signifies a shift from unilateralism to multilateralism in jus ad bellum considerations (Shany, 2009).

The purpose of this chapter is twofold. First, the chapter explores the notion of self-defense and how it is manifested in contemporary law of war. I am especially interested in the imminence criteria for deciding on self-defense and to what extent Kant completely ruled out any preventive use of force. Secondly, I will set out to use the EU as a practical example of the possible use of collective self-defense as a constitutional duty emanating from the Lisbon Treaty on the basis of solidarity. Moreover, the EU's fight against terrorism and security regulation in general have many parallels with the "war" model and security enforcement practices and preventive justice model, which will be discussed in this chapter. While the EU does not have its own army despite a longstanding debate on this matter, the security mission of the EU is not only a crime fighting task but one which boils down to the uses of force against terrorism and security threats.[3] The chapter will try to highlight the EU security context and its similarities to war.

The chapter is structured as follows. First, some fundamentals of self-defense will be sketched out and I will link it to the manuscript on *Kant and the Law of War*. I will compare the just war theory to the extensive security regulation framework within the EU and address some pertinent constitutional questions. Subsequently I will try to discuss the EU solidarity and mutual assistance clauses in the context of collective self-defense. The EU represents an interesting case study as the constant response to security concerns questions from border control and migration management to counterterrorism strategies means that the EU is currently very active in this area.[4] Yet this empirical dimension might seem strange, given Kant's view on empiricism (Ripstein: chap. 1, section I). As Ripstein explains in *Force and Freedom*, Kant's ground for rejecting empiricism and anthropological starting point rest on the normative idea that as a matter of right each person is entitled to be his or her own master (Ripstein, 2009:21). And for Kant war is not merely empirical but instead he insists on the possibility of finding right in a condition of war. Be that as it may, it is arguably still interesting to look at the question as to whether the EU's current security practices are compatible with a Kantian approach.

In addition, some of the EU's security strategies to fight terrorism, and especially the solidarity provision amounts to that of collective self-defense and therefore, as noted, represent an interesting case study.

[3] *European Parliament Brief: "Defence: Is the EU creating a European Army?"*

[4] On other areas of EU law including the migration crisis, see B Wolthuis & L Corrias in this volume.

2. Self-Defense and Preventive Force

While the notion of self-defense constitutes one of the legitimate grounds for using force against another state, for Kant preventive force is generally ruled out, but could be used under certain circumstances (Ripstein 2021:chap. 3, section VI). For example, preventive force could be morally superior to wars of national defense if fighting early is less costly on both sides (Benbaji, 2014:179). Specifically, for Kant legitimate war does not show that preventive war is defensible as such as potentially all nations are possible threat to each other (Ripstein, 2009:81). And secondly that where each can only do what seems right and good to them, two states might be entitled to attack each other. Ripstein adds to this that according to the Kantian model, neither perfidious deceit such as for example false surrender, nor targeting of noncombatants, are available as a means through which a nation is entitled to defend itself neither as pure self-defense or preventive force (ibid.110). As is clear from *Kant and the Law of War*, for Kant, defensive war is not a procedure at all (Ripstein 2021: chap. 2).[5] Instead the right self-defense, as Ripstein explains, is closer to an innate right of protecting oneself. Consequently, on this account Kant's theory is narrower than just war theory and regular war traditions which generally have more expansive grounds of war with regard to preventive force also (ibid.).[6] Hence, Kant's view of preventive force is extremely restricted, while just war theory in general would not rule it out (Meisels, 2018).

Unlike some just war theorists however, Kant disregarded the successfulness test as to whether for example the use of self-defense needs to be successful in order to be justified (Ripstein, 2021:chap. 3, section VI).[7] The contemporary law on the right to defense under the international legal framework follows a similar pattern and does not attach much weight to the successfulness criteria but rather focuses on the proportionality and necessity tests (Meisels, 2018:76–77). But the legal framework for regulating the parameters of warfare has been criticized for, inter alia, in essence allowing the aggressor to have better cards in its hands at the start of the war (Beer, 2018:chap. 1). Indeed, from the perspective of fairness, it seems odd that the aggressor is rewarded at the law—abiding self-defendant's expense. According to some scholars instead of deterring aggression, the current law seems to incentivize it. For this reason, Yishai Beer (ibid.182), for example, has suggested that in an international arena in which the central law enforcement apparatus of the UN does not function (at least not properly), there should be an effective shield for law-abiding states. Moreover, according to Beer, preventive force including that of defensive retaliation may be justified when a state has good reason to expect a series of attacks from the same source.

While a Kantian theory in general would not allow preventive force or at least adopts a restrictive view of it, Kant would not rule it out in specific circumstances however (Ripstein 2021:chap. 3, section VI). For him defensive war is just the upholding of the respective independence of nations; by contrast, preventive war is inconsistent with any such independence. It is connected to the fact that for Kant, war

[5] See also the chapter by Malcolm Thorburn in this volume, discussing the idea of authority and war in Ripstein's *Kant and the Law of War*.

[6] See also the chapter by Massimo Renzo in this volume

[7] See also the discussion by Johan Olsthoorn in this volume.

is not a procedure for establishing rights. Instead, as Ripstein (2021: chap. 3, section IV) explains, while a nation is entitled to defend itself, it does not settle who has right. More specifically, in the *Perpetual Peace*, Kant argues that preventive war could never be rightful, and likewise a state's right to security must be the same as every other state's right to security. The rationale is that all nations are always potential threats to each other and therefore preventive war cannot be rightful. As Ripstein (ibid.) explains, however, in the *Doctrine of Rights*, Kant also links the idea of preventive war to that of breach of contract, that is, an unjust enemy. The line between defense and prevention are not easy to draw though. Michael Walzer, for example, argued that preventive strikes are justifiable if three conditions exist: an obvious intent to injure, a degree of active preparation, and a risk to a state's territorial integrity or political independence if no action is taken (Walzer, 1977; Harris & Monaghan, 2018). On the Kantian account, the second criteria regarding the degree of active preparation would be nearly impossible to meet (Ripstein, 2021: chap. 2).

Moreover, in just war theory (Kretzmer, 2013; Tan, 2017) proportionality has long been regarded as one of the key elements for determining whether resort to war is justified. Proportionality means that the destructiveness of war must not be excessive in relation to the relevant good the war will achieve. But only a legitimate war, as Tamar Meisels claims, can pass the proportionality test to count as a "just war" (Meisels, 2018). In just war theory, proportionality is not measured by the number of civilians killed but is both forward looking and retrospective with regard to the risk posed in any given case. From an empirical point of view, proportionality has been criticized for leaving too much uncertainty on the battlefield, literary, and that it is not an empirically grounded concept as it leaves a lot of uncertainty in practice.[8] While, Walzer famously said that "war is always judged twice first with reference to the reasons state have for fighting, secondly with reference to the means they adopt," the *jus ad bellum* and *jus in bello* have been debated further (Walzer, 1977).[9] For example, Eyal Benvenisti has added an extra dimension to the proportionality assessment in this regard (Benvenisti, 2009). Specifically, he has suggested that *ad bellum* considerations do matter for the *in bello* proportionality analysis. This is because widespread reactions to conflicts and civil society, should have a say in these matters and from a fairness perspective jus in bello and jus ad bellum are intermixed. Others (Christodoulidou & Chainoglou, 2015) have argued that, for example, in the scenario where the use of force is considered legal in terms of its proportionality under *jus ad bellum* but illegal in terms of its proportionality under *jus in bello* (or vice versa), the assessment of the proportionality of the actions in question will not be distorted by the applicability of the other legal regime. On this account, the legality of actions in *jus in bello* cannot mitigate their illegality under *jus ad bellum*. In any case, the increased participation of non-state actors in armed conflicts complicates the contemporary law of war as they do not follow the traditional state-bound pattern of just war. More specifically, Benvenisti has suggested that while Walzer famously referred to jus

[8] Daniel Statman, presentation at authors workshop February 1–2, 2018. "Unreliable Protection: Proportionality Judgments in War" (unpublished manuscript by Daniel Statman, Ra'anan Sulitzeanu-Kenan, Micha Mandel, Michael Skerker, & Stephen de Wijze).

[9] See also the chapter by Alon Harel in this volume.

ad bellum and jus in bello as "logically independent," there appear to be good reasons to question this distinction because if each side view itself as just the two camps could descend to ruthless brutality.

Ripstein (Ripstein 2021: chap. 3 and chap. 5) also tackles the question of *ad bellum* and *jus in bello* in his discussion of Kant who argued that the rationale for that distinction is that while one side in a war is fighting defensively, both sides in the war can still violate the in bello rules. The in bello norms governing the conduct or war are restrictions that govern the use of defensive force, as national defense is the only legitimate reason for waging war. As Ripstein puts it: the norms apply to both sides in a war because both sides do wrong if they violate them, even if one side was already doing wrong by going to war at all.

In positive international law, self-defense under Article 51 UN Charter must also be proportionate and necessary. In the legal context (Kretzmer, 2013), proportionality is an umbrella term and casts a wide net that includes both harm of violating territorial integrity, damage to infrastructure, and effects on third parties. Thus, in the well-known *Nicaragua* case, the International Court of Justice (ICJ) held that self-defense must be proportionate and thereby not excessive.[10] The ICJ has stipulated that there is a "specific rule whereby self-defense would warrant only measures which are proportional to the armed attack and necessary to respond to it, a rule well established in customary international law" a use of force that is proportionate under the law of self-defense, must, in order to be lawful, also meet the requirements of the law applicable in armed conflict which comprise in particular the principles and rules of humanitarian law.

Moreover, in *Nicarague*, non-state actors were liable of an armed attack, but a line was drawn between state sponsored terrorism and terrorism that is not state sponsored. Yet in the advisory opinion on the construction of the wall in the Palestinian territories the ICJ held that an armed attack can only be done by a state against another.[11] The case concerned the boundaries of Article 51 UN Charter. Judges Higgins and Kooijmans in their conflicting opinions pointed out that there was nothing in the text of Article 51 that stipulated that self-defense is available only when an armed attack is made by a state.[12] As will be discussed, in the framework of the suppression of terrorism, the division between non-state actors and state actors has been significantly blurred in the last couple of decades.

But perhaps it should be asked whether independence, which is a central question for Kant in the context of self-defense, if Kant ruled out preventive attacks against non-state actors. It is interesting to pose the issue as a general query in the context of the imminence criteria and when thinking of how self-defense has developed in contemporary international law. Central to this question seems to be what the imminence criteria really means as well as the role of the state. The next section will explore the meaning of imminence in further detail.

[10] *Nicaragua v United States of America, International Court of Justice 27 June 1986*. See also the discussion in Williamson (2009).

[11] *Legal Consequences of the Construction of a Wall in the Occupied Palestinian Territory, Advisory Opinion, ICJ Rep 2004*.

[12] Yet both dissenting opinions questioned the proportionality of the wall.

2.A. The Imminence Criteria and Self-Defense

In international law, and in following the old *Caroline* case, the notion of "imminence" means the following: "necessity of self-defense, instant, overwhelming, leaving no choice of means and no moment of deliberation."[13] But a problem when gauging the likelihood of an attack is what level of evidence is required in order to make an informed decision (Guiora, 2013). On the one hand absolute certainty is nearly impossible in these cases; on the other hand a lax standard which allows preventive strikes and anticipatory self-defense may be very dangerous.

The idea of preventive war is of course to avert an imminent harm or harm before it occurs. The reason to have a high threshold as to when preemptive war is necessary is obviously the danger of an overuse of attacks. As Meisels explains, adopting a consequentialist approach one could argue that it would permit a general rule permitting prevention and secondly "bad practice" objections do not in themselves rule out any possibility of justified prevention in exceptional circumstances (Meisels, 2018:112–113). Moreover, David Luban (Luban, 2018), Meisels (Meisels, 2018) and others have called for more reliable intelligence information before any use of preventive force is admissible. We seem to encounter here, according to Lubell (Lubell, 2015) the territory of risk regulation and emergency-driven regulation posing the question whether there can ever be certainty in these matters anyway (Guiora, 2013).

As Yitzhak Benbaji (Benbaji, 2014) puts it, Article 51 UN Charter copies to a great extent the domestic criminal law system of self-defense, to prevent violence. While self-defense in war is often likened to the law of self-defense in criminal law there are some differences.[14] For example, unlike self-defense in criminal law, the idea of targeted killing as part of a preemptive strategy reflect a planned military operation representing the convergence of intelligence gathering that law enforcement has been tracing. (Guiora, 2013). Self-defense in criminal law, however, usually requires a decision made on the spot and this kind of preventive strategy would in most cases fall outside the criminal law of defense, with the exception of (for instance) the debate on the law on battered women which also cover non "imminent" situations (McColgan, 1993). Yet it is unclear how the assessment of self-defense for states can have the same imminence criteria as an individual in criminal law context. The issue seems inexorably linked to the idea of prevention.

Consequently, we may ask what exactly is meant by "preventive" self-defense? A preventive justice model, beyond the law of war is most evident in, for example, the fight against terrorism, the combating of the financing of terrorism, anti-money laundering legislation, and measures to fight organized crime (Ashworth & Zedner, 2014; Carvalho, 2017). Prevention in this regard is often taken to mean more efficiency and a risk based approach (Herlin-Karnell, 2012: chap. 5). In an interesting article on the contours of self-defense in the law of war and self-defense in the national criminal law setting, respectively, Shany has highlighted the limits of any analogy when "context" is missing (Shany, 2009). The key question in this "context," that is, the

[13] US Secretary of State Daniel Webster dated Apr. 24, 1841, in Caroline Case, 29 British and Foreign State Papers (1841) 1137–1138.

[14] On criminal law, see Thorburn in this volume.

imminence criteria for self-defense seems still to be that if one could equate states with individuals? As mentioned earlier, Kant saw war as essentially a public endeavor. The revisionists think otherwise, and privatize the decision-making on the use of force to all situations, and not only to war crime scenarios as in contemporary international humanitarian law where soldiers cannot commit war crimes regardless of an order to do so. In defending this view, Jeff McMahan stated that military decision-makers are not chosen by the people (McMahen, 2009:92–93). In his view it remains doubtful why soldiers can be forbidden to engage in unjust war but at the same time be denied to fight just wars when they would fight in opposition. For a revisionist, imminence as such does not seem an important question. The important question for them is instead the risk of limited information and the knowledge of the soldiers participating and not knowing what is "just" (McMahan, ibid.184).

How then are the imminence criteria used when more than one state is concerned? Does it apply to each of the collective actors in self-defense or only to the actor who asks for help? While it is clear from a legal perspective (e.g., Article 5 North Atlantic Treaty) that it only applies to the state that is attacked from a theoretical view point it remains somewhat unclear why that is the case. In criminal law, self-defense applies to a broader category of people who are present at the time of the crime being committed. For sure states cannot be "present" in the same way as individuals and it does not matter if it is the neighboring state or an ally from far away that comes to help. But what does the imminence criteria mean when applied in a complex EU security law context within a constitutional structure of EU law one may legitimately ask. And where is the distinction between self-defense and law enforcement action? It is often said (cf. Uniacke, 2015) that terrorism particularly with regard to terrorist action which directly contravenes domestic criminal and counterterrorism laws also activate central elements of the laws of war embodied in international humanitarian law, the most salient being the *jus in bello* principle.

The chapter will now turn to the idea of security regulation in the EU before addressing the question of collective self-defense based on EU solidarity. The purpose is to highlight the security context and its similarities to war. For example, does solidarity and self-defense include retaliation measures?

3. The EU Security Context and the Vocabulary of War

For Kant (Ripstein 2021:chap. 8), peace requires international institution to give effect to the law of peace and states must exist the non-rightful condition and enter into a league of nations in accordance with the original social contract. The EU is a lot more than a federation of states but in relation to states outside its borders, it functions a lot like a state when it comes to external actions and military strategy. For example, one of EU policies, *The EU Global Strategy*, is to demonstrate an ongoing commitment to the deepening of multilateralism ranging from new trade rules, climate deals, to security cooperation with third states.[15]

[15] https://eeas.europa.eu/topics/eu-global-strategy_en

While the vocabulary of just war theory may be dormant in the EU context, the EU has nonetheless an extensive security discourse not totally unrelated to that of war. Security in the EU context is a highly multifaceted concept and has been influenced by the US war on terror, with a number of EU measures enacted (Bossung, 2012; Kaunert & Zwolski, 2015). The measures for combating terrorism in the EU has had a revolutionary impact on the member states, partly because of the very short time limit to transpose it into national law, and partly because it was no longer regarded as sufficient to address the threat of terrorism through common offences (e.g., Nuotio, 2006; Noorda, 2020). The EU's main tactics in the war on terror finance is mainly criminal law as opposed to military force, in contrast to the United States.[16] The EU has also a wide-ranging regime for restrictive measures through administrative law means that have been frequently criticized for being overly coercive without sufficient due process guarantees.

The EU security mission follows from the ambitions set out in the Treaty of Lisbon, inter alia, the EU shall "endeavor to ensure a high level of security." Security, however, is rarely defined in EU law. Still the Area of Freedom, Security and Justice (AFSJ) sphere is becoming ever-more securitized, with an increasing mixing and mingling of the internal and external. The external dimension of security in an EU context (the Common Foreign Security Policy) is also very central to the EU's security ambitions, and concerns the EU security towards third countries. Here, the EU acts as one legal entity, with legal personality in its relationship with the rest of the world.[17] The constant response to security concerns from border control and migration management to counterterrorism strategies means that the EU is currently very active in this area and is largely following the international law norms. Yet, in AFSJ matters and security regulation, such as Schengen and the European border-free zone, it is paved by ad hoc approaches to cooperation, in which not all the member states are moving in the same direction (the so-called multi-speed phenomenon in which not all the member states participate in legislative measures or cooperation).[18]

While the EU's reliance on security as a justificatory tool for the EU's presence both on the internal and the external stage has been criticized by academics for a long time, it still plays an important role as the main driving force for the furthering of the EU security agenda (e.g., Douglas-Scott, 2012). It confirms a precautionary approach to the fight against crime and terrorism. It could be argued that due to the strong security focus in the EU, the status quo in AFSJ law has become one of a "too preventive" regime, in which the safeguards of the individual are lost, despite the grand treaty values. It has also resulted in an increase of EU competences in the area at the possible expense or gain of individual member states as matters of security has a clear cross-border dimension.

[16] See, e.g. ,*Directive 2017/541 on Combating Terrorism, OJ L 88, 31.3.2017*, pp. 6–21.

[17] The main difference between the internal and external security agenda is that the AFSJ is subject to the same constitutional framework as the internal market both within the EU, with ordinary legislative procedure, and within the EU Court's jurisdiction, while the foreign policy area is still largely subject to intergovernmental co-operation. The EU regulates security through a mixture of risk and emergency approaches.

[18] For example, and much more trivial than war theory, the Brexit debate poses a challenge to the Northern Ireland question and whether the "old" troubles security threat will reappear.

Why then is the EU security endeavor interesting in the context of war theory? As Walzer points out, the threat of terrorism is rooted partly in its unpredictability of its source (Walzer 1977:266–267). More specifically, in the aftermath of 9/11, and in light of more recent terrorist attacks in Europe such as the Paris and Brussels attacks in 2015 and 2016, by using the emergency provisions to fight terrorism, the "normal" criminal law framework has generally been set aside. Moreover, measures are frequently adopted under administrative procedures without the full package of procedural safeguards (Herlin-Karnell, 2016) and this is seen as a way of ensuring effectiveness of the law. Interestingly, the EU Counter Terrorism Directive adopted in 2017 mainly uses the criminal law framework, and not the administrative one, which would mean that countries cannot derogate from their obligations to ensure a fair trial, and have, in any case, to comply with proportionality.[19] Almost all security situations these days have a cross-border impact. Therefore, one could argue that, although the EU has no specific regulation for a state of emergency, thus leaving it to the member states, the values set out in the Lisbon Treaty and the EU Charter of Fundamental Rights, to wit, the rule of law, dignity, and human rights protection, still need to be respected. As was recently observed in the context of the French counterterrorism strategies and the state of emergency that was recently brought to an end (in force since the Paris attacks in November 2015), the French Intelligence Act raises the difficult constitutional question of balance between the respect for freedom and the safeguarding of public order (Mastor, 2017:707). The new French counterterrorism bill now absorbs the exceptional powers used under the state of emergency into ordinary law.

In fact, the whole EU and the AFSJ is underpinned by a claimed solidarity principle. The EU values and objectives tell us that not only shall the AFSJ and its policies be based on solidarity and trust in the EU but also the EU's response to any armed attack. EU member states owe solidarity toward one another, if taking the Treaty seriously. It is codified the EU grand values, Article 2 Treaty of the European Union (TEU), and across the Treaty. It is also part of social justice question in the EU Charter. Solidarity as a constitutional question in EU law is usually framed as a loyalty question. Interestingly, the war dimension in the EU context is intermixed of internal and external. The 2017 Counter Terrorism Directive,[20] mentioned before, is interesting here. In para. 37 of the recital it stipulates, inter alia, that the Directive should not have the effect of altering the rights, obligations and responsibilities of the member states under international law, including under international humanitarian law. In addition, in recital 38, the Directive states that "The provision of humanitarian activities by impartial humanitarian organisations recognised by international law, including international humanitarian law, do not fall within the scope of this Directive, while taking into account the case-law of the Court of Justice of the European Union." While it could be argued that recitals are not part of the core legal text anyway, it is nonetheless very interesting to ask if there can be such a distinction of EU vis-à-vis international

[19] *Directive 2017/541 on combating terrorism and replacing Council Framework Decision 2002/475/JHA and amending Council Decision 2005/671/JHA.*

[20] Ibid.

law at all? For example, can the EU have its own idea of collective self-defense and how could it be formulated? And why is solidarity so important in EU law?

To a great extent, it has to do with competence allocation in the EU, which has been restricted when it comes to national security. So while security is predominantly a national competence, Article 4 TEU which states that national security remains the responsibility of each member state, the EU has, in fact, adopted a number of security measures that severely blurs the distinction between the EU and national security, and even global security.[21] Many of the EU's policy agendas especially in the area of migration focuses on solidarity and humanity as the key words here. Those policy agenda do not deal with the classic EU law idea of subsidiarity, that is, local level first, but instead highlights the importance of solidarity and the need for joint action when fighting terrorism and other transboundary crimes. The policy documents of the EU Security Agenda 2015–2020 and the new EU Security strategy of 2020 also highlight the importance of solidarity among the member states in order to support one another when states are attacked or subject to a security threat, as is also a constitutional duty as set out in Article 222 Treaty of the Functioning of the European Union (TFEU).[22]

The idea of collective self-defense in EU security regulation then is connected to solidarity in an EU context. But what is solidarity in EU law? Most scholars analyzing this question focus on distributive justice or justice and social welfare questions. Indeed, for Andrea Sangiovanni (Sangiovanni, 2013), for example, the principle of solidarity in EU law is primarily a social justice question. The concept of solidarity for him could be divided in roughly two parts. The first part defines principles for relations between member states (member state solidarity), and the second part principles governing our relations qua European citizens and residents (transnational solidarity). But solidarity in the EU context is not only about distributive justice in the EU. There are a number of intriguing provisions in the EU Treaty that links solidarity to that of collective self-defense. Beyond the traditional debate as to whether the fight against terrorism, for example, should be fought through the "war on terror" slogan, or if the criminal model is a better one, it asks some fundamental questions about the idea of solidarity.

Although Kant did not directly discuss the obligation for states to assist one another or whether federations have a duty to assist members in need, Pauline Kleingeld (Kleingeld, 2004:304–325) has argued that he discussed it in the context of individuals and that states have a duty to maintain those who are unable to provide for their own needs.[23] This seems perhaps more akin to the global justice debate than a question of the use of solidarity as self-defense. But it could also be seen as an expression of solidarity between states in an international framework.

[21] There are several references. See Directive of the European Parliament and of the Council on the Prevention of the Use of the Financial System for the Purpose of Money Laundering and Terrorist Financing, Directive (EU) 2015/849, Directive 2013/40/EU, L 218/8 Directive on attacks against information systems.

[22] European Agenda on Security COM(2015) 185 final, *EU Security Strategy* COM(2020) 605 final.

[23] See also the chapter by Peter Niesen in this volume, on assisting in the hour of need.

4. EU Solidarity as Collective Self-Defense?

Collective self-defense in the EU context is mainly a constitutional obligation originating from the EU Treaty. And when it is used the law of war applies to the member states as one EU entity (and most EU states are also members of NATO). The EU security agenda is clearly also external and it is fight against terrorism is both internal and external. The solidarity clause echoes that of collective self-defense and poses interesting questions about the self-defense imminence criteria.

The interesting question from the perspective of this chapter is the potential link, if any, between the solidarity provision in EU law and that of self-defense. The following section will try, in descriptive terms, to outline the law on collective self-defense at the EU level.

According to Article 222 TFEU and Article 47 TEU, the solidarity and mutual assistance provisions respectively EU member states are required to assist one another in connection with a terror attack or an armed conflict.[24] When for example Germany, Belgium, Denmark, the Netherlands, and the United Kingdom and other EU countries sent troops to fight in Syria to support France in the wake of the terrorist attacks in Paris in 2015, defense was invoked. Initially, France appealed to solidarity but opted for the use of Article 42 (7) TEU and mutual assistance.[25] Under Article 222 TFEU, arrangements for the implementation by the Union of the solidarity clause shall be defined by a decision adopted by the Council acting on a joint proposal by the Commission and the High Representative of the Union for Foreign Affairs and Security Policy. Article 42.7 TEU in turn is more intergovernmental and based on political willingness, simply stating that the area shall be consistent with commitments under NATO.

Similarly, NATO has a collective defense clause. The debate is linked to the question of self-defense against non-state actors and particularly the fight against terrorism. For example, in the aftermath of 9/11 NATO countries invoked Art 5 of the North Atlantic Treaty. The Security Council Resolutions 1368 and 1373, passed after the 9/11 attack on the United States by non-state actors, refers within the context of this attack to the inherent right of individual and collective self-defense in accordance with the UN Charter (Kreztmer, 2013). It confirms a widened definition of war. Consider, for example, Article 45 UN Charter which stipulates that: "In order to enable the UN to take urgent military measures, Members shall hold immediately available national air-force contingents for combined international enforcement action." To what extent is it reflected in EU law?

Self-defense is reflected in the solidarity provision of Article 222 TFEU: The Union and its member states shall act jointly in a spirit of solidarity if a member state is the object of a terrorist attack or the victim of a natural or man-made disaster. The Union shall mobilize all the instruments at its disposal, including the military resources made available by the member states, to:

[24] There is also the general EU law obligation of loyalty, article 4 (2) TEU.

[25] http://www.europarl.europa.eu/thinktank/en/document.html?reference=EPRS_BRI(2016)581408 and https://www.iss.europa.eu/sites/default/files/EUISSFiles/Brief_37_Article_222.pdf

— prevent the terrorist threat in the territory of the Member States;
— protect democratic institutions and the civilian population from any terrorist attack;
— assist a Member State in its territory, at the request of its political authorities, in the event of a terrorist attack.

Moreover, the EU has a mutual assistance clause in Article 42(7) TEU. In short it means that when an EU country is the target of armed aggression on its territory, the other EU countries must assist it by all the means in their power. Specifically, Article 42 (7) TEU states, inter alia, that if a member state is the victim of armed aggression on its territory, the other member states shall have toward it an obligation of aid and assistance by all the means in their power, in accordance with Article 51 of the United Nations Charter. This shall not prejudice the specific character of the security and defense policy of certain member states. Still the solidarity clause does not have a similar derogation but is much broader in scope.

Article 43 TEU in turn states that the inter alia "*Union may use civilian and military means … conflict prevention and peace-keeping tasks, tasks of combat forces in crisis management, including peace-making and post-conflict stabilisation. All these tasks may contribute to the fight against terrorism, including by supporting third countries in combating terrorism in their territories.*"[26]

The threat of terrorism networks poses a significant threat to the internal security of the Union and to the safety and livelihood of its citizens, according to the EU security policy agendas.[27] The threat assessments show that criminal and terrorist groups are becoming increasingly poly-criminal and cross-border in their activities. National law enforcement authorities therefore need to cooperate more closely with their counterparts in other member states. Therefore, the EU argues that it is necessary to equip Europol to support member states better in Union-wide crime prevention, analyses, and investigations.[28]

The EU's security mission equally follows from the ambitions set out in the Treaty since the EU promises to ensure a high level of security even where there is no clear division between internal and external security. The focus on security, and the desire to import norms from the global arena in the fight against terrorism and crime, has frequently overshadowed the need to ensure due process rights within the EU.[29] For Kaarlo Tuori, war and crime, and external and internal security are hard to keep separate in combating terrorism, which leads to the emergence of a security constitution in EU law (Tuori, 2014). This is similar to Bruce Ackerman's "The Emergency Constitution," in which he claimed that we need "new constitutional concepts" in order to avoid the downward spiral in the protection of civil liberties while we wait for

[26] The High Representative of the Union for Foreign Affairs and Security Policy, acting under the authority of the Council and in close and constant contact with the Political and Security Committee, shall ensure coordination of the civilian and military aspects of such tasks.

[27] *EU Security Strategy* COM(2020) 605 final.

[28] Regulation 2016/794 on the European Union Agency for Law Enforcement Cooperation (Europol) and replacing and repealing Council Decisions 2009/371/JHA, 2009/934/JHA, 2009/935/JHA, 2009/936/JHA and 2009/968/JHA, L135/53.

[29] See also E Herlin-Karnell, *The Constitutional Structure of Europe's Area of Freedom, Security and Justice and the Right to Justification* (Oxford: Hart publishing 2019), chap. 5.

politicians to respond to each new terror attack by enacting laws that become increasingly repressive with each attack (Ackerman, 2004). As David Dyzenhaus points out, a crucial part of meeting this test is the demonstration that judges can play a meaningful role in keeping legislatures and governments within the rule-of-law project (Dyzenhaus, 2006). Similarly, with Aharon Barak then judicial review is not necessarily undemocratic, but is, in fact, a democratic institution which consists of an extra check, and shows that the legislator has done its job (Barak 2012). For Dyzenhaus, if the judges fail to carry out their duty, they also fail to clarify to the people what constitutes responsible government—that is, government in compliance with the rule of law.

Yet collective self-defense as discussed earlier does not refer to the collective of the people, but collective of states. As noted above, after the attacks in Paris on November 13, 2015, the French government declared a formal "state of emergency." The emergency regime granted a broad array of powers, including to search houses day or night and issue assigned residence orders without prior judicial authorization. The state of emergency was extended several times.[30] In July 2016, following the attack in Nice, the state of emergency was extended for six months. A new bill providing for the fifth extension of the state of emergency was concluded in the National Assembly in December 2016.[31] Instead of ending the state of emergency, France tabled a new counterterrorism bill which absorbs the exceptional powers used under the state of emergency into ordinary law.[32] The French counterterrorism strategies have been described as the following:

> What is the criteria for an emergency to become an "absolute" emergency? Perhaps more than any other legislation, the Intelligence Act raises the difficult question of balance between the respect for freedom and safeguarding public order," (Mastor, 2017:710)

Measures to prevent the threat of terrorists raise questions of burden sharing and distribution of responsibility between the member states.[33] As noted before, in the EU context, it asks interesting questions about a solidarity obligation. It should be recalled that, in the wake of the Paris attack in November 2015, France initially invoked the solidarity clause Article 222 TFEU but in rhetorical terms than any concrete action plan. However, crucially, France also chose to invoke the mutual assistance clause in Article 42 (7) TEU. This allowed France to send a strong political signal that Europe stands united against a common threat to its territory and society, especially as it implies a common European response in internal and border security domains in which NATO could not take action (Feinberg, 2017). The Solidarity Clause would have activated a wider array of EU instruments, but these require complex inter-institutional cooperation and a longer time frame for their execution (e.g., Faleg, 2015).

[30] See http://www.assemblee-nationale.fr/14/rap-enq/r3922-t1.asp and the Financial Times, "France: The Permanent State of Emergency," October 2, 2017, by A.-S. Chassany.

[31] "Dangerously Disproportionate: The Ever Expanding National Security State in Europe," Amnesty International 2017 report.

[32] The Financial Times, "France: The Permanent State of Emergency," October 2, 2017, by A.-S. Chassany.

[33] See also Saul (2014) and Herlin-Karnell (2019).

The idea of solidarity here seems to be between states and collective self-defense at the EU level is more *NATO* style than EU cooperation in a traditional meaning. Article 222 TFEU is framed as a constitutional duty. But if the collective here are the member states acting together, how does this correspond to the EU law on the right to judicial review? There is still no conclusive answer here regarding what is the criteria for deciding on self-defense. And who is the "collective"? The state who invokes it decides or the collective of states that decide jointly or also the people?[34]

5. The Constitutional Element

The reach of solidarity as collective self-defense beyond the borders of Europe pose questions about legitimacy and the role of the state (Buchanan, 2008). Kant (Ripstein, 2021:chap. 7) in the Doctrine of Right, stated that the requirement of a lasting peace command also means that everyone everywhere on this globe of earth and water is secured under laws. Most important for Kant, according to Pauline Kleingeld, is that peace has to be backed up by a normative foundation, not just that it is in everyone's interest (Kleingeld, 2004). There is no global demos although Kant talked of world citizenship (Ripstein, 2021: chap. 9). In democratic theory, of course, cosmopolitan democracy is often discussed as a way out of the dilemma of no demos, as state borders are becoming less central in our connected world and as many domestic laws have global reach. As Kok-Chor Tan (Tan, 2017:134–135) put it, the all affected principle commonly used in cosmopolitan theory is difficult to demarcate these days with so much activity online (even warfare). This is interesting in the context of the EU and transnational law in general, as the EU promises to promote its values abroad and many of its laws have effects on third states. But if the all-effected principle applies and the outcome is extremely high, it probably runs the risk that no one is affected in the end.

Consider the constitutionalism and justice debate, for example. What justifies the EU's expanding action and is it better suited as a collective—unilateral—actor, compared to the member states acting for themselves? Recently, the EU Commissioner for security stated that, "When it comes to security, national authorities remain the first line of responsibility, and the first line of defence."[35] The EU Commission has also stated that, "Toughened EU anti-terror laws mean those travelling to Syria to fight for Isis or those receiving terrorist training face criminal prosecution."[36] These EU policy documents form an important role in scrutinizing the underlying politics and background to the EU's motivation. This can be viewed in the sense of an individual rights-oriented attitude that justifies a restriction of state sovereignty. The idea of judicial review is commonly seen as the hallmark of a constitutionalist way of thinking about law (Harel 2014; Kumm, 2010). A common republican argument in favor of courts taking account of power differentials in certain contexts (Daly, 2015), would remain skeptical as to whether judicial power can reduce power differentials in society at

[34] See also chapter by Massimo Renzo in this volume, discussing autonomy.
[35] EU Commission for security, Julian King, available at: https://euobserver.com/opinion/138785
[36] Ibid.

large. Rainer Forst, for example, argues for a context-based approach in which it is necessary to pay attention to the perceived experience of freedom and un-freedom on the part of citizens (Forst, 2012). An international or domestic court could rule that a decision going to war was unconstitutional, but in most cases for obvious reasons it happens in retrospect.

Yet with regard to collective self-defense provision, the role of the individual here is almost nonexistent. This then raises interesting questions about the impact of constitutionalism in this area. There is a debate in international law scholarship on the "civilianization of the laws of war" and involvement civil society the recognition that civilian interests matter just as much as military interests (Benvenisti & Cohen, 2013). This would open up to a similar debate in an EU context. On the one hand the EU has to safeguard human rights, security, freedom, and justice of EU citizens (Article 2 TEU and Article 67 TFEU) and the Union's aim is to promote peace, its values and the well-being of its peoples (Article 3 TEU). On the other hand, the EU must also ensure that everyone has the right to effective judicial protection (member states shall provide remedies sufficient to ensure effective legal protection in the fields covered by Union law, Article 19 TEU) and that there is transparency of EU action (inter alia Article 11 TEU). This means that at least in theory a collective self-defense decision, or conversely a decision not to using it, could potentially be challenged in an EU court. Any adjudication of solidarity might be very difficult though and probably more akin to a proportionality assessment.

In any case, one could imagine that these questions could be questioned as regard the proportionality of collective self-defense as well as the democratic aspect of it. The question of judicial review in this context is a very relevant question that deserves exploring in further detail in future research.[37] In short, the idea of collective self-defense at the EU level would presuppose an idea of EU democracy proper.

6. Conclusion

This chapter has tried to discuss the idea of self-defense in the specific context of collective self-defense by discussing it in the light of Ripstein's manuscript on *Kant and the Law of War* and by linking it to the broader EU security regulation debate. I have tried to test the solidarity clause in the EU as a manifestation of collective self-defense. While the Lisbon Treaty refers to both mutual assistance and solidarity in cases of armed aggression and terrorist attacks respectively, the division of security cooperation and resort to collective self-defense seems at best a blurred one. It confirms a very dynamic relationship between the EU security framework and that of EU external relations amounting to a war model.

In addition to that, the chapter sought to dissect the imminence criteria as to when force is justified and tentatively looked at the EU constitutional landscape and the fight against terrorism as well as the AFSJ mission when exploring this question.

[37] See unpublished conference presentation by Katja Ziegler, workshop 1 *Kant and the Law of War*, Amsterdam 1–2 February 2018.

The constitutional dimension is highly dynamic here and relevant to any war theory. It gives rise to a whole host of questions that ought to be asked: inter alia the role of courts and who can challenge a decision to go to war. From Kant we learn that the reason to go to war is always only self-defense and that law restricts the use of force. The means to go to war is a whole different story. But, as Kleingeld reminds us, if smaller states would be dominated by bigger states, justice and right would be endangered (Kleingeld, 2004). An international legal regime is therefore very much at the center of a Kantian theory of peace. The EU in this regard is a state among states.

References

Ackerman, B. (2004). "The Emergency Constitution." *Yale Law Journal* Vol. 113(1): 1029–1091.

Ahsworth, A., & Zedner, L. (2014). *Preventive Justice*. Oxford: Oxford University Press.

Barak, A. (2012). *Proportionality* (Cambridge: Cambridge University Press).

Beer, Y. (2018). *Military Professionalism and Humanitarian Law. The Struggle to Reduce the Hazard of War.* Oxford: Oxford University Press.

Benbaji, Y. (2014). "Distributive Justice, Human Rights and Territorial Integrity: A Contractarian Account of the Crime of Aggression." In Farbre, C., & Lazard, S. (eds.), *The Morality of Defensive War.* Oxford: Oxford University Press.

Benbaji, Y., & Statman, D. (2019). *War by Agreement*. Oxford: Oxford University Press.

Benvenisti, E. (2009). "Rethinking the Divide between Jus ad Bellum and Jus in Bello in Warfare against Nonstate Actors." *Yale Journal of International Law* Vol. 34(2): 541–548

Benvenisti, E., & Cohen, A. (2013). "War Is Governance: Explaining the Logic of the Laws of War from a Principal-Agent Perspective." *Michigan Law Review* Vol. 112(8): 1363–1415.

Buchanan, A. (2008). "Human Rights and the Legitimacy of the International Order." *Legal Theory* Vol. 14(1): 39–70.

Bossung, R. (2012). *The Evolution of EU Counter-Terrorism: European Security Policy after 9/11.* Abingdon: Routledge.

Carvalho, H. (2017). *The Preventive Turn in Criminal Law*. Oxford: Oxford University Press.

Christodoulidou, T., & Chainoglou, K. (2015). "The Principle of Proportionality from a Jus Ad Bellum Perspective." In Weller, M. (ed.), *The Oxford Handbook of the Use of Force in International Law.* Oxford: Oxford University Press, DOI:10.1093/law/9780199673049.003.0056

Daly, E. (2015). "Freedom as Non-domination in the Jurisprudence of Constitutional Rights." *Canadian Journal of Law and Jurisprudence* Vol. 28(1): 289–316.

Douglas-Scott, S. (2012). "The Problem of Justice in the European Union." In Dickson, J., & Eleftheriadis, P. (eds.), *Philosophical Foundations of EU Law.* Oxford: Oxford University Press, chap. 16.

Dyzenhaus D. (2016). *The Constitution of Law: Legality in a Time of Emergency*. Cambridge: Cambridge University Press.

Faleg, G. (2015). "European Security after the Paris Attack." CEPS commentary, available at: https://www.ceps.eu/system/files/GF%20European%20Security%20after%20Paris.pdf.

Feinberg, M. (2017). "Terrorism Inside Out: Legislating for Humanity to Cooperate Against Terrorism." *North Carolina Journal of International Law* Vol. 42(1): 505–543.

Forst, R. (2012). *The Right to Justification: Elements of a Constructivist Theory of Justice.* Columbia, NY: Columbia University Press.

Forteau, M. (2015). "Rescuing National Abroad." In Weller, M. (ed.), *The Use of Force in International Law.* Oxford: Oxford University Press, pp. 947–961.

Guiora, A. (2013). *Legitimate Target, A Criteria-Based Approach to Targeted Killing.* Oxford: Oxford University Press.

Harel, A. (2014). *Why Law Matters.* Oxford: Oxford University Press.

Harris, L., & Monaghan, R. (2018). "Ethics and Human Rights in Counterterrorism." In Silke, A. (ed.), *Routledge Handbook of Terrorism and Counter Terrorism.* Routledge, ch 42.

Herlin-Karnell, E. (2012). *The Constitutional Dimension of European Criminal Law.* Oxford: Hart Publishing.

Herlin-Karnell, E. (2016). "The EU as a Promoter of Preventive Criminal Justice and the Internal Security Context." *European Politics and Society* Vol. 17(1): 215–225.

Herlin-Karnell, E. (2019). *The Constitutional Structure of Europe's Area of Freedom, Security and Justice and the Right to Justification.* Oxford: Hart Publishing.

McMahen J (2009). *Killing in War.* Oxford: Oxford University Press.

Kaunert, C., & Zwolski, K. (2015). *The EU as a Global Security Actor.* London: Palgrave.

Kleingeld P. (2004). "Approaching Perpetual Peace: Kant's Defence of a League of States and His Ideal of a World Federation." *European Journal of Philosophy* Vol. 12(3): 304–325.

Kreztmer D. (2013). "The Inherent Right to Self-Defence and Proportionality in Jus Ad Bellum." *European Journal of International Law* Vol. 24(1): 235–282.

Kumm, M. (2010). "The Idea of Socratic Contestation and the Right to Justification: The Point of Rights-based Proportionality Review." *Law & Ethics of Human Rights* Vol. 4: 141–75.

Luban D (2018). "Knowing When Not to Fight." In Lazar, S., & Frowe, H. (eds.), *The Oxford Handbook of the Ethics of War.* Oxford: Oxford University Press, online version. DOI:10.1093/oxfordhb/9780199943418.013.20

Lubell N. (2015), "The Problem of Imminence in an Uncertain World." In Weller, M. (ed.), *The Oxford Handbook of the Use of Force in International Law Oxford University Press*, online version. DOI: 0.1093/law/9780199673049.003.0032

Mastor, W. (2017). "The French Intelligence Act: 'The French Surveillance State?'" *European Public Law* Vol. 23(4): 707–722.

Meisels, T. (2018). *Contemporary Just War, Theory and Practice.* London: Routledge.

McColgan, A. (1993). "In Defence of Battered Women Who Kill." *Oxford Journal of Legal Studies* Vol. 13(4): 508–529.

Noorda, H. (2020). "Law Reform as a Response to Terrorist Threats." *New Criminal Law Review* Vol. 23(2): 271–299.

Nuotio, K. (2006). "Terrorism as a Catalyst for the Emergence, Harmonization and Reform of Criminal Law," 4 *JICJ* 998

Ripstein, A. (2009). *Force and Freedom, Kant's Legal and Political Philosophy.* Cambridge, MA: Harvard University Press.

Ripstein, A. (2021). *Kant and the Law of War.* Oxford: Oxford University Press.

Sangiovanni, A. (2013). "Solidarity in the European Union." *Oxford Journal of Legal Studies* Vol. 33(2): 213–241.

Saul, B. (2014). "Terrorism as a Transnational Crime." Sidney Law School (working paper). Available at: http://ssrn.com/abstract=2386462.

Shany, Y. (2009). "The Analogy's Limit." *Journal of International Criminal Law Justice* Vol. 7(3): 541–553.

Statman, D., Sulitzeanu-Kenan, R., Mandel, M., Skerker, M., & de Wijze, S. (2018). "Unreliable Protection: Proportionality Judgments in War" (unpublished manuscript, workshop Amsterdam, 1–2 February, 2018).

Tan, K-C. (2017). *What Is This Thing Called Global Justice?* London: Routledge.

Tuori, K. (2014). "A European Security Constitution?" In David Jenkins, D., Jacobsen, A., & Henriksen, A. (eds.), *The Long Decade: How 9/11 Changed the Law.* Oxford, Oxford University Press.

Uniacke, S. (2015). "Terrorism." In Lazar, S., & Frowe, H. (eds.), *The Oxford Handbook of Ethics of War.* Oxford: Oxford University Press.

Vidmar, J. (2017). "The Use of Force as a Plea of Necessity." *American Journal of International Law Unbound* Vol. 111(1): 301–306.

Walzer, M. (1977). *Just and Unjust Wars.* New York: Basic Books.

Williamson, M. (2009). *Terrorism, War and International law.* Farnham: Ashgate.

Other References

Legal Consequences of the Construction of a Wall in the Occupied Palestinian Territory, Advisory Opinion, ICJ. Rep 2004, Separate Opinion of Judge Higgins and Judge Kooijmans, paras. 35 and 34 respectively.

European Parliament brief "Defence: Is the EU creating a European Army?" available at: https://www.europarl.europa.eu/news/en/headlines/security/20190612STO54310/eu-army-myth-what-is-europé really-doing-to-boost-defence

US Secretary of State Daniel Webster dated Apr. 24, 1841, in Caroline Case, 29 British and Foreign State Papers (1841) 1137–1138 The EU Global Strategy, available at: https://eeas.europa.eu/sites/eeas/files/eu_global_strategy_2019.pdf

Directive 2017/541 on combating terrorism, replacing Council Framework Decision 2002/475/JHA and amending Council Decision 2005/671/JHA.

Directive 2017/541 on combating terrorism and replacing Council Framework Decision 2002/475/JHA and amending Council Decision 2005/671/JHA.

Directive of the European Parliament and of the Council on the Prevention of the Use of the Financial System for the Purpose of Money Laundering and Terrorist Financing, Directive (EU) 2015/849, Directive 2013/40/EU, L 218/8 Directive on attacks against information systems.

European Agenda on Migration, COM (2015) 240 final

EU Security Strategy 2020, COM/2020/605 final

Regulation 2016/794 on the European Union Agency for Law Enforcement Cooperation (Europol) and replacing and repealing Council Decisions 2009/371/JHA, 2009/934/JHA, 2009/935/JHA, 2009/936/JHA and 2009/968/JHA, L135/53.

15
Europe's Cosmopolitan Union
A Kantian Reading of EU Internal Market Law and the Refugee Crisis

Bertjan Wolthuis and Luigi Corrias

1. Introduction

Cosmopolitan law, in Kant's theory, is restricted to rules of universal hospitality. A person who comes in peace to a foreign state has the right to be received in peace by that state. In this chapter we claim that European Union (EU) law is best viewed as an instance of cosmopolitan law in Kant's sense. The reason is that Kant's universal hospitality right and EU law both concern the relation between a state and a foreign person—in the EU case: between an EU member state and a person from another EU member state. One might object that Kant's right is quite narrow whereas EU law is extensive. Our response is that Kant's right has to be limited only because it is natural; EU law can be extensive because it is positive. Another objection is that the EU is a superstate and, as Kant has explained, several such superstates are necessarily in a state of war with one another. Our response here is that the EU is no state but a union of states, of the cosmopolitan kind, which entails that these states have brought the relation between their citizens and the participating states under law, to advance peace in this way. The question in this chapter is whether EU law viewed as positive cosmopolitan law can be consistent with state law and international law, especially with the freedom of EU member states. Our answer is affirmative. It rests on Kant's idea of the civil or legal condition (section 3), on the constitution of EU internal market law (section 4), and on EU external cosmopolitan law (section 5). First we introduce the EU as a cosmopolitan union.

2. Europe's Cosmopolitan Union

The claim in this chapter is that the EU is a union of cosmopolitan law,[1] in Kant's sense of cosmopolitan law, with EU internal market law as presumably the most extensive

[1] We approach the EU here from the point of view of legal philosophy. This means that we are not primarily interested in international relations or politics within the EU. See Middelaar (2013) for an example of an IR approach. We are interested here in the EU as a legal order. The difficulty is to uncover that order. For not all that is promulgated as EU law can be qualified as law in a Kantian sense. It is sometimes only a proclamation of what would be desirable for member states to achieve together. Member states do not bring

Bertjan Wolthuis and Luigi Corrias, *Europe's Cosmopolitan Union* In: *The Public Uses of Coercion and Force.* Edited by: Ester Herlin-Karnell and Enzo Rossi, Oxford University Press. © Oxford University Press 2021.
DOI: 10.1093/oso/9780197519103.003.0015

system of positive cosmopolitan law in history. In Kant's legal theory, public law consists of three parts: state law, international law, and cosmopolitan law.[2] The first concerns the relation between persons (within a state), the second applies to the relation between states, and the third to the relation between persons and foreign states (and, through this, to the relation between persons from different states). In a world of persons and states these three types of law are required to bring all possible relations between persons and states under public law, to create a universal civil condition.[3] In the vocabulary of Kant's theory, EU law has to be classified as an instance of cosmopolitan law, we argue, because this law grants the citizen of an EU member state rights in foreign states that are members of the EU. The citizen of an EU member state has the right to freely bring goods, services, and capital to foreign EU member states, for example. Trade tariffs or similar obstacles are prohibited. The citizen of an EU member state also has the right to travel, work, study, and retire in the foreign state that is a member of the EU. EU internal market law is cosmopolitan law in Kant's sense because it creates rights and duties that bear on the relation between each member state citizen on the one hand and each member state on the other. These cosmopolitan rights aim to establish an internal market and an ever closer union between Europe's peoples—which echoes Kant's (1977: para. 62) emphasis on "*Verkehr*" or "*commercium*."

The EU is, as far as we know, not yet regarded by Kant scholars as a possible contribution to a universal civil condition.[4] An explanation is that the EU is often viewed as a continental state;[5] and that Kant wrote that such "Korporationen" necessarily stand in a state of war with each other.[6] But this qualification of the EU fails to appreciate the cosmopolitan way in which the member states of this union seek to establish peace between them. The peace logic of the superstate, and that of the world state, is that the independent states that preceded it are discontinued by it, so that the peoples that have become part of it no longer have the right to wage war on each other. The EU does not follow this scenario. Its member states aim to establish peace between them

themselves under law with only such proclamations. Yet these find their way into EU law handbooks. So positivist EU lawyers perhaps find our cosmopolitan union and its law a stripped-down version of what they have come to know as the EU and its law through EU law handbooks. Our claim that the EU legal order is cosmopolitan is not a political proposal (about what the EU should be) but a philosophical analysis (of what the EU is).

[2] Our translations of Kant's vocabulary do not always follow those proposed in the Cambridge edition of Kant's *Metaphysics of Morals*. We write "law" and "civil/legal" condition, for instance, whereas Gregor writes "right" and "rightful" condition (Kant, 2017: xvi). On the topic of terminology, see the contribution by Mertens in this volume.

[3] See on these categories Ripstein (2021: chap. 8, section I). This contribution builds upon Ripstein's work on Kant and aims to add to it this Kantian analysis of EU law and the refugee crisis. We thank Arthur Ripstein for his comments and the editors of this volume, Ester Herlin-Karnell and Enzo Rossi, for hosting two authors' conferences at which we discussed contributions against the background of themes from Ripstein's work. We thank all participants for their comments. We thank Jelena Belic, Dion Kramer, Irina Leichenko, Peter Niesen, and the editors for their comments on an earlier version of the paper.

[4] Höffe (2007) is no exception because he thinks he has to distance himself from Kant to include the EU, which he sees as a continental state. For an alternative line of reasoning, see the argument in this paragraph.

[5] See the discussion in Eberl and Niesen (2011: 349–350).

[6] Kant, *Metaphysik der Sitten*, Rechtslehre, para. 61. Notice, however, that Kant writes about the situation in which there is no positive international law between such "Korporationen," which means that his argument does not apply to the situation in which the EU finds itself, even if the EU were a superstate.

by granting cosmopolitan rights to each other's citizens, so by bringing the relation between citizens and participating states under positive law. It is too hasty, therefore, to dismiss the EU as "unkantisch."[7] A world of cosmopolitan unions need not be a world at war. What matters in Kant's theory is that there is positive state law, that there is positive international law (in the form of a voluntary federative league of nations), and that there is positive cosmopolitan law. A set of cosmopolitan unions could be a legitimate temporary form of cosmopolitan law; as long as the more extensive cosmopolitan rights enjoyed within these legal orders are consistent with the more restricted universal hospitality right given shape in migration and refugee law. We discuss the latter in section 5. Our basic point here is that the EU should not be categorized as a continental state or a superstate but as a cosmopolitan union.

The EU is not commonly viewed as a cosmopolitan union of states, however.[8] An explanation is that the differences between Kant's cosmopolitan law and EU law are easier to notice than the similarities. Kant's *Weltbürgerrecht* is restricted to the right of a person to make an offer, in a peaceful manner, to a foreign state, concerning this person's visit; and to be heard by this state, again in peace.[9] There is no unqualified right to visit a nation; the foreigner's offer need not be accepted.[10] The provisions of EU internal market law are numerous and far-reaching compared to Kant's "narrow" (Ripstein 2021: chap. 9, section VII) cosmopolitan right. However, what matters is that Kant's cosmopolitan law and EU internal market law both concern the rights of persons vis-à-vis foreign states. The structure is the same. Some deny this and say that EU internal market law cannot be cosmopolitan law because it is more or less equivalent to state law and has erased internal borders.[11] We think this is incorrect. The member states are united by treaties through which each contracting state has granted the citizens of the other contracting states the same set of rights to travel, study, work, conduct business, retire, etc., on its territory. The result is one body of positive cosmopolitan law which is precisely the "new legal order [...] the subjects of which comprise

[7] For this qualification of continental states such as the EU, see Eberl and Niesen (2007): "Charakteristisch für heutige, sich an Kant anlehnende Weltstaatskonzepte ist, daβ sie eine ganz und gar unkantische Zwischenebene zwischen die staatliche und die globale Ebene schieben, die mit "kontinentale[n] oder subkontinentale[n] Zwischeneinheiten" bzw. "kontinentaalstaatlichen" Akteuren bevölkert ist. Im Hintergrund stehen regionale Groβregime nach dem Vorbild der Europäische Union, die nun für den ganzen Erdkreis verbindlich werden sollen. Kant lehnt solche partikularen staatlichen Zusammenschlüsse ab, da sich zwischen ihnen unweigerlich ein machtspolitisches Rivalitätsverhältnis herausschält, das 'einen Kriegszustand herbeiführt.'"

[8] Ferry (2009) and Hoogenboom (2016: 81–88) view the EU as a cosmopolitan union. There are several publications about Europe and cosmopolitanism, but these do not analyze EU law as cosmopolitan law in Kant's sense. Eriksen (2006: 252–269), discusses EU foreign policy and not from Kant's point of view. Beck and Grande's *Cosmopolitan Europe* (2007) concerns the spirit of toleration in Europe. Alec Sweet Stone and Clare Ryan view the legal order erected by the European Convention on Human Rights as a cosmopolitan order in Kant's sense. See their *A Cosmopolitan Legal Order: Kant, Constitutional Justice, and the European Convention on Human Rights* (2018). Their interpretation of Kant's cosmopolitan law differs remarkably from ours, however: they do not structure this law in terms of the rights of citizens vis-à-vis a foreign state (contra Kant, we think); they also view cosmopolitan law neither as narrow in Kant's writings (contra Ripstein) nor as natural (contra Niesen).

[9] Kant, *Zum ewigen Frieden*, Dritter Definitivartikel. Kant writes that both parties, foreigner and state, have to be "friedlich," "nicht feindselig."

[10] Idem: a foreigner can only be rejected if that can be done "ohne seinen Untergang."

[11] This remark was made at the second authors' conference leading up to this volume, at the University of Amsterdam, on June 7, 2019.

not only member states but also their nationals" (Court of Justice, Case 26/62, *Van Gend en Loos v Nederlandse Administratie der Belastingen*, 1963).[12] So at the border of an EU member state, citizens from other EU member states no longer need to make an individual offer about the terms of their visit. The terms of visits for all these citizens are equal; they are now laid down in EU law. The promulgation of these terms makes the crossing of internal borders clearly much less complicated than before but it does not erase the member states or the borders that come with states. Without these borders EU law would make no sense; internal borders are the raison d'être of EU law.

That Kant's *Weltbürgerrecht* is narrow and EU internal market law is not does not mean that EU internal market law cannot be cosmopolitan law, in terms of Kant's theory of law. The reason that Kant's cosmopolitan right *has to be* limited while EU cosmopolitan law *can be* wider, is that Kant's right to a hospitable reception is *natural* while EU cosmopolitan law is *positive*.[13] In the absence of positive cosmopolitan law, persons only have the limited natural right to make the foreign state an offer in peace.[14] A less narrow natural right would subordinate the foreign state's freedom to this too wide natural cosmopolitan right of persons. A narrower natural right, however, would no longer protect the freedom of the person in question. However, less narrow *positive* cosmopolitan law, for example, EU internal market law, need not necessarily subordinate the participating states' freedoms to the cosmopolitan union because these states have *chosen* to give each other's citizens these rights; and they remain free to make a different choice; they are free to leave the union.

Kant's theory of law requires the three parts of public law to be "coordinate" with each other in this way, in order to establish a *legal condition* in which all persons and states can be free.[15] It is against the background of this idea of a legal condition, in which the equal freedom of persons and states is possible, and the requirement of consistency between the three parts of public law, that we analyze EU law in the remainder of this chapter. Kant's understanding of law as "the conditions under which the choice of one can be united with the choice of another in accordance with a universal law of freedom" already makes clear that persons or states, in order to be equally free, have to be subjected to law.[16] That is only possible if the executive and judicial authorities themselves also stand "under law" and are not in the hands of legislators.

[12] What is "new" about this order, we propose, is exactly the cosmopolitan way in which member states and their citizens are tied together as subjects to this law. Technically speaking, the EU is an international governmental organization (because it is treaty-based and member states are free to leave) of a cosmopolitan kind: the material law the states have agreed upon by treaty is cosmopolitan.

[13] We follow here the reading of Kant's *Zum ewigen Frieden* and his *Rechtslehre* as defended by Eberl and Niesen, "Kommentar." This reading is crucial for our argument in this paper. See on the topic of natural and positive law in Kant: Niesen (2017: 83–106; 2014: 170–196).

[14] Kant's *natural* cosmopolitan right is, in a sense, the right to the peaceful conditions under which communication about *positive* cosmopolitan law (an individual agreement between a person and a state about a visit; EU internal market law is the result of an agreement between states about visits of their subjects) has to take place. Some authors appear inclined to think that cosmopolitan law in Kant has to remain limited to this, perhaps driven to this thought by the title of the third article of *Zum ewigen Frieden*. We think this does not make sense; clearly the *natural* right has to be limited to this; but notice that this natural right concerns the conditions under which cosmopolitan law should be *made*.

[15] Ripstein (2021: chap. 9, sections I and VII) examines the consistency between the three parts of law systematically. What is meant by coordinate or consistent in this context is discussed in section 3 of our chapter.

[16] Kant, *Metaphysics of Morals*, 6: 230.

A legal condition in which all three types of relations between persons and states are brought under law in a consistent way makes the equal freedom of states and persons possible. We discuss this in section 3.

So, in this contribution, with the EU viewed now as a cosmopolitan union we ask whether EU law, understood as positive cosmopolitan law, can indeed be qualified as an extension of the legal condition, and whether it can be viewed as consistent with the other two parts of public law, especially with the freedom of EU member states (section 4). The answers to these questions also depend on the possible connection to global, much less extensive, systems of positive cosmopolitan law such as migration law (section 5). The rest of this chapter is structured as follows. First we briefly review the three parts of public right, the requirement of coordination and the civil condition (section 3), to prepare the discussion of the issues just distinguished (sections 4 and 5).

3. The Idea of a Legal Condition

In this section, we discuss the three parts of public law and the legal condition these coordinate parts are meant to establish.

State law. According to Kant (*Metaphysics of Morals*), every human being has the innate right to be independent from the constraining choice of another.[17] But this natural innate right, together with natural acquired rights such as the right to property, can only be enjoyed if they are specified in a system of positive law and maintained by state institutions. The reason for this is ultimately not that man is a wolf to man in the state of nature (Hobbes, 1997: 3–4) or that natural law, while clear, is not safe in the hands of ignorant and self-interested private persons (Locke, 1980: par. 124). Even if human beings were "well-disposed and right-loving" (Kant, *Metaphysics of Morals*, 6: 312), public law is necessary. The reason is contained in the idea of the state of nature itself. What natural law prescribes in concrete cases is something persons might reasonably disagree about (contra Locke).[18] Hence, in the state of nature you ought to rely necessarily on your own view of what is right. If you follow another person's view, you are dependent upon that person, which frustrates your innate right to independence from the start. However, if all persons rely, as they should, on their own view of what is right, each person's right to be independent from the constraining choice of others *still* depends upon the constraining choice of others, who (in the best of cases) make their choices according to their own view of what is right. So the state of nature is a state in which a plurality of persons can never be free and equal, no matter how good or righteous they are. Persons can only be independent from each other if the equal freedom of persons is secured through a system of positive law through which a collective of persons ultimately governs itself (cf. Ripstein: chap. 2, section I). Then all persons can be independent because they can depend upon the positive law that is administered by independent state institutions in the name of them all.

[17] See Ripstein (2009a: chap. 2) for an interpretation of this right.
[18] For a detailed discussion see Ripstein (2009a: 190).

International law. States, understood as such systems of positive law, themselves stand in a state of nature relation to each other, in the absence of a system of positive international law. The equivalent of the innate, natural right of a person to independence is the right of a state to decide how to act in its relations to other states, which is constrained only by what a state ought to be, that is, a civil condition for its subjects, who have, as a people, "an original right to unite itself" (Kant, *Metaphysics of Morals*, 6: 349). This limit means that a state cannot abandon its task to establish a legal condition for those subjected to its system of positive law. So we see here how international law has to be coordinate with state law. This view of the relation between state law and international law also implies that states cannot legitimately be forced into a world state (in which states cease to be), while persons have the duty to enter a civil condition with persons with which they cannot avoid interaction. States can freely enter a *federative* union, if "that [union] can be renounced at any time" (Kant, *Metaphysics of Morals*, 6: 344). Positive international law governing their relations, administered by an international court, is a form of peace, a civil condition, between them. They are subordinated to rules given to each other by treaty, administered by treaty-based judicial and administrative authorities in the name of all of the participating states.

Cosmopolitan law. Cosmopolitan law is the third part of public law. Kant's natural cosmopolitan right states that a person who wants to enter a foreign land in peace has the right to make an offer to the foreign state about the terms of this person's visit and to be received in peace by the foreign state while the offer is being made. Kant uses the words "Wechselwirkung (*commercium*)" and "Verkehr" (Kant, *Metaphysik der Sitten*: para. 62) to refer to the exchange or, more broadly, interaction between world citizens that is indirectly at stake here (the content of the offer). Eberl and Niesen ("Kommentar," 251) argue that Kant's *commercium* has to be viewed broadly as including economic, political, and cultural interaction. The natural cosmopolitan right formulated by Kant only prohibits the addressee of the offer to treat the foreigner with hostility; and, vice versa, the foreigner has to come in peace. The reason for this restriction is that a broader understanding of the natural right (a foreigner's right to settle, for instance) would subordinate the freedom of the state in question to the freedom of the globetrotter (or a movement of settlers: colonialism) while an even narrower view (a hostile reception?) would of course no longer protect the freedom of the globetrotter.[19]

Legal condition. With these three kinds of public law now distinguished, we move on to what public law itself is and what it is meant to establish. In Kant's theory, public law is defined as law that has to be positivized to create a legal condition. (cf. Kant, *Metaphysics of Morals*, 6: 311) This normative understanding makes it an open question whether EU law is public law. Does EU law extend the legal condition on the territory of the EU member states? The answer will depend on what a legal condition is and how it can be brought about. From the *Rechtslehre* and *Zum ewigen Frieden* we learn that not all kinds of rule necessarily constitute a legal condition. A despotic regime is not rule of law at all but rule of men. Kant distinguishes despotism from republicanism; republicanism is for Kant the doctrine of the separation of powers.

[19] See the thorough discussion in Eberl and Niesen (2007: para. 3.3.3).

Law should ideally be *given* by all persons who are also *subjected* to the law, since *volenti non fit iniuria* (cf. Kant, *Metaphysics of Morals*, 6: 313–314). However, these persons can only be subjected to *law* (as opposed to *men*) if they do not also *execute* the law, because then the executive power is not "under" or "bound by" law, since it is in the hands of those who already stand "above" law, as legislators. Then a majority of "men" would rule over a minority of men. This makes such a direct democracy an "Unform";[20] it does not constitute a legal condition at all; it may have the appearance of a constitution and it may produce laws but the executive power is never under law.[21]

Coordinate. A recurring theme in Ripstein's *Kant and the Law of War* is that the three parts of public law have to be "coordinate," a Ripsteinian term of art we borrow here. They have to be coordinate in the obvious sense that they have to bring all possible relations between persons, between states, and between persons and states under law. Ripstein (2021: chap. 9, section X) uses the term coordinate in this sense in a discussion of "legal black holes," which designate relations not covered by public law. It has to be distinguished from the second meaning of coordinate in Ripstein, which concerns the possibility of equal freedom of persons and states in one system of public law. So the requirement is not only that there has to be public law to cover all relations, the relations also have to be shaped through public law in such a way that each person's and each state's equality and freedom is made possible.

It is the latter understanding of coordinate that implies that the three parts of public law are *not* necessarily coordinate with each other in the sense that the three parts have to be of equal rank. The concern is that the equal freedom of *persons* and *states* is protected.

In this regard the *sequence* of Kant's line of argument, from innate human right to private right, to state law, to international law and, finally, to cosmopolitan law, appears to be relevant. In the above overview of the three parts of public law, we emphasized, following Ripstein, that state law (whereby a civil condition between persons is provided) limits international law (treaties whereby this condition is threatened are invalid) and that international law (the equal freedom or independence of states) limits cosmopolitan law (natural cosmopolitan right has to be narrow because of states' equal freedom). An argument in reverse order would generate a different kind of public law, if it would generate anything comprehensible at all.

The normative implications of this sequence need not necessarily be interpreted as conservative, as protecting the status quo of a world of states.[22] After all, the crucial point is that the equal freedom of persons is only conceivable in the form of state law, which makes the state the correct place to start. A similar argument is made by Rawls, who takes the "basic structure" of a society as the first subject of justice, for the reason that "the freedom and equality of moral persons require some public form" ("at some level there must exist a closed background system") (Rawls, 1993).[23] The sequence in

[20] Kant (*Zum ewigen Frieden*, Erster Definitivartikel).

[21] These insights about law, the legal condition, and the republic can already be found in Rousseau (1997, Book II).

[22] That opinion is prevalent among contemporary cosmopolitans (not, we think, in Kant's sense). For example: Valentini (2011).

[23] Lecture VII, 281 and 272 footnote 9. See for Kantian readings of Rawls's theory: Rainer Forst (2017: 123–144) and Arthur Ripstein (2006: 1391–1438).

Rawls's constructivism is also one-directional: in his *Law of Peoples* the principles of justice chosen by parties representing citizens in the first original position determine the interests ascribed to peoples represented by parties in the second original position. These interests in turn shape the law of peoples.[24]

Against the background of this understanding of the legal condition and the requirement that the equal freedoms of persons and states have to be coordinate, our question in the next section is whether EU law really extends the legal condition and whether it is coordinate with the freedom of the European states who are united by the cosmopolitan rights they have reciprocally given to each other's citizens.

To clarify: this question involves a critical assessment of a given, "empirical" system of positive cosmopolitan law against the background of Kant's view of law "as idea" or "norm."[25] This approach need not be alien to the Kant scholarship as proposed by Niesen. Niesen draws from his reading of Kant's cosmopolitan right as a natural right the conclusion that "it makes more sense to understand the positions "developed" from Kant in the secondary literature not as rivaling interpretations of natural cosmopolitan law, but as arguments to be considered in its transformation toward a global public legal condition" (Niesen, 2017: 100). We agree. We take an instance of positive cosmopolitan law and ask whether it contributes to the global legal condition. EU internal market law raises interesting issues with respect to Kant's theory because it is an *extensive* system of positive cosmopolitan law (section 4) between only a *limited* group of states (section 5).

4. EU Internal Market Law and the EMU

In the second section we introduced the EU as a cosmopolitan union. In the third section we examined Kant's notion of the legal condition and the requirement that the three parts of public law are coordinate. In this section and the next we answer the question whether EU law contributes to the establishment of a global legal condition, for EU member states and their citizens (this section) and for non-EU citizens (next section).

We do not concentrate on individual EU rules (individual EU treaty articles or pieces of EU legislation) or policies but on the structure or "constitution" of EU law, because we are interested in the legal condition (4.A), which concerns the separation of powers, as we have seen, and because we are also interested in the way EU law coordinates with international law (4.B), especially with the equal freedom or independence of its member states.

[24] See for an analysis of Rawls's "sequential" constructivism: Wolthuis (2017: 454–466). Rawls's justice as fairness offers no cosmopolitan principles. In that article Wolthuis proposes a third-level original position, to make justice as fairness applicable to the European Union. In light of this chapter (see also the third section), we think the title of Wolthuis's article should have read: "Principles of Cosmopolitan Union," since the principles do not concern the participating states' household management (which is the topic of the EMU) but *commercium*.

[25] We adopt here phrases used in paras. 45 and 51–52 of Kant's *Rechtslehre*; on this topic of noumenal and phenomenal state/law: Ludwig (1999: 173–194).

4.A. Legal Condition

The answer to the main question is not a straightforward "yes" or "no" because there are multiple parts of EU law, which are structured rather differently.[26] Here we zoom in, to keep it manageable, only on EU internal market law on the one hand and on EU economic and monetary union (EMU) law on the other. The first extends the legal condition, we argue, and is coordinate with the freedom of states, whereas the second does not extend the civil condition. Moreover, we submit that EMU law as it is cannot be coordinate with the freedom of states. We have selected EMU law only because it stands in clear contrast to the way EU internal market law is constituted.[27]

Internal market law. Kant's cosmopolitan law and EU internal market law are both concerned with "Verkehr," with exchange or interaction between persons. The German "freie *Verkehr* von Waren, Personen, Dienstenleistungen und Kapital" (art. 26 TFEU) captures this better than the English "free *movement* of goods, persons, services and capital." These four freedoms define the rights of citizens of EU member states and prohibit EU member states to restrict the border crossing transport of goods, services and capital and to hinder the free border crossing exchange of workers and students, irrespective of the member states of which they are citizens.

These rights and prohibitions fall under titles I, II, and IV of part three of the Treaty on the Functioning of the European Union. These provisions are maintained by the Commission, which is an executive authority independent from the member states. They can be invoked by persons and states before national courts of law, who can refer questions about them to the European Court of Justice. Legal conflict about what this law entails need not be resolved through force, and need not depend upon certain persons, but can be settled in a court of law. The legislative, executive, and judicial powers are separated within internal market law, which does not conflict with but instead appears to extend the legal condition for citizens of EU member states and these member states with respect to border crossing *commercium* within the EU.[28]

EMU. Several EU member states have a common currency and aim to create an economic and monetary union (EMU).[29] The member states who have chosen to cooperate with each other within the EMU regime have subjected themselves to economic standards: they have to avoid excessive government deficits (126 (1) TFEU); they have to coordinate their economic policies with the Council (121 (1) TFEU); and their economic policies have to be guided by certain "principles: stable prices, sound

[26] EU member states cooperate in a many different ways, on various policy areas, and sometimes they codify their common aims. This does not place them under law but is intended to foster international solidarity. EU internal market law is different. See on this topic, Wolthuis (2021).

[27] So we ignore here that the character of EMU law is not cosmopolitan; we discuss it only to illustrate how EU law can fail to establish, from a Kantian point of view, a legal condition.

[28] The implication is that a state that decides to leave the EU internal market willingly takes away some of its citizens' cosmopolitan rights.

[29] So we stress the distinction between an economic union and what could be called a "commercial" union (in the Latin sense) or, better, "cosmopolitan" union. While "commerce" and "economy" may be thought to refer to identical phenomena, for present purposes we suggest they should be sharply distinguished. "Economy" refers here and in the EMU to the classical Greek "oikos," to household management. It makes sense to talk about a family's economy, a business's or a state's. The unit in question has to decide how to conduct its business, how to invest, how much to save, etc.

public finances and monetary conditions and a sustainable balance of payments" (128 (3) TFEU). If a member state has an "excessive deficit," the Commission is obligated to write a report. If the Council decides that such a deficit exists (126 TFEU), it adopts a recommendation. If the member state fails or refuses to comply with the recommendation, sanctions must be applied by the Council.

Some EU lawyers doubt whether these EMU rules really constitute law. These provisions are given by treaty and are in this sense obviously promulgated, positive norms. However, they do not specify rights but are mere "guards of economic wisdom" (Herdegen, 1998: 9–32). It is this character of the rules that may explain that they are not executed *ultimately* by an independent authority (such as the Commission) and cannot be appealed to in a court of law. They are executed by the member states themselves, in their capacity as members of the Council, to themselves, in their capacity as subjects of EMU law. This constitution of the EMU resembles the structure of the direct democracy that Kant rejects in *Zum ewigen Frieden* as an "Unform" (see previous section). The implication of its constitutional structure is that EMU "law" cannot be viewed as an extension of the legal condition. EMU "law" may have the appearance of law but does not establish the rule of law. Member states do not stand under law, as they do in internal market law, but above it, because they have in their own hands legislative, executive, and (quasi-)judicial powers. That establishes no legal condition; it is a form of despotic rule.[30] Stronger member states (or the majority of member states) may overrule weaker member states (or a minority of member states). The conditions under which the Eurogroup (minus Greece) was prepared to lend support to Greece, as an answer to the Greek debt crisis, illustrated according to some commentators that the Eurogroup is a form of rule by these states over Greece. In the aftermath of the sovereign debt crisis, several measures were taken to prevent new crises of this kind (Chalmers et al., 2014: 46). We need not discuss these here; sufficient for our purposes is that even EU law handbooks disqualify these measures in terms of the relations of dependency and hierarchy they generate between member states (ibid.).[31] We conclude that Kant's theory of law can supply the arguments on which such judgments can be based.

4.B. Coordination

Internal market law. By itself the creation of positive cosmopolitan law is not necessarily contrary to the freedom of persons and states as protected by state and international law. A person has no natural right to travel or settle in a foreign country. Those rights can only be created.[32] We regard EU internal market law as positive cosmopolitan law. By treaty the member states have given each other's citizens the border crossing four freedoms referred to earlier and specified in many regulations and EU

[30] See also the analysis of the UN Security Council as a form of despotic rule in Eberl and Niesen (2007: 326).

[31] "The European Stability Mechanism generates new forms of dependency and hierarchy, whereby, through the conditions established in the Memoranda of Understanding, donor states have the opportunity to impose their own vision of economic life on recipient states."

[32] See the discussion in sections 2 and 3. Kant's natural right refers to positive cosmopolitan law.

case law. There are no obvious reasons why states are not free to make treaties of this kind. As discussed in the previous section, the freedom to make treaties is only restricted by what states ought to be or provide, that is, legal conditions in which their citizens can enjoy their rights. States also have to be independent, in the sense that they have to be free to decide for themselves what is required for maintaining the civil condition for its citizens. EU internal market law leaves this freedom of EU member states intact. These states may even have succeeded, by creating by treaty positive cosmopolitan law, in *extending* the civil condition for their citizens (4.A).

Perhaps this could have been different, if EU member states had reciprocally granted their citizens an unqualified right to settle in the member state of their choice.[33] Large-scale migration (also by persons not able or willing to provide for themselves; sometimes called "welfare tourism") may frustrate a state's capacity to preserve itself as a civil condition.[34] Is it also problematic if a state subjects itself to rules of positive cosmopolitan law but has no authority to change the rules? Boris Johnson wrote that a state who finds itself in that situation—which he feared would be the outcome of a Brexit with "backstop"—has the "status of colony" (@BorisJohnson, July 9, 2018). Because the element of force is absent here, we believe this term is not accurate.

EMU. Now that the economic and monetary union is unmasked as an "Unform," whereby no legal condition is established (4.A), the issue of coordination (coordinate or not coordinate?) may no longer seem to apply, since there is no law (at least in Kant's sense) to begin with, for other parts of law to be coordinate with or not. However, recall that we interpreted the coordination requirement in terms of the possibility of equal freedom of persons and states. Coordination in this substantial sense is still relevant here. The rule of states in EMU "law," with dependency and hierarchy between donor and recipient states (section 4.A) is clearly in conflict with the equal freedom of states. The Greek debt crisis is for this reason better classified as a crisis of international justice—whereby the independence of Greece is at stake—than as a crisis of international solidarity (Wolthuis, 2021). That Greece found itself in a situation in which it had to depend upon the solidarity of other member states, already implies that its independence, its freedom, and equal position was at stake. The conditions under which it received help were of such a nature that it no longer could decide how to manage its own household.

This "rule of states" character of EMU raises the issue whether a hypothetical "rule of law" type EMU *could* be coordinate with the equality and independence of participating states.[35] The topic of household management may be thought to be too intimately connected to a state's freedom to decide for itself how to preserve itself as a legal condition for its citizens. However, if a state can enter a rule of law type EMU freely and has the freedom to exit it, its independence need not be threatened by

[33] See for the qualifications Kramer (2017: 172–188).

[34] See on this topic of a state's mandate to preserve itself: Ripstein (2009a: chap. 8).

[35] It is crucial to see that "supranational" (as used by EU lawyers) does not imply "civil condition" (as used by Kantians). The first notion is already used when the Council decides with qualitative majority. See the discussion in the previous section.

membership of such a union. Its membership of such a union may be, according to this state, the best way to preserve itself as a legal condition, at least for the time being.

So the provisional conclusion is that the EU is, as it is, best approached as a cosmopolitan union, as soon as it is viewed from the point of view of Kant's theory of law. The states that are part of this union have used their freedom of choice—and they remain free to choose differently in the future—to bring themselves and their citizens under cosmopolitan positive law, enacted by themselves by treaty, but administered by the Commission and the Court of Justice, authorities independent from them.

5. EU Migration Law

The analysis in this section is driven by two main questions. The first question concerns whether the system of EU migration law constitutes a civil or legal condition in the sense envisioned by Kant (see section 3 above). In order to answer this question, we will first sketch the constitutional outline of this area of EU law (5.A). The second question is how this part of EU law—understood as a form of *positive* cosmopolitan law—relates to universal cosmopolitan law. In other words: is EU migration law coordinate with universal cosmopolitan standards (5.B)? This latter question contains two sub-questions: first, whether coordination is in principle possible; second, whether the factual situation is coordinate.

5.A. Legal Condition

EU migration law is part and parcel of the so-called Area of Freedom, Security and Justice (AFSJ). Article 3 (2) TEU offers the legal basis for the AFSJ: "The Union shall offer its citizens an area of freedom, security and justice without internal frontiers, in which the free movement of persons is ensured in conjunction with appropriate measures with respect to external border controls, asylum, immigration and the prevention and combating of crime." The objectives of the AFSJ that are relevant for our purposes are enumerated in the first two paragraphs of Article 67 TFEU: "1. The Union shall constitute an area of freedom, security and justice with respect for fundamental rights and the different legal systems and traditions of the Member States. 2. It shall ensure the absence of internal border controls for persons and shall frame a common policy on asylum, immigration and external border control, based on solidarity between Member States, which is fair towards third-country nationals. (. . .)"

Institutionally, the Lisbon Treaty has brought about a number of important changes in this area. First, the European Commission has been given a more important role within the AFSJ. The Commission now holds the power to monitor the application of the law of the AFSJ and to start proceedings against those member states that fail to comply with their duties. Second, the European Court of Justice has obtained more powers. The Court has attained the competence to deliver preliminary rulings on all aspects of the AFSJ. Third, several special agencies have been established that are tasked with monitoring policies that are part of the AFSJ. For the purposes of this contribution, the European Border and Coast Guard Agency (Frontex) is the most relevant one: the agency is charged with the coordination of external border control. Nevertheless, fourth, in chapters 1 and 2 of Title V of the Treaty on the Functioning of the European Union the treaty-making states have granted themselves, within their role as members of the Council, an important role, along with the European Parliament, in the development and execution of policies in the AFSJ.[36]

Nowadays, EU law contains a comprehensive system of migration law consisting on rules of entry, residence, and return of migrants (cf. Boeles, den Heijer, Lodder, & Wouters, 2014: 37) The rationale of the system as a whole resides in the need to regulate the *external* borders of the EU, now that the *internal* border controls are abolished.[37] Central here is the distinction between EU citizens and third-country nationals. As the Treaty also establishes, EU citizens are all persons with the nationality of a member state. (cf. Article 20 (1) TFEU) The group of third-country nationals consists of those persons who are not EU citizens, including stateless persons. Furthermore, European migration law distinguishes between forced and voluntary migration (cf. Boeles, Den Heijer, Lodder, & Wouters, 2014). Starting with the latter: voluntary migration may have different reasons, including economic activity. In case of voluntary migration, migrants may enjoy a number of rights under EU migration law. The precise content and scope of these rights depend mainly on the reason why a person wants to migrate to the EU and for how long (s)he plans to stay.

Forced migration refers to people fleeing their country and asking for a form of protection (asylum). While traditionally perceived as a direct repercussion of a state's sovereignty, the right of asylum is nowadays understood as a human right under what lawyers refer to as international law (cf. ibid.: 234), and Kant scholars as positive cosmopolitan law. Central is the right of *non-refoulement*: the protection of a refugee against "being returned to his or her country of origin, or any other country for that matter, where he or she is at risk of being subjected to serious harm" (ibid.: 244). It is codified in the 1951 Refugee Convention in article 33 (1). Besides this Convention, human rights law forms a second fundament underlying the international law of asylum and both are central to an emerging EU asylum law, or Common European Asylum System. (cf. ibid.: 245).

An important instrument is the Dublin Regulation (604/2013), which aims at reducing the secondary movements of asylum seekers within the EU. (cf. ibid.: 257) It determines that only one member state is in principle responsible for an application, usually the one through which the applicant has entered the EU (either regularly or irregularly), provided that the applicant was registered there. (cf. ibid.: 258–259)

[36] See, TFEU, Title V, chapters 1 and 2. This is an executive authority, in terms of Kant's theory of law; that the treaty and EU lawyers refer in some of these matters to "legislation" does not make this different. Within Kant's theory of law the treaty makers are the legislators and the treaty is the legislation. See also sections 2 and 3.

[37] "It should not be forgotten that the main reason why the EU stepped in to implement the rules on asylum, whether those of the Refugee Convention or of the ECHR, as interpreted by the ECtHR, was the need to regulate asylum in a uniform manner. This was a necessary step towards establishing a common juridical space in which, among other things, the freedom of movement of European citizens is assured as this would be hindered by the presence of internal border controls. The origin of the Schengen Agreements illustrates very clearly the link between the desire to create that space and the harmonisation of legislation on asylum and immigration in general" (Cherubini 2014: 254).

Among other things, the Regulation gives the applicant the right to a personal interview (article 5), to appeal against a transfer decision and to stay while the appeal is pending (article 27).

The Reception Conditions Directive (2013/ 33/EU) gives the regulatory framework for the reception of asylum seekers within member states (cf. ibid.: 267). Important for our purposes: it grants asylum seekers the freedom of movement and residence, yet under certain conditions (article 7). The Qualification Directive (2011/95/EU) has as its aim to ensure that the member states apply the 1951 Refugee Convention in a uniform way (cf. ibid.: 293). The Directive gives refugees a number of rights, such as protection from *refoulement* (article 21), a residence permit (article 24), access to employment (article 26), and the freedom of movement (article 33) (cf. ibid.: 338–339). Furthermore, it offers so-called subsidiary protection to forced migrants who do not qualify as a refugee but "face a real risk of suffering serious harm" (ibid.: 347) in their country of origin. Here too the protection against *refoulement* (e.g., when execution, torture, or inhuman treatment looms) forms the cornerstone of the protection. (cf. ibid.: 334–346). Just like the protection given to refugees, subsidiary protection cannot be invoked by EU citizens (yet, they do possess the right of *non-refoulement* as it derives from both the ECHR and the EU Charter of Fundamental Rights) (cf. ibid.: 348). The prohibition of *refoulement* as it derives from article 3 ECHR is absolute and *notstandfest* (meaning that not even in times of war or other emergencies may it be set aside) (cf. ibid.: 366). Regarding the content of the protection: the Qualification Directive also envisions such rights as access to employment (article 26), initial residence of one year to start with (article 24), and free movement within the member state (article 33) (cf. ibid.: 370–371).

Let us now return to the main question of this subsection: Does the AFSJ, particularly regarding EU migration law, establish a legal condition in the sense of Kant?[38] It appears to approach it. Just as is the case with internal market law, the rights and duties regarding European migration law are *monitored* by the European Commission. In this way, an authority independent from the member states is responsible for their correct execution. Furthermore, the national judge and, ultimately, the European Court of Justice may be called upon to settle conflicts. However, the Council has crucial executive powers, which means that within AFSJ powers are not as clearly separated as in internal market law. Hence, our provisional conclusion is that EU migration law as discussed earlier may be understood as approaching but still falling short of a civil or legal condition in the Kantian sense.

5.B. Coordination

Now, we will analyze EU migration law from a Kantian perspective. More specifically, the question is whether EU migration law is consistent with universal cosmopolitan

[38] Whereas it falls outside the scope of this chapter, we acknowledge the important role played by the case law of the European Court of Justice in interpreting the relevant directives and pointing at the limits of mutual recognition. Important judgements include Case C-163/17, Jawo v. Germany of March 19, 2019; Joined Cases C-411/10 and C-493/10, N.S. and Others of December 21, 2011; and Case C-578/16 PPU, C. K. and Others of February 16, 2017.

law. This analysis is driven by two major lines of thought, both constitutive of the Kantian framework. The first submits that every individual ought to reside within a state of law: every person is a member of a legal order. The second says that legal orders ought to respect each other's independence. As we have discussed above (sections 2 and 3), cosmopolitan right forms the final step in Kant's architecture of perpetual peace. As a project to bring the whole globe under the rule of law—indeed, to make the world into a *legal* world—the relations of states with people who are not members of those states must also "be brought under law: there must be both a rule that governs them and a rule identifying the officials charged with its application" (Ripstein 2021: chap. 9, section IV). So, using the terminology of H.L.A. Hart, cosmopolitan right consists of both the primary and secondary rules on this theme. Ripstein stresses that Kant's starting point in conceptualizing cosmopolitan right is "the presumption that everyone is always already a member of some state, and the correlative presumption that every inhabited region of the earth's surface is—as against both other nations and their individual members—always already a rightful condition and entitled to occupy the space that it does" (ibid.). Hence, "individual visitors always bring with them their membership in some nation, and always visit a rightful condition over which the visitor has no authority" (ibid.).

This last aspect is important in order to understand the narrow content of the right of the visitor: as a nonmember, (s)he has the right to visit but not the right to stay and settle. This implies a number of negative duties, both for the visitor and the receiving state. For the state, most importantly, it entails the duty to abstain from any form of hostility vis-à-vis the visitor (cf. ibid.) For the visitor, the duty consists in not importing the rules of her home legal order into the state she visits. In this context, Ripstein (ibid.) gives slavery and polygamy as examples. But what exactly are the rights of the visitor? Formally, these "are the rights that every human being has as a member of the global public" (ibid.). Materially, the right to visit grants you the right to travel and bring possessions with you (even if the state you enter may tax them) without losing any rightful property you might have in your home state. (Ripstein 2021: chap. 9, section II) Furthermore, the right to visit includes the right to peacefully interact with the citizens of the host country: the right to *commercium*. (cf. ibid.) This narrow content of cosmopolitan right is ultimately grounded in the principle of non-domination underlying the whole edifice of peace: coordinating the relations between states and people, guaranteeing the independence of each nation's legal order by making sure "that relations between individual human beings do not become a means of subordinating one legal order to another" (ibid.).

The EU is viewed here as a league of states, which are united by the cosmopolitan law that regulates the relation between each of them and the citizens of the other states—as, in short, a cosmopolitan union—and this union's positive cosmopolitan law can be seen as an extension of the civil condition for its member states and these member states' citizens. So far, so good. However, the EU is not a *universal* or *global* cosmopolitan union of state: far from it. This raises the question whether a *limited* union's positive cosmopolitan law can be consistent with universal cosmopolitan law, such as Kant's natural universal hospitality right or the (nearly) worldwide systems of positive cosmopolitan law we know as refugee and migration law. To be more precise: we want to address the following two questions. The first, by way of introduction, is whether a limited cosmopolitan union can in principle be consistent with universal cosmopolitan law. The second, critical question is whether the actual institutions and practices of the EU and/or its member states with regard to migrants and refugees are in fact consistent with positive refugee and migration law.

Regarding the first question: we do not see why a limited cosmopolitan union has to be in conflict with universal cosmopolitan law, natural or positive. That the civil condition for its member states and citizens is extended need not imply that the rights of others, outside of this union, are thereby threatened. To recall: Kant's natural right only includes a non-hostile reception. A state or union of states may have its reasons to reject migrants and it is allowed to return refugees if their homeland is safe again. That the rights of some have been promulgated and have become more extensive does not mean that others have lost rights. Other states can follow the example and form cosmopolitan unions themselves. To bring relations between states and foreigners under law by itself brings the aim of eternal peace closer, we think.

Does it make a difference whether the *union* pledges to respect the standards of the universal migration and refugee treaties and has, with the promulgation of *union* migration and refugee law, subjected itself to a worldwide civil condition with respect to migration and refugee law or whether *each of its member states* is party to these treaties and has its own rules and procedures of migration and refugee law? We think not; we find that in both cases the more extensive but limited system of EU migration law can be consistent with universal refugee and migration law. The exact difference depends of course on the content of what would then be union migration and refugee law. That law could state that refugees and migrants can reside wherever they want in the union, but it could also say that the refugee or migrant has to reside in the member state with which it came into contact with first. (Or a system of distribution could be agreed upon.) In all these possible cases we assume that legal solutions are arrived at: that legal conditions as specified in the second section are established. We now turn to the solution that the EU and/or its member states have actually have agreed upon and we analyze whether that solution conflicts with universal cosmopolitan law, natural or positive.

The recent migration crisis has shown the failure of the system to distribute asylum seekers fairly over the member states (Boeles, Den Heijer, Lodder, and Wouters 2014: 265). Trying to aid Greece and Italy in the summer of 2015 when these two countries were dealing with a huge inflow of migrants, the *Council* adopted a majority decision that envisioned the reallocation of 120,000 asylum seekers (Council Decision (EU) 2015/1601 of 22 September 2015 establishing provisional measures in the area of international protection for the benefit of Italy and Greece (OJ 2015 L 248, p. 80)) Yet, several countries refused to cooperate and Slovakia and Hungary asked the European Court of Justice to annul the decision. The Court, however, ruled against the two countries, thus establishing that the decision was taken on the basis of the appropriate legal basis, according to the right procedure and was a suitable instrument to attain the purpose.[39] Nevertheless, what we see here is the discrepancy between what member states are legally obliged to do and what they politically are willing to do. In terms of Kant's understanding of the legal condition the problem is that the

[39] Judgment in Joined Cases C-643/15 and C-647/15, *Slovakia and Hungary v Council* of September 6, 2017.

main executive authority is in the hands of a *non-representative* body (the Council); it is, in this sense, rule over a minority of states by a (qualitative) majority of states. The *Unform* that we introduced in section 3 and that we identified in EMU (section 4), also lurks here. That the proper procedure was followed, is one thing; that the procedure itself is proper, is another. This situation also partly explains (but certainly does not excuse) Matteo Salvini's, Italy's former Minister of Internal Affairs, blunt refusal to let asylum seekers arriving by boat to Italian ports to leave their ship and set foot on Italian soil as this would make Italy responsible for handling their claim to asylum (following the Dublin Regulation).

One of the developments in EU migration law that also needs to be mentioned in this context is the externalization of migration policies. As Afailal and Fernandez (2018: 217) put it, "to externalize means to delocalize the limits of the control of a sovereign country through the implication and accountability of other countries." This process has been going on for a while and is at odds with the rule of law.[40] Migration, according to Afailal and Fernandez (2018: 217, 221), "is understood as part of the safeguarding of European security," "of controlling access to 'fortress Europe.'" In this process, some scholars have argued, the EU acts as a neocolonial power by trying to influence the sovereignty of its neighboring countries in interfering with their migration policies. The example of this externalization that has recently received most attention is the deal between Turkey and the EU (cf. European Council, EU-Turkey statement, March 18, 2016) on the reallocation of migrants, which was signed on March 18, 2016. According to the agreement (cf. Haferlach & Kurban, 2017: 87), as of March 20, 2016, all new irregular migrants who pass from Turkey to Greece are to be returned. For every Syrian migrant who is returned to Turkey, another Syrian migrant will be resettled in the EU. Turkey and the EU have pledged to improve humanitarian conditions within Syria. In return for its efforts, Turkey will be financially compensated by the EU, the lifting of visa requirements for Turkish citizens traveling to the Union will be accelerated and the process of Turkish accession to the EU will be reinvigorated.

There are several problems with this agreement. This type of "out-sourcing stands in stark contrast to values and norms that the EU member states purport to promote. Moreover, the extra-territorialisation of migration control transfers and diffuses political responsibility, making it difficult to identify the bodies and states responsible for the breaches of international law and hold them accountable" (Haferlach & Kurban, 2017: 86) Most importantly, Turkey has made a restriction to the 1951 Refugee Convention, implying that only persons from Europe can obtain the status of a refugee (cf. Afailal & Fernandez (2018: 218–219). Indeed, the agreement itself does not contain a definition of a refugee, thus giving way to arbitrary treatment. (cf. Afailal & Fernandez, 2018: 220) The agreement falsely regards Turkey as a "safe third country," despite its ambiguous relationship with regard to the 1951 Refugee Convention and the situation on the ground with its recurrent reports of violence against and maltreatment of refugees by Turkish border guards and police officers

[40] Compare Jorrit J. Rijpma and Marise Cremona, "The extra-territorialisation of EU migration policies and the rule of law," EUI Working Papers, Law 2007/01, https://cadmus.eui.eu/bitstream/handle/1814/6690/LAW_2007_01.pdf?seque.

(cf. Haferlach & Kurban, 2017: 88–89). Ever since the failed coup d'état of July 15, 2016, the general human rights situation in Turkey has also deteriorated (cf. Haferlach & Kurban, 2017: 88). Yet, because of this deal, the EU has become dependent on Turkey. So, the agreement has decreased "the Union's leverage to push for positive structural reforms in Turkey in the years to come" (Haferlach & Kurban, 2017: 85). Even now, there remains debate whether the agreement has actually been effective in bringing the number of migrants down (cf. Haferlach & Kurban, 2017: 85, 90).

The agreement is a showcase for how the EU has failed as a normative power, betraying its principles and values and, most importantly, the fundamental rights of refugees, in order to defend the walls of the continent (cf. Haferlach & Kurban, 2017: 85). It leaves those who are persecuted without protection, sending them the message that they belong elsewhere, not in the EU: "Despite affirming that "human dignity is at the core of our common endeavor," Europe continued to give priority to applying problem-oriented policing to an emergency humanitarian situation at its doorstep, treating the forcibly displaced as irregular/illegal migrants that need to be stopped and prevented from entering the fortress" (Afailal & Fernandez, 2018: 220).

When we cast the problem with the EU's agreement with Turkey in Kantian terms, we could say that the EU does not take the right to hospitality of this group of migrants seriously. It denies them the right to present themselves at the EU border and engage in peaceful interaction. This interaction is precluded by their forceful removal to Turkey. From a Kantian point of view, this is all the more problematic for the group that qualifies as refugees.[41] Toward this group, as Ripstein argues, states bear a special responsibility to protect and welcome it.

Ultimately, including the stateless person into the legal order stems from the idea of independence. As Ripstein also stresses, independence is a relational concept that is developed in the three stages of innate right, private right, and public right. Formulated positively, independence consists in the ability to decide on the purposes which are to be pursued by your means (Ripstein, 2009b: 161–166). Formulated negatively, independence stipulates that one is not dependent on the choice of another person. It is this relational conception of independence, a notion that goes beyond modern autonomy understood as the ability of an individual to identify with her choice from a range of meaningful possibilities, which also acts as the yardstick for the legitimacy of institutions. Furthermore, understood in relation to other persons, independence points at the reciprocal relationships between persons as persons.

Now, the question is whether the EU institutions live up to this norm. What we can conclude from the overview of EU migration law is that the EU is doing its utmost to make it difficult for asylum seekers to present themselves at the border and plead their case for protection. One could argue that in this sense their independence is not respected. The innate right of humanity is a right to independence from and equality with others. As such, it establishes the fundamental norm of reciprocity at the heart of interpersonal relationships. EU migration law fails to live up to independence, for individuals seeking protection from persecution are not regarded as equals with legal entitlements but are seen as rightless. Even in those cases where they are welcomed

[41] We acknowledge that one of the challenges for EU migration law is the phenomenon of mixed migration. For more on this and its relationship to the AFSJ's claim to justice, see Herlin-Karnell (2019: 113–119).

("Wir schaffen das"), this is done on political and humanitarian but not on legal grounds. Yet, while our pity may grant them access, it also means that these asylum seekers are, when push comes to shove, not equal to us: they remain dependent on us, so no reciprocity can exist between us and them.

It is exactly this condition of rightlessness that is central to the analysis of Hannah Arendt. From his Kantian perspective, Arthur Ripstein also has a sharp eye for the vulnerable position refugees often find themselves in. Those without citizenship—stateless persons—ought to be taken in, since those without membership in a legal order are "not subject to law, but only to force, occupying a legal black hole" (Ripstein 2021: chap. 9, section X). Recalling a famous example of Hannah Arendt (1968: 296–302), in order to leave such a predicament, it may even be advantageous for a refugee to break the law, since that would make her into criminal with all the legal rights that come with such a position. Indeed, Ripstein stresses, also a refugee possesses an "innate right of humanity," which means that she "must be taken in on a longer term basis, permanently if necessary" (Ripstein 2021: section VIII). In a phrase that resonates with (some interpretations of) Arendt's "right to have rights," Ripstein contends that "Everyone is entitled to be a member of a rightful condition, a being with rights" (ibid.).[42] One could thus read Ripstein as defending an interpretation of Kant that would favor a legal reading of Arendt's "right to have rights": states are under a legal duty to welcome stateless persons to become a member of their legal order. This duty follows from the necessity that every individual leaves the state of nature and enters a state of law.

6. Conclusion

We have adopted in this chapter Kant's theory of law and confronted it with the EU and EU law. The EU has revealed itself in that encounter as a league of states united by the rights these states have reciprocally granted to each other's citizens. That makes this union's law cosmopolitan. EU internal market law establishes a civil condition: it places EU member states and their citizens under cosmopolitan law, administered by the Commission and the Court of Justice, institutions that are independent from the member states who have created them and who have subjected themselves to their rule, for as long as they want. EU internal market law can be consistent with international law and state law; member states do not abandon their task of maintaining a civil condition for their citizens and they remain independent in the sense that they can choose to leave the union. EU member states obviously do more together than maintain an internal market where their citizens can enjoy the four freedoms. But from Kant's point of view the EMU, for example, does not establish a legal condition. EMU is *not* consistent with international law: the states that participate in it come to depend on each other in a way that conflicts with their status as equal and independent states; in its current *Unform* the EMU cannot be part of a system of public law that protects the equality and independence of states.

[42] Ripstein, *Kant and the Law of War* (2021 forthcoming), chap. 9.

The cosmopolitan law of the EU is cosmopolitan in structure but not universal in scope. By itself that does not imply that it cannot be consistent with other, much less extensive but universal systems of positive cosmopolitan law. However, the way in which the EU has constituted its migration and refugee regulations and policies still falls short of establishing a civil condition. The EU refugee crisis shows that certain EU practices come in conflict with such universal systems. In this regard, the refusal of certain member states to let asylum seekers cross their borders, or let people off their ships denies them the opportunity to ask for asylum and is accordingly in conflict with international refugee law. The same can be said about the so-called externalization of EU migration policy of which the Turkey deal is an example.

So we can conclude that with Kant's legal doctrine, fundamental aspects and problems of EU law can come into sharp focus. But we think this chapter has also shown that there is feedback from the EU to Kant scholarship. The EU is an innovation to Kantian doctrine in the sense that its member states have tried to find peace *indirectly*, not in the international way, by creating a continental state or something similar, but in the cosmopolitan way, by granting rights to each other's citizens. Kantian legal thought has to make sense of this innovative step; it has to address this innovation in the best way possible, as potentially a new route toward eternal peace.

References

Afailal, H., & Fernandez, M. (2018). "The Externalization of European Borders: The Other Face of Coloniality Turkey as a Case Study." *Athens Journal of Mediterranean Studies* Vol 4(3): 215–223.

Arendt, H. (1968). *The Origins of Totalitarianism*. San Diego: Harcourt.

Beck, U., & Grande, E. (2007). *Cosmopolitan Europe*. Cambridge: Polity Press.

Boeles, P., den Heijer, M., Lodder, G., & Wouters, C. (2014). *European Migration Law*. Cambridge: Intersentia.

Chalmers, D., Davies, G., & Monti, G. (2014). *European Union Law: Texts and Materials*. Cambridge: Cambridge University Press.

Cherubini, F. (2014). *Asylum Law in the European Union*. Abingdon: Routledge.

Eberl O., & Niesen, P. (2011). "Kommentar." In Kant, I., *Zum ewigen Frieden und Auszüge aus der Rechtslehre: Kommentar von Oliver Eberl und Peter Niesen*. Berlin: Suhrkamp, 2011, pp. 349–350.

Eriksen, E. (2006). "The EU—A Cosmopolitan Polity?" *Journal of European Public Policy* Vol. 13(2): 252–269.

European Court of Justice, Judgment in Joined Cases C-643/15 and C-647/15, Slovakia and Hungary v. Council of September 6, 2017.

Ferry, J.-M. (2009). "European Integration and the Cosmopolitan Way." In Telò, M. (ed.), *The European Union and Global Governance*. London: Routledge, pp. 329–345.

Forst, R. (2017). "Political Liberalism: A Kantian View." *Ethics* Vol. 128: 123–144.

Haferlach, L., & Kurban, D. (2017). "Lessons Learnt from the EU-Turkey Refugee Agreement in Guiding EU Migration Partnerships with Origin and Transit Countries." *Global Policy* Vol. 8(4): 85–93.

Hart, H.L.A. (1997). *The Concept of Law*. Oxford: Oxford University Press.

Heinbach, G. (2013). *Ein Kosmopolitismus-Ideal für die Europäische Union*. Darmstadt: Techn. Univ. Diss.

Herdegen, M. (1998). "Price Stability and Budgetary Restraints in the Economic and Monetary Union: The Law as Guardian of Economic Wisdom." *Common Market Law Review* Vol. 35(1): 9–32.

Herlin-Karnell, E. (2019). *The Constitutional Structure of Europe's Area of "Freedom, Security and Justice" and the Right to Justification*. Oxford: Hart Publishing.

Hobbes, T. (1997). *On the Citizen*. Cambridge: Cambridge University Press.

Höffe, O. (2007). *Democracy in an Age of Globalisation*. Dordrecht: Springer.

Hoogenboom, A. (2016). *Balancing Student Mobility Rights and National Higher Education Autonomy*. Oisterwijk: Wolf Legal Publishers.

Kant, I. (1977). *Die Metaphysik der Sitten*. Frankfurt am Main: Suhrkamp.

Kant, I. (2017). *The Metaphysics of Morals: Revised Edition*. Cambridge: Cambridge University Press.

Kant, I. (2011). *Zum ewigen Frieden und Auszüge aus der Rechtslehre: Kommentar von Oliver Eberl und Peter Niesen*. Berlin: Suhrkamp-.

Kramer, D. (2017). "From Worker to Self-Entrepreneur: The Transformation of the Homo Economicus and the Freedom of Movement in the European Union." *European Law Journal* Vol. 23(1): 172–188.

Locke, J. (1980). *Second Treatise of Government*. Indianapolis: Hackett.

Ludwig, B. (1999). "Kommentar zum Staatsrecht (II)." In Höffe, O. (ed.), *Immanuel Kant: Metaphysische Anfangsgründe der Rechtslehre*. Berlin: Akademie Verlag, pp. 173–194.

Middelaar, L. (2013) *The Passage to Europe*. New Haven, CT: Yale University Press.

Niesen, P. (2017). "What Kant Would Have Said in the Refugee Crisis." *Danish Yearbook of Philosophy* Vol. 50(1): 83–106.

Niesen, P. (2014). "Restorative Justice in International and Cosmopolitan Law," In Flinkschuh, K., & Ypi, L. (eds.), *Kant and Colonialism: Historical and Critical Perspectives*. Oxford: Oxford: University Press, pp. 170–196.

Rawls, J. (1993). *Political Liberalism*. New York: Columbia University Press.

Rijpma, J., & Cremona, M. (1997). "The Extra-Territorialisation of EU Migration Policies and the Rule of Law." EUI Working Papers, Law 2007/01, http://hdl.handle.net/1814/6690. Oxford: Oxford University Press.

Ripstein, A. (2006). "Private Order and Public Justice: Kant and Rawls." *Virginia Law Review* Vol. 92(7): 1391–1438.

Ripstein, A. (2009a). *Force and Freedom: Kant's Legal and Political Philosophy*. Cambridge, MA: Harvard University Press.

Ripstein, A. (2009b). "Kant on Law and Justice." In Hill, T. (ed.), *The Blackwell Guide to Kant's Ethics*. Chichester: Blackwell Publishing, pp. 161–178.

Ripstein, A. (forthcoming). *Kant and the Law of War*. Oxford: Oxford University Press.

Rousseau, J. (1997). *The Social Contract and Other Political Writings*. Cambridge: University Press.

Stone Sweet, A., Ryan, C. (2018). *A Cosmopolitan Legal Order: Kant, Constitutional Justice, and the European Convention on Human Rights*. Oxford: Oxford University Press.

Valentini, L. (2011). *Justice in a Globalized World: A Normative Framework*. Oxford: Oxford University Press.

Wolthuis, B. (2017). "Principles of Economic Union." *European Law Journal* Vol. 23(1): 454–466.

Wolthuis, B. (2021). "The European Union between Solidarity and Justice," *Acta Politica* Vol. 57: 261–275.

Other References

Associated Press, "Migrant ship with 64 people denied safe port by Italy and Malta," *The Guardian*, April 4, 2019, https://www.theguardian.com/world/2019/apr/04/migrant-ship-storm-64-people-denied-safe-port-harbour-italy-malta.

Council Decision (EU) 2015/1601 September 22, 2015, establishing provisional measures in the area of international protection for the benefit of Italy and Greece (OJ 2015 L 248, p. 80).

European Council, EU-Turkey statement, March 18, 2016, available at https://www.consilium.europa.eu/en/press/press-releases/2016/03/18/eu-turkey-statement/.

European Court of Justice, Judgment in Case C-163/17, Jawo v. Germany of March 19, 2019.

European Court of Justice, Judgment in Joined Cases C-411/10 and C-493/10, N.S. and Others of December 21, 2011.

European Court of Justice, Judgment in Case C-578/16 PPU, C. K. and Others of February 16, 2017.

European Court of Justice, Judgement in Case 26/62, Van Gend en Loos v Nederlandse Administratie der Belastingen, 1963.

PART THREE
REPLIES

16

From Constitutionalism to War—and Back Again

A Reply

Arthur Ripstein

I am grateful to each of the contributors for their careful and sustained engagement with my work.[1] I am delighted that so many distinguished scholars found my work, and even more so, its topic, to be of sufficient interest to write about it. I am also grateful to Ester Herlin-Karnell and Enzo Rossi for organizing two workshops, one on the manuscript of *Kant and the Law of War* and the other on the draft papers for this volume. The chapters in the volume take up arguments in my book *Kant and the Law of War*, which Oxford University Press will publish at the same time. As a result, the comments engage with an earlier draft. Since I circulated that draft, I have made minor revisions to the book, and added an additional section, but have not made any significant changes that modify the lines of argument that the contributors to this volume take up, sometimes to criticize, other times to extend or to propose alternatives.

In responding to the contributions, I will work my way through them in the order in which they appear in Part I of this volume. There are some areas of overlap—in particular, although the specifics of their arguments differ, several contributors raised questions about my claim that the possibility of a future peace requires treating the past as settled. In order to fully engage with each chapter but at the same time avoid repetition, I have adopted the practice of multiple cross references; the cost of doing so is that I do not always take up all aspects of each of the contributors' arguments in one place.

The papers in Part I of this volume engage in different ways with my characterization of the ways in which facts matter to right, and the relation between the flawed world in which we find ourselves and the ideal case that Kant contemplates. Yitzhak Benbaji and Rainer Forst both challenge my claim that past wrongdoing is irrelevant to the right against invasion. Benbaji proposes that something more is required in order to justify a right to territorial integrity even to past wrongdoers; Forst worries that the account I provide is insufficiently sensitive to the effects of past wrongdoing. Thomas Mertens comes at the same set of issues from the opposite direction, suggesting that Kant is not offering an account of rights at all, but rather trying to explain the legal practice of his day. On this reading, then, there is no Kantian justification to be found, only a practice in which that idea figures. Anna Stilz raises a more general

[1] I am also grateful to Michael Law-Smith for careful and detailed comments on a draft of this reply.

Arthur Ripstein, *From Constitutionalism to War—and Back Again* In: *The Public Uses of Coercion and Force*.
Edited by: Ester Herlin-Karnell and Enzo Rossi, Oxford University Press. © Oxford University Press 2021.
DOI: 10.1093/oso/9780197519103.003.0016

question of the bounds of legitimacy, and therefore of the occasions in which a state has a right to political independence and territorial integrity. Peter Niesen takes up the relevance of contemporary boundaries in a different way, suggesting that the cosmopolitan right that Kant describes in the *Doctrine of Right* and *Towards Perpetual Peace* is an incomplete and provisional version of it. Alice Pinheiro Walla extends this line of argument, seeking to show that not only cosmopolitan right but the very idea of a territorial right must be understood as a temporary provision awaiting a properly cosmopolitan ruled order. Coming from a very different direction, the papers by Alon Harel and Massimo Renzo question Kant's conclusion that states always have territorial rights. Harel argues that Kantian premises lead to the conclusion that remedial war is sometimes permissible, and Renzo doubts that political independence has any value outside the narrow domain of democratic states.

The papers in Part II deal with more specific issues in the Kantian theory of war and peace. Katrin Flikschuh challenges my claim that anything so specific can even be generated from the materials Kant provides, and indeed whether the ambition of Kantian political philosophy is to generate specificity. Johann Olsthoorn engages with the Kantian opposition to redistribution as a ground of war, asking whether the obstacles to what he describes as "resource wars" are a priori or merely pragmatic. Aravind Ganesh draws out the implications of the obligations of those exercising political power in cases in which that exercise is itself illegitimate. Malcolm Thorburn proposes a way of deepening my account of the prohibition on punishing ordinary combatants who fight in a war of aggression. The papers by Ester Herlin-Karnell and Bertjan Wolthuis and Luigi Corrias draw on my development of the Kantian theory of public right to examine the juridical status of the European Union (EU), from both internal perspectives—is it a local version of cosmopolitan right?—and external ones—is it a state in relation to other states?

I take up these challenges in the order in which they are raised.

1. Part I: Kantian Philosophy and the Actual World

1.A. The Right to Territorial Integrity (Reply to Yitzhak Benbaji)

Yitzhak Benbaji's chapter approaches the question of territorial right against the background of the idea that the moral nature of relations between states is changed by the existence of international practices and agreements. He develops a more systematic and detailed account of this approach in his recent coauthored book *War by Agreement*.[2] His chapter brings that approach to bear on the question of how wrongful acquisition (for example, through conquest) can give rise to rights that bind others. In place of the Kantian account I defend, he offers what he calls a "quasi-Kantian" account. Where Renzo, Stilz, and Pinheiro Walla seek to provide an account of legitimacy in terms of the internal structure of the state, Benbaji proposes to specify the content of legitimacy in terms of recognition by other states; he contends that

[2] Yitzhak Benbaji and Daniel Statman, *War by Agreement* (Oxford: Oxford University Press, 2019).

only such recognition can make a territorial claim binding, regardless of its historical origin.

The Kantian view sets a much lower standard for legitimacy as between nations. As I will explain in more detail in my response to Pinheiro Walla, Kant's critique of colonialism rests on the wrongfulness of the colonial power imposing its will on the people—the collective—of the colonized nation; even if they lack developed legal institutions and do not participate in international legal institutions, they have a right against invasion. Perhaps Benbaji would want to say that there is more to legitimacy than a right against invasion; in understanding the wrong of wars of aggression, I am interested only in the right against invasion. Kant's view, which has much to recommend it, is that invasion is wrongful even against gravely defective social orders. Recent historical and anthropological evidence has made it clear that the Indigenous peoples of the Americas and Africa had much more sophisticated legal and political institutions than were acknowledged in the secondhand reports from travelers on which Kant relied (and who Kant himself conceded were unreliable). Kant opposes the colonial conquest of what he considered pastoral peoples for the same reason that he characterizes colonialism in the case of the British conquest of Ireland or Athenian rule over Greek islands as wrongful. In each case, the wrong lies in the subordination of a people to another nation's will.

Like Renzo and Stilz, Benbaji seeks to restrict the right against invasion to what he characterizes as legitimate states. Pointing to the striking disanalogies between a "state's (disputable) territorial right and the (indisputable) bodily rights of a natural person," he suggests that "a state is entitled to rule over a territory if and only if (first) it secures the private freedom and political participation of its inhabitants and (second) a treaty to which other states voluntarily entered obligates all actors not to use force in resolving future disputes over this territory" (Benbaji p __). All three attempts to narrow the range of states that are entitled to be free of aggression from others share a common problem. I will identify it before going into the particulars of the alternative sketched by Benbaji, turning to those of Renzo, and Stilz in my discussion of their chapters.

The fundamental difficulty with all of these proposals is that they run together two separate issues: how things stand between a state (considered as a set of institutions) and those over whom it exercises power, and how things stand between such a state and other states. That difference maps onto the Kantian distinction between the idea of the original contract and the postulate of public right. The idea of the original contract is the pure case of a system of republican government, one that properly secures participation, so that those subject to its laws rule themselves through those laws and institutions. Kant characterizes the idea of the original contract as an "idea of reason" because no actual society is, or indeed could be, fully adequate to it.[3] Choose whatever legitimate state you like—many American academics wistfully imagine my own nation, Canada, to be something like the ideal case, while others think about some hybrid of Scandinavian countries. None of these will be fully adequate, because the political institutions will still treat some, such as children, as merely passive citizens; the

[3] 6:372.

protection of private rights will be defective because of the cumbersome machinery of dispute resolution; and, as it turns out, the distributions of both private rights and political power, as well as the state's boundaries, will reflect a history of past injustice and war. Those are the general considerations; if you think about Canada, the treatment and condition of its Indigenous peoples is enough to show that it is not the ideal case. I mention these familiar points of history and political life as a reminder that a higher bar could never be satisfied.

The second general concern is that even if the bar of legitimacy is set lower, so as to accommodate contemporary Canada and other well-functioning democracies, almost no other legal orders in human history will turn out to have satisfied it. The problem of war to which Kant points was a problem in his own day even though states then behaved patrimonially; what they did was wrong in the highest degree, even when they invaded other patrimonial states. Something is always wrong with war, and it is always the same thing: it is the condition in which force decides. Territorial wars against societies of which that attackers had only recently become aware, and who they refused to recognize—the colonial conquests of much of the non-European world by European powers—was wrong in the absence of any shared recognitional framework. An uneasy peace—the situation that satisfies the preliminary but not the definitive articles of *Towards Perpetual Peace*—is unstable, but at least potentially allows a range of human affairs to be governed by law, or something close to law, rather than force. War is wrong because it replaces law (however defective) with force alone, which is worse.

Unlike the approaches of many of the other contributors, Benbaji's strategy is not to propose a higher bar, but rather to offer an alternative account of why a low bar—as low as Kant sets—is sufficient for territorial rights. On Benbaji's view, recognition by other states is sufficient to underwrite a right against invasion, even for states that have no genuine pre-recognitional right to their territory. I see two general difficulties with this proposed solution; I will explore these in relation to each of his three hypotheticals.

The first and most important point is that although a treaty outlawing war is morally and legally required—Kant writes of Practical Reason's "irresistible *veto*: there shall be no war"[4]—such a requirement cannot be coherently represented as the product of a voluntary arrangement as between the parties to it. To represent it in this way is morally and conceptually incoherent. Parties can arrange relations between them, varying the rights that they have as against each other, only against the background of a system of rights that they have antecedently. You and I can make an arrangement that permits you to do to me things that would otherwise be wrongful, or that obligates me to do things for you that I was under no antecedent duty to do. We can only do so because we are already bearers of rights, and so prohibited from doing certain things to each other without proper authorization. The moral possibility of varying relations between us, however, is not so broad as to make our background rights against interference contingent on the agreements into which we enter. Instead, the starting point must be our claims against interference, however precisely those are specified.

[4] 6:354.

In the individual case, I do not need to get your agreement in order to be entitled that you not attack me; the idea that everyone has, in Hobbes's vocabulary, a "right to all things," is a view that can be coherently stated, but from which there is simply no way out. Hobbes seeks to show that it is to everyone's advantage to exit the state of nature, and further, to put themselves under an absolute sovereign. But even if that argument is successful, it shows only the prudence of the exit, not that it is morally mandatory.

If the agreement that Benbaji contemplates is not to be redundant, it would have to only be binding once it has been entered into. Moreover, for the fact of agreement to carry the justificatory weight, it would need to be something that could have been binding even if it had had different terms. It is difficult, however, to see what those different but still acceptable terms might have been. Rather than a contractual right (*pacto*), the right against aggression is required by law (*lege*). That does not make the agreement irrelevant, but it makes it more like an oath of office by a public official (or an express undertaking of loyalty by a fiduciary) than a contract. In both of those examples, the duty in question is binding apart from contract because of the relation between the parties; the express agreement is expressive and cautionary, but its terms cannot be modified by the parties. The agreement isn't entirely redundant (in the way in which, for example, a gangster's promise to stop killing people is) but its terms are not binding just because they are the ones that have been accepted by the parties. I want to make a parallel claim about the right against aggression.

Benbaji's remarks about the differences between territory and human bodies indicate that he would probably accept something like this structure of mandatory precontractual rights in the case of individual human beings, but in the case of nations he thinks that the analogous point does not hold. As he puts it, a state's claim to its territory is disputable, while each person's claim to their own body is not. But even this idea is ambiguous, and once the ambiguity is uncovered, it undermines the disanalogies that Benbaji seeks to establish. Your right to your own body does not require an affirmative act to establish it, only the fact of you being identifiable with it. But the claim of a legal order as against other nations does not require a particular affirmative act to establish it either. As I explain in my responses to Flikschuh and Harel, the creation of a legal order, and so its standing against other nations, is by its nature an extralegal act. Once global institutions are in place, there can be orderly procedures for secession and perhaps the amalgamation of legal orders, but either of those presuppose already individuated legal orders. Indeed, when Benbaji points to illegitimate states, he presupposes this kind of apparatus of individuation; to call a state illegitimate requires identifying it as a state. Boundaries may well be disputed, as may histories, but the proposed solution of a shared framework of recognition could only solve the problem with respect to something that is already in place to recognize. Indeed, the puzzle with which Benbaji introduces his account—a state's claim to territorial integrity with respect to territory that it acquired through wrongful conquest in the past—already supposes that the factual question—who governs this territory—can be answered without answering the question of whether it came to govern in the right way. The idea that boundaries are disputable then must be about whether others are entitled to dispute those boundaries, the very question at issue.

Perhaps Benbaji means to point to a different, epistemic sense in which territory is disputable. On this version of the contrast, the contours of a person's body are clear, and what counts as part of a body is a matter of growth or perhaps in some cases grafting or transplanting. Once the body part grows (or the graft takes) it is part of the body. Other things, such as prostheses and clothes accede to a person's body in a similarly clear and orderly way. Territory, by contrast, is acquired through exercising jurisdiction over it—making "an actual deed (taking control) the condition and the basis for a right."[5] As such, there could be unclarity about who is exercising jurisdiction, and, Benbaji might suggest, about whether the acquisition was legitimate.

I do not mean to deny that territory subject to dispute in the sense that one state might refuse to accept another state's claim to territory, or even its legitimacy. Historically, such claims were often stated in the language of corrective justice; in chapter 3 of *Kant and the Law of War*, I explain why states cannot be entitled to enforce claims of corrective justice. That they assert such claims does not show they are valid. Nor is the lack of clarity about borders sufficient to entitle states to resolve such unclarity to their advantage by attacking others.

I am not sure why Benbaji supposes that the Kantian view is committed to the claim that even what he describes as "failed states," that is, those that fail to provide a legal order and leave their inhabitants in a condition of barbarism, possess a right against aggression. Part of the difficulty with the vocabulary of failed states is that it is not clear whether a failed state is a type of state, or rather a characterization of why there is no state where there had once been one. I take Benbaji to mean the latter, given his reference to leaving inhabitants in a condition of barbarism. In such a situation, outsiders can intervene, though it is not clear that doing so would count as aggression; there is nothing against which to aggress in the relevant sense. At the same time, any license to intervene is extremely limited. If they have no defensive ground for entering the region of the earth's surface ("territory" is probably the wrong word) that no longer has any legal order in place, they can assist in institution building, and provide infrastructure for peaceful interaction, enabling the inhabitants to regain the rightful condition that they lost through the failure of what had been their state. The rationale for belligerent occupation in the course, or even in the aftermath, of the war—the Allied occupation of the former Nazi Germany, for example—is precisely that the end of the war has led a state to fail. Prior to its military defeat, Nazi Germany was already a failure along many other dimensions already, and arguably a condition of barbarism.[6] But at the end of the war, the rationale for occupation and institution building was to see to it that its inhabitants were in a rightful condition. Given the breakdown of the existing institutions, new ones could be put in its place, including new, less warlike ones.

In the version of the manuscript Benbaji read, I only talked in passing about the possibility of humanitarian intervention; here I can say a little bit more. To prevent barbarism, or to enable the return from a complete breakdown, a state may be entitled to intervene on territory not its own. But it cannot make a unilateral judgment about its entitlement to do so; that is a question for international institutions. More than that, such an authorization rules out annexing the broken-down state or the territory on which it had operated.

[5] 6:371.

[6] I take this issue up in the final chapter of *Force and Freedom* (Cambridge MA: Harvard University Press, 2009).

This narrow scope of dealing with failed states is in part the Kantian response to the ambitious claims of European colonial powers with respect to what they claimed was their "civilizing mission." Nations are not entitled to make this their mission, let alone in a way that would entitle them to make unilateral judgments about it.

Let me make this point in terms of each of Benbaji's three examples, which he structures through a contrast and tension between what he calls "force" and "legitimacy," both of which he takes the Kantian account to be committed to. As Benbaji describes the Kantian position, "States are legitimate in virtue of the legal order that they provide," and "one state uses force against another, if and only if it violates its territorial integrity." The tension is said to arise because legitimacy is an inward-looking and functional matter: whether a state provides the requisite service to its inhabitants is a question about the relation between it and its inhabitants, while force looks to how things stand between states and is indifferent to the quality or, on Benbaji's reading, even the provision of legal order.

Before turning to the examples in detail, I should be explicit that the reason I see no such tension is that one state is not permitted to take it upon itself to determine whether another is doing a good enough job of providing a legal order. Kantian political philosophy operates with relational concepts, asking how things stand between one legal person and another; how things stand between the state and its inhabitants is the first question of public right, but how things stand between two states is not fixed by the estimation of one of how things stand between either state and its inhabitants.

The example of Well-Ordered's annexation of Warlord's territory from Failed is presented as a case in which there is some doubt about Failed's claim against Well-Ordered, on the grounds that Failed is not providing a rightful condition on the territory controlled by Warlord. Meanwhile, Warlord is permitting incursions into the territory of Well-Ordered. Failed is unable to stop those incursions; Warlord is unwilling to stop them. Benbaji concludes that the only way to stop them is through annexation of the territory controlled by Warlord.

In that circumstance, it may well be that the only thing to make of the horrible situation is that Well-Ordered may, with appropriate international authorization perhaps, intervene to protect both itself and the inhabitants of the Warlord-dominated territory. The example is not described with enough specificity for this to be obviously correct, but I will suppose that it could be filled out in a way that it was. In such a circumstance, it does not follow that Well-Ordered is entitled to annex the territory. Benbaji adds one detail to reach this conclusion, specifying that the annexation takes place only because the Warlord's gang will otherwise reoccupy the territory. Perhaps in such a situation, the only thing to do is to decisively defeat Warlord; looking at recent interventions of this kind, that may not be possible, and indeed, it may be that an invasion to stamp out warlordism has no real prospect of success. But I will follow Benbaji in supposing that, at least in the example as described, such success is possible, but can only be achieved through annexation, rather than shorter-term military action and temporary occupation. The Kantian idea, however, is that annexation (as opposed to, for example, extended occupation by a multilateral peacekeeping force) could still never be authorized. The difference rests on the fundamentally different normative structure of annexation and belligerent occupation. An occupying power must exercise all of its power on behalf of the occupants of the territory in question.

It must not change the demographic character of the territory, either by moving its civilian population in or removing the prior occupants. If the disputed area becomes part of Well-Ordered's territory, it becomes part of Well-Ordered's legal order, to be governed on behalf of its entire population, rather than just that of the disputed region. It can, among other things, tax it for national purposes, move its civilian population in, and so on. By contrast, it can do no such thing as a belligerent occupier. If it is occupying the territory as a result of a defensive war, it must administer it on behalf of its inhabitants, and, further, standing in place of, and in preparation for the future handover to, the government of Failed. These contrasts explain why Benbaji might suppose that annexation is the only thing that could remove Warlord by changing the demographics in which we are supposed to assume warlordism is able to survive. That nonetheless does not make doing so rightful.

In his discussion of the conflict between Failed and Well-Ordered, Benbaji puts this in terms of the distinction between force and legitimacy, suggesting that Well-Ordered's actions are justified by the idea of legitimacy but prohibited by the idea of force. But in order to make the distinction apply in this case, he needs to suppose that Failed's officials have a continuing right to rule in the absence of any effective institutions or even access to the territory. That is, it appears to rest on an ambiguity between a defective state and no state at all. Now there is one sense in which the government continues to have a right to rule even when ousted from its territory: a government in exile after an enemy invasion remains entitled to rule, but its rule is not effective. So, too, if there really are favelas in São Paulo that the police and other government officials are afraid to enter, that are run by organized criminal gangs, the Brazilian government is still entitled to rule over those gangs. That means that, if Well-Ordered were to invade those neighborhoods in São Paulo, it could only do so in order to restore order on behalf of the Brazilian government. Well-Ordered would not be entitled to invade and take the disputed territory for itself, even with the consent of its inhabitants. There is thus no tension in the Kantian account. Benbaji sets up a conflict where there is none by assuming that Failed is a permanent and irreparable failure—that it could not have power returned to it because it is incapable of ruling—while at the same time assuming that it nonetheless retains its right to rule. If it is able to control other territory, just not where Warlord is, then it is not a complete and permanent failure, and Well-Ordered can assist it in restoring order, but not oust it from the territory. If it really is a permanent and total failure, then Well-Ordered can only protect itself in ways consistent with restoring what had once been its territory to political independence and territorial integrity. Benbaji moves between talking about the provision of legal order and annexation, but these are fundamentally different.

Benbaji's further example of Hong Kong is complicated by the fact that the treaty that ended in 1997 was patrimonial: political authority followed territorial control which was ceded by a (coerced) lease, as though the land and its inhabitants were property that could be passed from one state to another. As Benbaji notes, this was the last major colonial possession of the British Empire, and, for reasons that I detail in chapter 8 of *Kant and the Law of War*, colonialism is an inherently defective form. Thus the inhabitants of the territory retain their right to govern themselves. Although, as Kant puts it, the undoing of the residual effects of acquisition by conquest, gift, or purchase (but also lease) can be postponed, it cannot be postponed indefinitely. Geopolitical realities led the United Kingdom to make an arrangement with China rather than with the inhabitants of Hong Kong, as it had done with many of its other colonies.

Legitimacy, as Benbaji conceptualizes it, leaves the question open of whether, on the alternative history in which Britain granted sovereignty to Hong Kong as an independent nation, China would have done wrong in immediately annexing it. But it seems to me that the answer that he says is given by Force—that Hong Kong is entitled to sovereignty—in this case is superior. It differs from the Warlord example because granting Hong Kong independence would have been a transition from one lawful condition to another, less defective one, not an instance of a lawless condition. Had China invaded and annexed Hong Kong, it would not have been a case of filling a legal void. A more powerful nation is not entitled to annex a weaker one; it has neither a juridical nor a historical claim to do so.

In his discussion of the example of Holy, Benbaji mischaracterizes the commitments of force, as he had described it. Force rests on the idea that the creation of the legal order is extralegal, but that it is still wrong for another legal order to invade it; by contrast, legitimacy, as he describes it, sees the creation of a legal order as conditional on further criteria.

With this in mind, the idea that "any action that substitutes right with force is impermissible" is indeed what force says. Applied to particulars, however, as in his example of Holy, in which the departing colonial power decides to cut and run without leaving functioning institutions in its place, the problem is substituting right for force in such a situation, and in their attempts to do so, given the absence of functioning institutions, each side can do no better than what seems good and right to it. As Benbaji describes the dynamics, each side has minimalists who are willing to make territorial compromises for the sake of peace (which for the Kantian must also be understood as doing so for the sake of right, that is bringing force under law) and maximalists who want entire territory of Holy for themselves and will accept no compromises. In such a situation, the best that either side can do is to defend its claims against the maximalist on the other side, and trying to reach a compromise with minimalists. But in the case "in which implementing the best arrangement" involves extralegal force, unilateral action is not ruled out if, as in Benbaji's description of the case, there is no way of having legal officials resolve it. So if the colonial power's departure begins a civil war among the subject peoples, all force will be extralegal until some kind of order arises, or international institutions impose one. This concession about the unilateral and extralegal nature of all force in the situation does not license the use of force by maximalists on either side of the conflict in Holy or suggest that it is in any way morally unproblematic. It means instead that in such a situation, the absence of institutions entails there can only be force, and those in it must exit the state of nature, and so must exit using force. If confronted with maximalists who will kill them or expel them into a world where no one will take them in (which seems to be the situation with one side but not the other in the hypothetical) they can rightfully resist that force. Force, as Benbaji formulates it, prohibits substituting force for law, but it does not, and could not, prohibit responding defensively to aggression.

Further, those who find themselves setting up a rightful condition, by means of force where necessary, are charged with setting one up, not with setting up a situation that is by its nature unstable, plagued with violence, and certain to fail.

Having addressed the three hypotheticals, I now want to come back to Benbaji's positive account of legitimacy.

There was no treaty outlawing war before 1928, but that is only to say, as Lauterpacht puts it, that there was an incoherence running through all of international law until 1928.[7] To put it in terms of an oxymoron, a legal system in which one legal person can unilaterally deprive another of its legal personality is not a legal order at all. It is a condition of barbarism. A condition in which war is outlawed may still be a condition of anarchy—an ideal case of a state of nature, in which there is in fact no war, but also no legal institutions to guarantee peace. Such a situation remains unstable because each can do no better than what seems good and right to it, and in such a condition some states may continue choose to go to war when they think they can get away with it. But for all that, it has some element of legality in it. The situation in which states are entitled to use force whenever they see fit, by contrast, let alone one that, like the pre-Kellogg-Briand situation, permitted one nation to forcibly compel another to resolve a dispute through force, leaves everything always potentially subject to force because any claim can be asserted through the use of force. Further, such nations, imagined as independent but entitled to use force at their own initiative, are not independent at all; their independence is entirely conditional on the goodwill of others.

The entitlement to be free of aggression, like the capacity to enter into binding agreements, is not itself the product of a binding agreement. It is not that the people or states have agreed to be bound by their agreements, and therefore they are binding; the binding nature of agreements is presupposed by any such agreement. The fact that the bindingness of agreements is not itself the product of an agreement is conceptual and entirely familiar. The parallel normative point also applies: the right against aggression is presupposed by any agreement. Your agreement is only needed for others to do things to you that it is not already permissible for them to do to you. You and others do not need to enter into an arrangement to refrain from attacking each other; an agreement to comply with the prior obligation is no agreement at all. That is why a peace treaty is not like any other treaty or contract. That is also why, conversely, a peace treaty is binding even though its specific terms were the result of duress, and why it cannot include terms in violation of *ius cogens* norms.

More generally, the suggestion that the moral right to territorial integrity is a result of contract runs together two importantly different claims. The first is the claim that states are under an affirmative obligation to enter into an arrangement whereby institutions are set up for the peaceful resolution of disputes, that is, to exit a condition in which each can do what seems right and good to it. They can only exit this condition by making some such arrangement. The claim that the wrongfulness of aggressive

[7] In *War by Agreement* Benbaji and Statman concede that on their view an agreement to permit interpersonal disputes to be resolved by dueling could be binding. Such an account would seem to allow any party to it to invoke the "procedure" entirely at its own initiative, and so to make the use of force always available. This is just the difficulty that Lauterpacht identified with the pre-Kellogg-Briand international situation: a "system" in which one party can unilaterally assert a claim and enforce it by right is the absence of law.

war depends on an agreement turns on the very different claim that absent such an agreement, states do no wrong when they resort to war for any reason whatsoever. As Benbaji notes, every rightful condition is internally insecure if it is subject to the aggression of others; in order to bring its own laws fully under law, it needs to set up institutions that will do so. But that is just to say that it can only enjoy its rights in a legal order. It is not to say that how things stand between states is somehow changed by entering such a legal order. Kant's characterization is more plausible: entry into a legal order generates "no further or other duties among" nations than can be thought in their prior state.[8]

The claim that the right to territorial integrity and political independence is established *pacto* (by agreement) not only supposes, as noted before, that contractual power is virtually unlimited; more significantly, if the idea is that of a voluntary arrangement between states—rather than an arrangement *lege*, that is, one with mandatory terms required by the nature of the situation—then it must be that a different set of terms is possible. This point is already apparent in the difficulties with the suggestion that contractual obligation is itself the product of a contract. It is not as though the most basic principles—*pacta sunt servunda* (agreements must be kept)—could be any different. Of course, many particulars can be. Perhaps Benbaji's concern is really with the particulars of a possible prohibition of aggression; he describes the Kellogg-Briand pact as "incomplete and imperfect." But its incompleteness and flaws can only be judged in terms of its point; it fails to completely and flawlessly protect the political independence and territorial integrity of states conceived as independent of each other. The idea that such a pact should be maximally inclusive goes not just to its scope, but in some sense carries with it the central idea of a plurality of independent states.

The idea that rights are only conclusive if guaranteed through institutions is not the same as the claim that those rights are constituted by those institutions. Indeed, when Benbaji says "The ideal treaty dictates a further instruction," I wonder whether the actuality of the treaty is doing any normative work in his account. I do not deny that the 1928 treaty marks "the beginning of a process in which a society of states" was formed. I just deny that states were morally entitled to attack each other at will before this process began. The duty to provide a self-certifying legal order not only "entails a duty to enter a maximally inclusive peace treaty," it entails the duty to refrain from interfering in other states that are at peace.

Benbaji points to particular treaties, noting that they do not have mandatory terms, but have terms that must rather be worked out by the parties. While I agree that this is true of many treaties, I deny that it is true in general—*ius cogens* norms prohibiting slavery, genocide, and the refusal of consular services[9] restrict the range of permissible terms, because a state cannot contract out of providing a rightful condition to its inhabitants. Peace treaties have mandatory terms, formal requirements of being a peace treaty at all. To make this point, I want to draw attention to a mandatory term that any treaty to resolve any of his three examples would need to have, and then argue that it is the mandatory term of any kind of actual, positive treaty as between nations. The basic term is the one that Kant identifies as the first preliminary article

[8] 6:307.

[9] That is, of a nation's protection of innate and acquired private rights of its citizens.

of perpetual peace: the treaty must provide closure, that is, it must leave no residual grounds of war as between the disputants. The prohibition of aggressive war marked by the Kellogg-Briand pact of 1928 had a mandatory term also. But, as I suggested, following Lauterpacht, the mandatory term was in some sense already implicit in the possibility of international interactions being brought under law. Nations had duties not to attack each other. In the absence of a treaty, (and in the decades immediately following 1928, even in its presence) they disregarded those duties. Further, in the absence of competent legal institutions, no other mode of dispute resolution was available in a systematic or dependable way. But the fact that war was difficult to avoid does not show that aggressive war was ever acceptable. The basic term of international interaction must be that interactions are to be governed, and disputes are to be resolved, by words rather than force. As such, no nation can take it upon itself to attack another. This structure is the precondition of the possibility of entering into agreements. Just as no agreement is required in order to prohibit perfidy—unless perfidy is prohibited, agreement is impossible—so, too, no agreement is required in order for each state to have a right, as against each other nation, to its own political independence and territorial integrity.

1.B. Dissolving the Paradox of Peace (Reply to Rainer Forst)

Rainer Forst's paper characteristically goes to the heart of the matter, raising a challenge to what he calls the paradox of peace,

> peace is supposed to avoid, end or overcome war as the practice of might making right; but the principle of peace itself at crucial points seems to require us to accept that might makes right, at least as far as the past is concerned. But can this be reconciled with the principle of right that ought to guide our normative thinking about these matters? If we follow Kant, do we have to accept that the paradox of peace is unavoidable, as Ripstein implies?

Forst proposes to resolve this paradox in terms of the relevance of, and a different interpretation of, Kant's distinction between constitutive and regulative principles:

> This is an intriguing argument, yet it might puzzle some readers of Kant, since he does not—at least not explicitly—make the distinction in the exact way argued for by Ripstein. In fact, one might wonder what use the distinction between constitutive and regulative principles has at all in the practical realm, since Kant develops it in his theoretical philosophy and makes rather sparse use of it in his practical philosophy. And where he does use it, the regulative character of transcendental ideas of reason is most prominent, including his view of perpetual peace as a moral duty and an idea that guides political action, even though such a condition of peace may never be fully realized (MM VI 350; PP VIII 386). For many, this makes it into an idea that reasonably regulates our action in a similar way to others, such as

historical progress toward a republican state[10]—that is, a regulative idea. So what is peace: a constitutive principle, as Ripstein argues, or a regulative idea or both? (Forst p. 34).

Forst goes on to point out that in the *Critique of Pure Reason* Kant distinguishes the constitutive from the regulative *use* of ideas of reason. In the Transcendental Dialectic, Kant argues that the traditional arguments of metaphysics concerning the world (antinomy), the soul (paralogism), and God (idea) are fallacious, because all of them inappropriately extend rational inquiries that are acceptable within the bounds of experience beyond their proper use. Nonetheless, Kant grants these same ideas a regulative use: although they do not refer to objects, they guide our inquiries. Most strikingly, having criticized Leibnizian arguments, Kant goes on to insist that the progress of inquiry depends on the regulative supposition that the world as a whole (which itself must not be taken to be an object) is as it would be created by the Leibnizian God, with maximum variety and maximum unity. In its regulative use, reason figures in cognition without providing any content to it.

As Forst points out, Kant uses the idea of a constitutive principle differently in his practical philosophy. In the *Critique of Practical Reason* Kant writes that the ideas of reason have a constitutive use "inasmuch as they are grounds of the possibility of making real the necessary object of pure practical reason (the highest good)."[11] And Forst points out that Kant, in the *Critique of the Power of Judgement*, writes that the constitutive is also regulative.

The challenge, then, at a textual level, concerns the way in which peace can be the constitutive principle of war. But Forst is also concerned about the practical implications of this idea, in particular, whether the priority of peace, whether constitutive or regulative, really means that the past cannot be undone. As I will explain, the priority of peace entails that force cannot be used to undo past injustices; it does not entail that they are not to be undone, or even that there is no duty to undo them. In order to explain the practical point, however, I must first say more about the relation between constitutive and regulative uses of the concept of peace.

I now see that my choice of vocabulary was potentially misleading; I should have said that the concept of peace has both a constitutive and a regulative use. The issue is not just about wording, however. There is also a genuine disagreement between me and Forst about how to understand the Kantian conception of constitutive and regulative uses of rational principles, and its application to the specific case of the *Doctrine of Right*. On my reading, it differs in important ways from both its application to the case of theoretical reason discussed in the *Critique of Pure Reason* and to that of pure practical reason as discussed in the *Critique of Practical Reason*. But I fear my choice of words was misleading and gets in the way of getting to that point. As Forst points out, the distinction between constitutive and regulative principles is not Kant's preferred vocabulary; in each of the three *Critiques*, he distinguishes between constitutive as opposed to regulative *uses* of rational principles.

10 Pauline Kleingeld (2008), "Kant on historiography and the use of regulative ideas," in *Studies in History and Philosophy of Science* 39/4, pp. 523–528.

11 CPrR V 135.

Let me begin with my understanding of Kant's distinction between constitutive and regulative uses of reason. As Forst points out, in the *Critique of Pure Reason*, "Constitutive principles are used to construct objects of appearances of which we can have secure knowledge, such as in mathematical reasoning, whereas regulative principles are heuristic rules for seeking a well grounded unity of experience" (Forst p. 34). In theoretical reason, regulative principles are used only hypothetically, in order to assist reason in its task of unification. For example, we have, and indeed could have, no grounds to believe the Leibnizian idea that God would create a world with both maximum unity and maximum variety, and a finite intellect could never encounter that idea in experience, yet it can serve as a regulative principle to guide inquiry. Even the successful use of such principles does not show that they constitute their objects.

In the case of practical reason, by contrast, the same practical principle has a constitutive as well as a regulative use. In the *Critique of Practical Reason*, reason constitutes the ideas of God, freedom, and immortality, not as objects of theoretical knowledge but as "making real the necessary object of pure practical reason (the highest good)."[12] As Forst points out, this line of thought continues in the *Critique of the Power of Judgement*, with Kant's characterization of practical principles as both regulative and constitutive.

There is little point to defending my misleading vocabulary, so let me suggest this instead: in the case of practical principles, their application to particulars presupposes a separate way of identifying the objects of their application. This is not an unfamiliar thought in Kantian philosophy: you have no empirical basis for applying ideas of freedom to yourself, only a moral basis. Once we move, however, to relations between persons—from the categorical imperative as described in the *Groundwork* and *Critique of Practical Reason* to the universal principle of right—some principle of individuation of other rational beings is required. Kant says little about this, because he accepts the naïve but obviously correct principle of individuation, understanding the moral category of persons—beings to which actions can be imputed[13]—to be the category of human beings, which are in turn identified through nothing more than human form. I fear I caused unnecessary confusion, by seeming to claim some analog of the idea that human form is the constitutive principle for the which universal principle of right is the regulative principle. The point is rather that concepts of right apply to beings that imperfectly realize them. I take this to be Kant's point in introducing the universal principle of right as a "postulate incapable of further proof" and again in introducing the postulate of practical reason with regard to rights and the postulate of public right. In each case, the extension of moral concepts to a domain of objects is morally required but not provable through theoretical reason. The Universal Principle of Right extends the application of the concept of freedom, which is itself a rational and so non-empirical concept, to embodied rational beings, individuated by their occupation of space. Nothing could count as evidence in support of this extension. Nonetheless, it must have some way of identifying the range of space-occupying things as embodied rational beings.

[12] *Critique of Practical Reason* 5: 135.
[13] 6:223.

The same structure manifests itself in Kant's treatment of the idea of a legal order. Kant writes that

> "the original contract is only the idea of this act, in terms of which alone we can think of the legitimacy of a state. In accordance with the original contract, everyone (*omnes et singuli*) within a people gives up his external freedom in order to take it up again immediately as a member of a commonwealth, that is, of a people considered as a state (*universi*)."[14]
>
> Yet he also writes that the idea of a perfect constitution is something "to which no object given in experience can be adequate," and that
>
> "If then a people united by laws under an authority exists, it is given as an object of experience in conformity with the idea of the unity of a people as such under a powerful supreme will, though it is indeed given only in appearance, that is, a rightful constitution in the general sense of the term exists."[15]

Kant's solution is to understand every existing legal order as the defective version of a rightful condition, a condition in which "the principle of freedom under law" is satisfied, even if only to a small degree. That is, if something partially satisfies the principle, it must be understood as a genuine, albeit defective version of it. At the same time, the pure case provides the regulative standard against which every legal order is to be assessed, through which its defects are identified, and on the basis of which it is incumbent on the legal order to bring itself more fully into conformity with its own regulative principle. The Postulate of Public Right specifies the existence of a rightful condition, which must not be overthrown or subjected to force, and the Idea of the Original Contract provides the norm for its assessment.

My wording was insufficiently clear when I extended this form of reasoning to the case of war, and described peace as a constitutive principle in a way that suggested that there was, as a result, a different, regulative principle. Instead, a better way of making the point would have been to say that the concept of peace has both a constitutive and a regulative use. In its constitutive use, the concept of peace is the concept of fully realized rightful condition, the one that Kant goes on to call "the final end of the doctrine of right." It constitutes the object, of which actual peace treaties are defective versions. Just as "the state as idea" provides the standard for identifying something as a legal order and the norm for assessing the performance of things so identified, the idea of peace provides the basis for identifying something as a peaceful interaction. Peace thus constitutes the possibility of rightful relations between nations, and is regulative for all (defective) interactions between them.

I hope that this textual clarification goes some way to explaining why I read Kant in this way. At the same time, rather than defusing what Forst evocatively describes as the "paradox of peace" it seems instead to cast it into sharper relief. The paradox is that we must accept the results of war if we are to have peace between nations, just as, in the domestic case, we must refrain from asking "with practical intent" about the origins of a particular legal order, that is, whether it came into being through

[14] 6:315.
[15] 6:371.

law or whether instead force arrived first in law only later or, as Kant tellingly puts it, "whether they should have followed in this order." That is, there is a double structure here, in which the pure case of a rational concept (to which nothing encountered in experience can be adequate) is required to apply it to it objects encountered in experience, all of which will be defective in relation to the idea through which they are identified. Nonetheless, the fact of defective application is sufficient to preclude treating anything to which it defectively applies in certain ways. In the case of the right of the state this double structure leads to a prohibition on revolution, because even a defective legal order qualifies as a legal order. In the international case it underwrites Forst's paradox of peace.

The first thing to say about this double structure is that it is a familiar structuring feature of Kant's use of ideas in practical contexts. In the introduction to the Dialectic in the *Critique of Pure Reason*, Kant cautions against what he calls the "transcendental ideas," which involve an inappropriate application of a regulative principle for theoretical reason into a constitutive principle of the objects to which it seems to apply—the world as a whole, the soul, and God.

Before introducing the transcendental ideas, however, Kant discusses what he calls "the ideas in general," the first example of which is the idea of a sage of perfect virtue, and the second of which is the idea of republican government. Kant goes on to explain that the same structure applies in the way in which we think about living things, such as plants and animals. Here, ideas function differently: the possibility of identifying a particular as an instance of the kind horse, or leaf, requires that we consider it in relation to what is "most perfect in its kind." Nothing encountered in experience will be congruent to the idea. The only things we will ever encounter in experience are defective horses and defective leaves. So, too, with virtue and republican government. This seems to open up the question of how these ideas can be applied to objects of experience: to think of it as a whole, the parts of which are subordinated to their place in the whole, requires both an exercise of reason and, at the same time, the application of the appropriate rational concept to some objects but not others.

The same structure applies to the idea of republican government: only a certain class of things qualify as legal orders and so as legitimate. In describing the transition between a state of nature and a public legal order, Kant notes that there are forms of sociability "compatible with rights (e.g., conjugal, paternal, domestic societies in general, as well as many others)."[16] These forms cannot be understood through the idea of the original contract. These contrasts show that the application of the idea of republican government is limited in scope to the application of the idea of government. Conjugal, paternal, or domestic society, as well as a condition of anarchy, could not be understood as defective forms of a rightful condition. Nor could a condition of barbarism. There is thus a threshold condition for the application of the idea of republican government, although, as I explain in my response to Pinheiro Walla, visitors to a social formation must take it to have met that threshold. Nothing encountered in experience could be fully adequate to the idea of republican government; all existing governments are at best defective realizations of it. But not everything even meets the

[16] 6:306.

threshold for being a defective version of it; only something that satisfies the postulate of public right, that is something in which private persons have rights as against each other. Anything that meets that threshold is a rightful condition, a condition in which force is subordinated to law, however incompletely. That is why Kant is so adamantly opposed to revolution. He says that revolution "hands everything over to savage violence" because the revolutionary substitutes force for law, however noble the revolutionary's motives for doing so might be.

My proposal was to understand that concept of peace as occupying both the threshold condition and the regulative condition. The threshold condition is the existence of procedures whereby force is disciplined by law, and so the existence of a condition in which people can consistently enjoy their rights. The modality is significant here: a rightful condition is one in which human beings "can consistently enjoy their rights" rather than one in which they "do consistently enjoy their rights," precisely because something can satisfy this boundary condition while being despotic. The mark of despotism is the lack of a distinction between legislative and executive functions, but that, in turn, is itself a mark of the existence of private and/or prerogative power within the legal order, that is, one whereby an official can act on any grounds whatsoever. By contrast, a system of republican government is one in which the executive branch acts in accordance with the laws, acting only for the properly public purpose of providing a rightful condition for those living under those laws, as prescribed by properly enacted law. In such a system, there is discretion and room for judgment, but no prerogative.

This double structure tracks the distinction that Kant draws in *Towards Perpetual Peace* between the "preliminary" and "definitive" articles. The preliminary articles marked the transition from a condition of barbarism to one of anarchy. The definitive articles characterize the condition of peace. States could act in conformity with the preliminary articles of perpetual peace without satisfying the definitive ones, and, in particular, non-republican systems of government can enter into and abide by peace treaties, refrain from intervening in each other's internal affairs, and repudiate wrongful means and methods of warfare. That is, they can be at peace without being yet in a condition of perpetual peace.

The three definitive articles correspond to the three forms of public right identified in the *Doctrine of Right*: the Right of the State, the Right of Nations, and Cosmopolitan Right. As I explain in my response to Flikschuh, these are the three ways in which "a system of laws for a people" can relate people and institutions: the right of the state corresponds to republican government, its pure case, and governs how things stand between the state and those living under its laws; the right of nations governs how things stand between a plurality of states ("a multitude of peoples"), and cosmopolitan right governs how things stand between states and those who are not members of them.

Taken together these three forms of public right are not ways of achieving perpetual peace; instead, they constitute it. That is also why they are, taken together, identified as instances of "principle of outer freedom limited by law." If that principle "is lacking in any one of these three possible forms of rightful condition, the framework of all the others is unavoidably undermined and must finally collapse."[17] Unlike some other

[17] 6:311.

readers, I do not take this to show that Kant thinks that we must enter a world state in order for any rights to be finally secure; instead, I think his claim is that the principle of right governs each of these forms of relation, and so the structure of the state, the international order, and cosmopolitan right are each subject to them as their norm. As I explain in more detail in my response to Flikschuh, taken together they provide a complete determination of all human interaction under law. Taking republican government as a model, then, we must understand the requirements of the right of nations and cosmopolitan right—exiting the state of nature and creating a universal right to visit—are themselves ideas of reason, to which nothing in experience will be adequate, but that nonetheless provide the measure for a certain domain of interaction. The same ideas of reason serve to identify the interactions to which they apply, even though nothing to be encountered in that class is adequate to the idea of reason.

The three forms of public right, then, are the forms of human interaction which can and must be brought under law, so that force can be subordinated to law. Each of these has the precise double character to which Forst draws attention: the principle of freedom under law operates as a minimal condition; it is why you need to leave a state of nature for a legal order, even one that meets conditions of legality to a "small degree." It is also the norm for each form of public right. At the level of the right of the state, it forbids revolution while insisting that reform is necessary to bring a particular legal order more fully into conformity with the idea of human beings ruling themselves through their institutions. At the level of the right of nations it says that you can always move from a condition of war to a condition of peace, and so that a peace treaty is binding on the parties even if entered into under duress. But you cannot make the converse move; you cannot move from a condition of peace to one of war, not even by agreement. The miserable comforters—Grotius, Pufendorf, and Vattel—who Kant criticizes in *Towards Perpetual Peace*, understand war as a kind of voluntary arrangement between parties to a dispute. If "war is not the way in which everyone should seek his rights"[18] then a bilateral agreement to resolve the dispute through force is no less barbaric than a unilateral act of aggression.

If peace has this double structure, however, then the quote with which Forst begins really must be Kant's view. Peace cannot be replaced by war, not even to repair the effects of past war. To create a condition in which right disciplines might, we must accept that might made right in the past.

Forst raises two concerns: the first is that treating the past as settled would prohibit attaching conditions to a peace treaty, including limiting the potential warlike conduct of a defeated aggressor. I do not think that this follows from the idea that peace requires treating the past as settled. As he notes, my main focus, like Kant's, is on territorial claims, and I contend that these cannot be shifted by war. Beyond that, I want to defend two claims, both of which Kant makes: first, that victory in war does not generate a right to condemn or punish the aggressor, or, more generally, to hold the aggressor to account for having started the war. Instead, a rightful war is narrowly defensive, and its aim can only be to stop the aggressor. Second, Kant also claims that the defeat of an aggressor does not entitle those who fought against it to divide its

[18] 6:354.

territory among themselves and make the state disappear, but, at the same time, it can be made to "adopt a new constitution that by its nature will be unfavourable to the inclination of war."[19] These two claims are consistent, because ensuring a less warlike constitution is a way of giving effect to the first Preliminary Article of Perpetual Peace, making sure that the peace treaty does indeed impose closure. There is no inconsistency between accepting the results of past wars and negotiating specific terms in a peace treaty. A nation cannot initiate a war to correct a past wrong, but it can negotiate a peace aimed at preventing grounds of future war. Although Forst presents this as an empirical requirement, I understand it as a formal one: peace would be impossible if the peace treaty did not impose closure on all past and future grounds of war. This is just the paradox that Forst considers: even if the terms are unjust, peace requires that they be addressed in some way other than through war. That requires accepting the results of past wars so as to create peace.

Forst's other ground for skepticism is that "A fully public condition of international peace must aim to create a public order that does not condone rightless and unjust acts of war as constitutive of an order of peace" (Forst p. 40). The Kantian claim, however, was not that historical injustices must be condoned, but only that past wrongs do not give rise to grounds for war. There is never a justification for moving from a condition of law, however defective, to a condition in which force decides. That requires treating the past as settled with respect to the use of force. In nations such as my own, with troubling internal colonial legacies, that requires a process of reconciliation and legal changes to facilitate it, but it does not give rise to grounds of war. Other nations have external legacies of conquest and colonization; these, too, must be addressed, but they do not give rise to grounds of war. As Kant puts the point in his discussion of the second, third, and fourth "preliminary" articles of perpetual peace, putting them into effect can be delayed, but not indefinitely. The "prohibition presupposed is directed only to the future way of acquiring a right (e.g., by inheritance), whereas the exemption from this prohibition, i.e., the permission, is directed to the present status of possession."[20] Giving effect to the second preliminary article, which prohibits patrimonial acquisition of (what had been) had been a state through purchase, marriage, gift, or inheritance requires reversing those wrongful transactions. A ruler "may not postpone to a nonexistent date (*ad calandas graecas*, as Augustus used to promise) putting into effect the law, for example, to *restore* in accordance with number 2 the freedom of certain states deprived of it, and so not restore it; he is permitted only to delay doing so, lest implementing the law prematurely counteract its very purpose."[21] It does not authorize the use of force to reverse them.

The same point applies to the effects of past wrongful conquests; their reversal brings the situation more fully into conformity with perpetual peace, but does not authorize going to war to achieve it. Nothing could count as an authorization to enter a situation in which force decides; that is morally practical reason's "irresistible *veto*."[22] The fact that force cannot be authorized does not condone past injustices; it

[19] 6:349.
[20] 8:348.
[21] 8:347.
[22] 6:354.

only denies that they give rise to grounds of war, because corrective justice is never a ground of war.

The continual approximation of perpetual peace therefore requires undoing historical wrongs peacefully; it imposes duties of right without appointing anyone to enforce them. This too might seem paradoxical; Flikschuh characterizes it as "the paradoxical conclusion that rights relations both are and are not coercively enforceable." As I will explain in detail in my response to her chapter, there is no paradox here, because the possibility of institutions bringing all uses of force under law leads inevitably to public officials who are authorized to use force being subject to law without themselves being subject to force. The Kantian conception of the separation of powers between legislature, executive, and judiciary demands three coordinate branches, none of which is supreme in relation to the others; the legal order as a whole, which they jointly comprise, is supreme. One consequence of this structure is that the executive branch, charged with enforcement, is not itself subject to enforcement; any enforcer charged with overseeing the executive branch would either be itself not subject to enforcement, or else be subject to yet another and so on ad infinitum. The fundamental idea is bringing freedom under law; because the possibility of equal freedom requires reciprocal limits on the use of force, it requires institutions charged with the use of force, where all uses are subject to legal oversight, but, on pain of regress, not all uses are subject to enforcement.

1.C. The Distinctiveness of Kantian Right (Reply to Thomas Mertens)

Thomas Mertens argues that my account of the doctrine of right in general and the law and morality of war in particular depends too heavily on importing liberal legal concepts of subjective rights into what he regards as fundamentally a classical treatise on the law of nations. Where Flikschuh's chapter, to be discussed, argues that the focus of the *Rechtslehre* is politics rather than law, Mertens argues that it is about law in a narrower sense than I suggest. He introduces his objection through a discussion of an issue about translation: Is the English word "right" the appropriate translation of Kant's word "recht," which has no precise English equivalent? In both *Force and Freedom* and *Kant and the Law of War*, I followed Mary Gregor's usage. Mertens identifies some issues of nuance in the German, suggesting that "law" would be a better word, which would then make the correct translation of *Rechtslehre* "The Doctrine of Law." I will say something briefly about the translation issue, before turning to the substantive issue that Mertens suggests that it illuminates. In responding to the translation challenge, I am interested in distinctions that Kant draws, and the uses to which he puts them. My remarks in support of Gregor's choice of words are based on the importance of those distinctions, and the ways in which Kant's position requires contrastive terms in a way that is not permitted by Mertens's proposal to always translate "recht" as "law."

Gregor proposed two grounds for her judgment that "right" is right: first, it preserves the relation—a relation I sought to articulate and explain but also appealed to—between a system of right, that is, a legal order, and individual rights, such as the right that you have to your property, the security of your person, and to contractual performance. This is not just Gregor's interpretive insertion; in the "General Division of Rights," Kant distinguishes between his use of the concept of a right "As systematic *doctrines*, rights are divided into natural right, which rests only on a priori principles, and *positive* (statutory) right, which proceeds from the will of a legislator" and his use of the idea of rights as "moral *capacities* for putting others under obligation."[23] The second concept of a right as a right that some particular (natural or artificial) person has against some other person figures centrally in Kant's conception of both natural right and statutory right. So we might say that the doctrine of rights is a proper part of the doctrine of right, but does not exhaust it. Further, the idea of one person wronging another—an ancient legal idea already fully developed in Aristotle—figures centrally in Kant's treatment of both private right and the right of nations. The task for the interpreter is to keep track of these fundamental distinctions even when the vocabulary is similar, rather than collapsing them. Reading Kant as a natural lawyer by rendering "*recht*" as "law" makes this more rather than less difficult. In addition, Kant uses different words, one of which, *gezetz* is standardly translated throughout has writings as "law." As is often remarked, by Gregor among others, German has two words, *recht* and *gesetz*, as do other European languages other than English, such as *ius* and *lex* in Latin, and *droit* and *loi* in French. The opening questions of the *Rechtslehre*, concerning what "the doctrine of right is" and "what is right" could not be questions of what law is, particularly given Kant's answer to it, according to which natural right provides "the immutable principles" for "which an external lawgiving is possible," a distinction marked with the *recht/gesetz* distinction.[24] So too with the question "what is right," which Kant answers by contrasting what is laid down as right with the rational basis for lawgiving. Here Kant's vocabulary parallels that of Rousseau, who also speaks of *droit*, carefully distinguishing it from *loi*. One could perhaps paraphrase this and say that natural law provides the immutable principles for the giving of positive law, but to do so would treat Kant as a natural law writer and overlook his distinctive conception of freedom under law.

None of this is necessarily inconsistent with Mertens's proposal that "According to this reading, the 'Metaphysics of Morals' 'merely' adds to this that these moral laws lead to two kinds of duties: those duties the compliance of which can be enforced by external means, i.e. duties of law, and the duties that require internal enforcement, i.e. duties of virtue" (Mertens p. 45). To put the contrast in terms of addition, however, would overlook the way in which Kant demarcates the difference between those two types of duties, displaying the formal structure of laws governing external freedom, which are in turn to be explained in terms of the relation between the will of one person and another. The idea that duties of right can be enforced is not a claim about when enforcement will be effective in the sense of increasing the likelihood of the desired conduct; it is an implication of the way in which concepts of right involve the ability of one person's will to necessitate another's will.[25] On Kant's analysis, duties of

[23] 6:237.

[24] 6:229. "für welche eine äußere Gesetzgebung möglich ist, heißt die Rechtslehre (Ius)."

[25] I am grateful to a forthcoming piece by Micha Gläser for this formulation.

virtue involve the will's ability to necessitate itself. As such, they cannot be coerced or enforced by external means, because any external necessitation would not be a case of the will necessitating itself.[26] Duties of right, by contrast, can be coerced because they involve the relation between one will and another.

Mertens also suggests that, translations aside, the *Metaphysics of Morals* as a whole is concerned with duties, the doctrine of right is just concerned with those duties that may be enforced. Although Kant distinguishes duties of right from duties of virtue, and treats both within a single book called the *Metaphysics of Morals*, this seems to me to mis-frame the conceptual structure of Kant has in mind. The possibility of enforcement is a subject-matter criterion, not an empirical one. Kant insists that the relation between right and coercion is, in his preferred terminology, "analytic,"[27] that the possibility of enforcement is already contained in[28] the concept of the idea of a system of equal freedom. His claim about enforcement concerns the subject matter of legality, which, according to Kant's characterization of the a priori nature of the concepts of right (and the *Metaphysics of Morals* as a whole) could not be analytically connected to any empirical concept and so could not be connected to anything like an empirical concept of coercion. Once again, he is not making the positivist point that there are some duties that can be complied with only internally and others that can be complied with through force, with the difference between them being merely empirical and contingent. Instead, he elucidates a distinctively juridical concept of right, explicated through the universal principle of right, which enters as a "postulate incapable of further proof."[29] The universal principle of right thus enters as a normative-juridical principle of the entitlement to a system in which no person is subject to the determining choice of another; that system turns out to also be a system of enforcement—a system of equal freedom subject to reciprocal limits that can also be described as the legitimate occasions of enforcement. For Kant, then, right is enforceable in its subject matter because freedom forms a self-contained system.[30] There is therefore a distinctive morality governing both freedom and enforcement, and so, with it, a distinctive subject matter of right. The duties that can be enforced or identified through the idea of external freedom, rather than being just an empirically demarcated proper subset of duties in general.[31] That is why Kant insists that the universal principle of right is a postulate, for it is the application of ideas of freedom to spatially embodied rational

[26] 6:239, 6:383.

[27] 6:396.

[28] Kant's use of the vocabulary of conceptual containment is discussed in detail in R. Lanier Anderson *The Poverty of Conceptual Truth* (Oxford: Oxford University Press, 2015).

[29] 6:231.

[30] As I explain in my responses to Flikschuh and Forst, within a legal system there will be some officials charged with duties of enforcement whose compliance with those duties are subject to law but to no further coercive enforcement.

[31] These two ways of understanding the scope of right lead to very different conclusions with respect to your duties to yourself. The question of whether external coercion could prevent you from suicide, self-mutilation, or gluttony is empirical. Kant's discussion of these vices in the *Doctrine of Virtue* explains the defect in each of these actions in terms of the relation between means and ends, and so in terms of the will's necessitation of itself. They are outside the realm of external enforcement because of their subject matter, not because such enforcement is likely to fail. As he explains in the "Division" of the *Metaphysics of Morals*, "Duties of virtue cannot be subject to external lawgiving simply because they have to do with an end which (or the having of which) is also a duty." 6:239

beings.[32] Commenting on Cicero's characterization of morals as a doctrine of duties and not also of rights, Kant remarks that the concept of a moral imperative applies to duties of virtue as well as of right, and can be appealed to in order to explicate the concept of right, that is, of putting others under obligations.[33]

On this reading, then, the fact that the *Metaphysics of Morals* as a whole is a doctrine of duties does not entail that in engaging with questions of right, Kant is writing "a classical legal philosophy," the kind of book that Grotius or Pufendorf or Achenwall wrote, outlining the duties contained in the positive law of the day and rationalizing them piecemeal. It is a systematic and philosophical doctrine. It is not exclusively a doctrine of right*s*, because many of the principles that shape it are ones which no enforceable right applies. That is why, as Kant notes in his discussion of Hobbes in "On the Common Saying," the absence of conclusively enforceable rights against the sovereign does not mean that the sovereign owes no duties of right to subjects.[34] Nor does it mean that the sovereign is subject to no duties of right, only that (as I explain in my responses to Flikschuh and Forst) there will be a point at which those charged with enforcement are not directly subject to (further) enforcement. As such, there will be no individual private claims with respect to those rights, even though the duties in question are duties of right. But, for all that, all of those cases follow from the way in which rights of individual human beings must be protected through legal and political institutions. The rights that Kant assigns to the supreme commander in the "general remark" about public right are all "moral *capacities* for putting others under obligation." Each of them each can be enforced, and is subject to its own internal principle. Kant's account of these powers focuses on the way in which they are grounded in the needs of a public legal order and, therefore, subject to its rationale. That is why he insists that hereditary offices are inconsistent with right, even though he recognizes that they are a commonplace of legal ordering in his time, and that institutions are empirically capable of entrenching and upholding them. These are duties without (enforceable) correlative rights, but they are duties having to do with right.

So, too, with cosmopolitan right: the public legal order under cosmopolitan right does not have the authorization to enforce those rights, but cosmopolitan right remains, nonetheless, as Kant puts it, "not a philanthropic (ethical) principle but a principle *having to do with rights*."[35] Kant could not here be insisting that this is a principle of positive law, or working through the implications of an international legal order in the absence of law. Instead, his claim is that this is a principle of right, on which it is

[32] In the *Critique of Practical Reason* Kant explains "a postulate of pure practical reason (by which I understand a theoretical proposition, though one not demonstrable as such, insofar as it is attached inseparably to an a priori unconditionally valid practical law)." 5:122. In the introduction to the *Metaphysics of Morals* as a whole Kant offers this characterization of postulates:

> It then seems less strange to find that these laws, like mathematical postulates, are incapable of being proved and yet apodictic, but at the same time to see a whole field of practical cognition open up before one, where reason in its theoretical use, with the same idea of freedom or with any other of its ideas of the supersensible, must find everything closed tight against it. 6:225

[33] 6:239.

[34] 8:304.

[35] *rechtliches* 6:352.

incumbent on states to act, even though it is not yet instantiated and no mechanism exists for its enforcement.

Mertens's concern about translation is, as I suggested, the way in which he introduces the more general point that he hopes to make, which concerns what he sees as a kind of slippage between the idea of a particular subjective right and a system of right. A system of right is not exhausted by the particular private rights of individual human beings. They nonetheless share a structural feature: both are always concerned with how things stand between the choice of one juridical person and another, and so with which of them is in charge of a particular question. This is not to say that individual subjective rights are the same as the powers that public authority has. An individual rightholder is free to exercise its right (consistent with the rights of others) for any purpose whatsoever, while a public authority is permitted to act only for properly public purposes. Nonetheless, both are forms of authority, because in each case, the subject matter of the authority has one juridical person's say-so determine what the other may do. Relations of private right are often said to be horizontal, those of public right vertical. However one characterizes the specifics of that division (given that each of public and private right has three forms, corresponding to the three categories of relation, substance, causality, and community), the relational structure of the morality in question is the same. So, too, with things that are wrong in the highest degree, because they make rightful relations impossible.

I take it that Mertens's main concern is not about what words to use to render Kant's German into English, but rather with the relation between ideas of individual subjective rights, such as property rights and contract rights, and ideas about legality, and the relation between all of this and any kind of question of justice or legitimacy. Mertens is certainly correct that we can study what is laid down as right without supposing it to be right, just, or even legitimate. Kant is plainly attuned to this possibility; that is why he frequently contrasts his account and that of what he variously calls "the jurist" or the "mere jurist,"[36] the one who can dutifully report what is laid down as right here or there. His objection to the jurist's "confusion" or subreption,[37] that is, mistaking a principle that applies at one level for a principle that applies at another, reveals that Kant takes himself to be answering the question about which the mere jurist is confused, by distinguishing different levels at which legal norms operate. The ambition of the Kantian enterprise is to explain how something can be laid down as right, not merely as positive law, but rather how something could be, or purport to be, both positive and rightful. Hence Kant's search for the supreme principle of lawgiving. His investigation of the Universal Principle of Right and his specification of Ulpian's precepts, as well as his description of the innate right of humanity as the basis of all rights, is not a report on what is laid down as right in some particular legal order or in legal orders generally; so, too, with the postulates of private right and public right, or the narrow characterization of the grounds of war, or the limits on acceptable means

[36] 6:229, 6:236, 6:345.

[37] Kant uses the concept of subreption to explain the mistake of "the ascription of objective reality to an idea that merely serves as a rule," and of transforming "the distributive unity of the use of understanding and experience, into the collective unity of the whole of experience." See *Critique of Pure Reason* A 509/B537, A583/B661.

of defense, or the preconditions of a possible peace. Each of these is introduced as a requirement of right. The creation of a legal structure, the prohibition of revolution, aggressive war, and wrongful means of war are instances of what Kant calls the Postulate of Public Right which, as I explain in my response to Forst, is the boundary condition on something being a rightful condition at all.

Let me fill this point out further by noting that Mertens has set things up as though the choice is between reading Kant as some variety of legal positivist and reading him as a natural lawyer. So the question then becomes whether he is giving an account of the positive law in this or that legal system, or ignoring what is "laid down as right" and only offering an external, extralegal account of law that is morally justified. Those options, as described, make the Kant-as-positivist picture seem compelling: Kant plainly thinks that you can only understand right by understanding it through the creation of a rightful condition in which there are public officials charged with making, applying, and enforcing law. He also seems to set the bar rather low for what counts as a rightful condition, hence his much-noted (including by Mertens) opposition to revolution.

But those are not the only options. Here is a better way of understanding Kant's view: on the positivist reading, there is a moral problem, to which positive law is the solution, but any system of positive law will solve the problem equally well. Something like this is Jeremy Waldron's interpretation of Kant.[38] On this reading, all that is required is that there be a system of positive law that clarifies who has which things, who gets to resolve disputes, and who gets to make the rules about who has which things and who gets to solve the disputes. This is a familiar and plausible way of reading Hobbes's social contract argument: life is dreadful in the state of nature, and so we must set up a sovereign, but nothing of the moral structure of the state of nature survives sovereignty. Beguiling as that positivist picture might be, it is not Kant's picture. For Kant, the basic question of right—how things stand between one human being and another, as they stand in practical relation to each other, such that none has their choice determined by the choice of another—both explains the need for a state and so for positive law, but so, too, imposes limits on the content of that positive law. Yet it does not turn him into a natural lawyer of the kind that Mertens appears to regard as the only alternative to legal positivism, that is, someone who adds a moral filter to the positive law, an external, extra-positive test. Instead, Kant is the type of natural lawyer who thinks that law's positivity is the solution to a moral problem, but that the shape and nature of that moral problem constrains possible content of positive law that is to provide a solution. On this view, morality is radically incomplete in the absence of public legal institutions, but those institutions are therefore not merely an instrument for producing morally desirable results. As he remarks near the end of his discussion of private right, entering a rightful condition

> contains no further or other duties of human beings among themselves than can be conceived in the former state; the matter of private right is the same in both. The laws of the condition of public right, accordingly, have to do only with the rightful form

[38] Waldron, "Kant's Legal Positivism," *Harvard Law Review* Vol. 109, No. 7 (May 1996), pp. 1535–1566.

> of their association (constitution), in view of which these laws must necessarily be conceived as public.[39]

The positivist thread in Kant's thought, then, turns on the need for positive, that is (partly) chosen law in order for a plurality of human beings to enjoy their rights as against each other. Kant's argument for this conclusion turns on rejecting an implicit premise of the way in which the supposedly forced choice between natural law and positivism is framed, according to which morality is already complete in the absence of legal and political institutions, either as natural law, with dictates for everything anyone must do; or, alternatively, is available to fill gaps in the positive law that has been laid down. For Kant, by contrast, positive law is more determinate and precise than morality. Right is the part of morality that can only be adequate to its own principle if it is made into law, because the requirements of a system of equal freedom can only be realized through a shared public standpoint for making, applying, and enforcing positive law, and a doctrine of right is the specification of the subject matter of legitimate external lawgiving. Kant can thus explain why there is a space between what is right and what is laid down as right, while maintaining that right provides the standard appropriate to evaluating positive law, "so as to establish the basis for any possible giving of positive laws."[40]

This alternative to both natural law and positivism, as Mertens presents them, runs through Kant's work. In the case of the law of war, it provides an alternative to the early modern ways of combining them. Christian Wolff distinguished between the natural law of nations, which required that wars only be fought by those with a just cause; and what he called the voluntary law of nations, which, he contended, conferred exemptions from natural law, permitting nations to go to war whenever they had a grievance.

Mertens's attempt to present Kant as writing a traditional natural lawyer's treatise focuses on right of defense, seeking to ground it in the idea of self-preservation. He interprets national defense through the idea of individual self-defense, and then seeks to explain the latter in terms of the duty of self-preservation, which, he points out, is discussed by Kant not in the doctrine of right but in the doctrine of virtue. This is certainly a familiar view of self-defense; indeed, it is the view taken by Achenwall, from whose textbook Kant was required to teach natural right. But, as the notes of Kant's student Feyerabend reveal, by 1784 Kant had already rejected this idea and, indeed, he rejects the more general idea that the doctrine of right is founded on any idea of self-preservation. Achenwall's three-fold gloss on the Ulpian precepts makes self-preservation the first duty of right; Kant's appropriation of them focuses instead on rightful honor, the demand that you not allow others to treat you as a mere thing. Kant rejects Achenwall's account of self-defense because it is non-relational, remarking that another person's self-preservation is not, on its own, a source of obligation, because "In right I have to arrange everything according to laws of freedom, or else everything gets confused."[41] I have written elsewhere about how this "internal duty" fits into the doctrine of right; for now I only draw attention to the fact that it is very different from

[39] 6:306.
[40] 6:230.
[41] *Naturrecht Feyerabend* 17:1334; translation modified at Micha Gläser's suggestion.

the duty to preserve yourself, because it can only be understood in relation to the choice of another. Yet, aside from the fact that Kant expressly says that all duties of right are indirectly duties of virtue, rather than, as Mertens proposes in this case, the reverse, in his discussion of the plank of Carneades, Kant expressly repudiates self-preservation. It is difficult to see how you could have a duty to do something that you are not permitted to do. Further, self-preservation would seem to give rise to a Hobbesian "right to all things," entitling you to preserve yourself even at the expense of another, but when Kant talks about self-defense (briefly, and at more length in the right of nations) this possibility is expressly repudiated.

In the *Doctrine of Virtue*, Kant identifies mandatory ends—your own perfection and the happiness of others—from which other duties, such as the duty of self-preservation and more general care of the self are said to be derived. That duty of self-preservation, however, is always understood as a duty of moral self-preservation, a duty that is always subject to moral limits. Virtue faces no difficulties about a right to whatever a person needs for self-preservation, because virtue takes right as setting its limit; it thus provides an explanation of why when another attacks you, from the standpoint of right it is up to you whether to "show moderation."[42]

Mertens also writes that "In the *Rechtslehre*, Kant merely gives us a description of what the international realm would look like without positive international law" (Mertens p 50) I am not sure I fully understand this claim. Kant's discussion of the right of nations does not conform to the international legal structure of his day. His point is that in the absence of public legal institutions, each nation can each only "do what seems good and right to it," but this is not a description of either a factual situation or a positive legal norm. Nor is it even a description of a hypothetical situation in which there is no positive international law. Instead, it is the characterization of a predicament, the distinctive version of the problem of the state of nature between moral persons that are themselves also legal orders. His express rejection of punitive wars is already a reminder of that; if the international realm is understood as a state of nature, it is one in which nations took it upon themselves to punish each other for their misdeeds, and exact remedies. Kant's rejection of this idea—his insistence that the victor in a war must let the matter of wrongdoing "drop"—is directly contrary to the legal norm practiced in his day.

At the beginning of his discussion of public right, Kant distinguishes between the right of the state, the right of nations, and cosmopolitan right, writing that "if the principle of outer freedom limited by law is lacking in any one of these three possible forms of rightful condition, the framework of all the others is unavoidably undermined and must finally collapse."[43] This is not a description of the legal order he has encountered, because the principle of outer freedom limited by law is constantly lacking in the world in which he found himself. Instead, Kant is identifying the structure of the principle of freedom under law in each of three possible forms of relation. The right of the state, the right of nations, and cosmopolitan right are different from each other, but each has its own proper internal principle of freedom under law.

[42] 6:235.
[43] 6:311.

That is why Mertens is incorrect to describe the Kantian state as patrimonial. The sovereign is the supreme proprietor in relation to all landholders in the state, as guarantor of title and provider of the systematic requirements of right. The supreme proprietorship is internal; the sovereign is a rightholder but not a proprietor as against other sovereigns. The powers that the sovereign is entitled to exercise *in subsidium* of its role as supreme proprietor are nothing like the powers that individual proprietors are entitled to exercise as against each other. Such powers as the power to tax and administer the state's economy finances and police cannot be characterized as having private analogs. As Kant notes, the police power "provides for public security, convenience and decency."[44] I have discussed Kant's treatment of these subjects in *Force and Freedom*, and will not repeat that discussion here, except to note that these are all distinctively public purposes, rather than private ones. The state is charged with maintaining itself as a rightful condition, rather than having untrammeled discretion to determine the purposes its institutions or territory will be made to serve. In this it contrasts with the situation of an individual property owner who is entitled to determine the purposes for which the property will be used. That, indeed, is the contrast on which Kant insists between despotism and republicanism. Even a despotic state is entitled to be independent of other states, but it is defective in relation to its own idea. The people, as sovereign, is in charge of it in the sense that no one else is in charge of them. But it is not subject to their private purposes in the way that the patrimonial state embraced by the writers of the regular war tradition is.

Once again, the idea of independence is contrastive and relational, rather than monadic.

Mertens takes from the fact that Kant acknowledges that, in an international state of nature, the right to national defense will be broader than we might be comfortable with, including a nation attacking a neighbor by whom it expects to be attacked, that Kant is finally just a miserable comforter who permits too much war. The difficulty with the miserable comforters, however, is not that their theories are insufficient to prevent war, or that they are used opportunistically by rulers eager to find justifications for what they want to do anyway. Instead, like Job's miserable comforters to whom Kant means to compare them, they will find a justification for whatever happens, identifying a possible fault to justify any act of injustice or aggression. That is, when used as directed, their catalogs of the grounds of war will justify pretty much anything, but never prohibit anything.[45] Kant's view is fundamentally different: on his view, it is always wrong to move from a condition of peace to a condition of war, and war has a distinctive immorality because it hands everything over to savage violence. He therefore repudiates all of the positive grounds of war introduced by the miserable comforters. At the same time he acknowledges that an international order without public legal institutions is defective because each can do no better than "what seems right and good to it."[46] As a result, the narrower standard that limits the grounds of war to national defense—essentially the standard that prohibits war as both the just war tradition and the miserable comforters understood it—will be applied by

[44] 6:325.
[45] 8:355.
[46] 6:312.

belligerents themselves, in an open ended and, no doubt, often self-serving way. The normative structure of the situation demands that each nation respect the independence of the others, but also that each may resist those who are attacking or about to attack it. Because they can do no better than what seems right and good in their own eyes, acting preemptively is consistent with the normative structure of the situation. But just as the fact that private rights in a state of nature are defective does not mean they have no normative structure, so, too, the fact that the right of nations has not fully brought freedom under law does not mean that there is not already contained in the structure of the situation a normative demand that it be resolved in a particular way. That is why practical reason has its "irresistible veto. There shall be no war."[47] The repudiation of war is not merely the demand for positive law; it is the demand for a system of freedom under law.

Immediately following the paragraph from which Mertens quotes Kant about practical reason's veto, Kant goes on to write

> establishing universal and lasting peace constitutes not merely a part of the doctrine of right but the entire final end of the doctrine of right within the limits of mere reason; for the condition of peace is alone the condition in what is mine and what is yours for a multitude of human beings is secured under the laws living in proximity to one another ... the rule for this Constitution, as a norm for others, cannot be derived from the experience of those who have hitherto found it most of their advantage; it must, rather, be derived *a priori* by reason from the ideal of a rightful Association of human beings under public laws as such.[48]

This is not positivism in the sense the Mertens identifies. Nor is it the kind of traditional natural law with which he contrasts it, or a traditional treatise on the nature of law. It is the distinctively Kantian project of finding the rational principle through which the possibility of the consistent exercise of freedom can be elucidated.

1.D. The Right to Territorial Integrity: Collective or Reductive? (Reply to Anna Stilz)

Anna Stilz's chapter offers an alternative Kantian interpretation of a state's right to political independence. In setting up her positive proposal, she offers a critique of my account. She introduces that critique by considering two ways in which the Kantian account of each state's right to political independence and territorial integrity might be understood. She characterizes the first of these as reductivist, the second as collectivist. On the reductivist view, the claim of the state is the exhausted by the claims of individual members of it; aggression against a state is wrongful because of what it does to the members of that state. On the collectivist view, by contrast, the wrong is against the state itself, a violation of its right to govern itself. Her own proposed account is offered as reductive, but in a different way than she interprets my account as reductive.

[47] 6:354.
[48] 6:355.

As Stilz points out, I appear to embrace both the collectivist and reductivist accounts, insisting on Kant's individualism while at the same time representing aggression against the political order as something that one collective does to another. As I will explain, I do indeed endorse both. I see no tension between the versions of each that I mean to endorse. Stilz raises a pair of fundamental objections to what she takes to be the reductivist account that I endorse. First, she argues that if the wrong of aggression is to be analyzed in terms of wrongs to individuals, the aggressor would need to have deprived those individuals of something. But, she suggests, imagining an invasion of Canada by the United States, each of the elements of public right that Kant considers necessary could be preserved. Intuitively, it seems such a bloodless invasion would still be wrong, even if all of the functions of the state remained intact. Second, like Forst, she draws attention to the fact that I also argue that the rightful condition of which a particular person is a member is historically contingent, and can reflect the fact of past wrongful aggression. I do not claim that there is a permission to shift borders through force, but I do insist that the possibility of peaceful interaction in future requires accepting the effects of past wrongdoing. Stilz remarks that

> "his effectiveness standard is morally unsatisfying, since it allows might to make right. Whichever power can maintain physical control for a sufficiently long period of time gains title, even if it conquered the area through bloody conquest or imperial subordination, and even if it maintains its control through continued repression. Such a standard permits territorial disputes to be settled by sheer force. (Stilz, p. 55)

In addition to these arguments about fungibility and past injustice, she also suggests that attaching so much weight to whoever happens to have come to be in charge, in whatever way, fails to attach sufficient weight to the competing claims of individuals who are mistreated by the states exercising power over them.

Stilz's alternative account, building on ideas developed in her groundbreaking recent book *Territorial Sovereignty: A Philosophical Exploration*[49] introduces on a contrast between two ways in which citizens relate to the institutions that rule over them.

> As institutional "takers," individuals have interest in specification, adjudication, and enforcement of their private rights that only a state can provide. This "taker" dimension of evaluation is concerned with the objective qualities or features of state institutions, independently of anyone's relation to them. But the right not to be treated as "passive" suggests that to be fully independent, individuals must also relate to their state in a certain way: they should be authors of its political institutions. Individuals should be not only state "takers," but also state "makers." (Stilz p. 57)

In responding to the objections, I will explain why, although the ideal case of a rightful condition is one in which the citizens are indeed "makers" who rule themselves through their institutions, satisfaction of that ideal requirement sets too high a threshold for the right against aggression, that is, the right to political independence

[49] *Territorial Sovereignty: A Philosophical Exploration* (Oxford: Oxford University Press, 2019).

and territorial integrity. Although Stilz expresses concern that accepting the past as settled "permits territorial disputes to be settled by sheer force," I will suggest that accepting that the past is settled is the only way to prevent territorial disputes in the future from being settled by mere force.

Stilz frames her discussion in terms of a contrast between individualism and collectivism. Stilz writes that "On plausible accounts of value individualism, the importance of state independence should somehow derive from its contribution to individual independence, autonomy, or well-being." (Stilz p 55) I am committed to a different understanding of individualism. I do not suppose that states, understood as legal orders, must somehow matter apart from their relation to their members. The account I endorse differs from many contemporary formulations of individualism inasmuch as it does not trace everything to questions of individual well-being. Nor does it understand the relation between state independence and individual independence in terms of an idea of contribution to something that can be fully understood or realized in its absence. On the Kantian account, everything ultimately turns on the fact that human beings are rights-bearers, capable of standing in rightful relations to each other. Right constrains interactions in two ways: first, you must not violate another person's right; second, you must interact with others in ways that are consistent with sharing (one of the three forms of) a public rightful condition with them. Wrongs against individual human beings are therefore not the only possible wrongs; war is wrong "in the highest degree"[50] because the way in which entering or willingly remaining in a condition in which might makes right it is inconsistent with the right of human beings, even if it were to somehow (as in Stilz's imagined US invasions of Canada) avoid violating any right of any individual human being in a particular case. The distinctive wrongfulness of violating public rights, either by remaining outside of a rightful condition or, however briefly, forcing someone who is in a rightful condition out of it (as in the example of bloodless annexation) remains even where no individual rights are violated. I share Kant's own assessment of the situation in which relations between nations are subject to force:

> This nonrightful condition is a condition of war (of the right of the stronger), even if it is not a condition of actual war and actual attacks being constantly made (hostilities). Although no state is wronged by another in this condition (insofar as neither wants anything better), this condition is in itself still wrong in the highest degree.[51]

Such a situation is wrongful even if no state is wronged, and even if no individual is. It is nonetheless wrongful because it is inconsistent with individual human beings enjoying their rights, since it creates a situation in which all interactions are subject to force.

The account I propose is individualistic but not reductively so. The thing that makes moral individualism seem compelling—the moral significance of each human being—is better accounted for by thinking of every human being as a bearer of a right to independence of another's determining choice, rather than by thinking of each

[50] 6:308, 6:343.
[51] 6:344.

person as the repository of well-being or as a thing that is vulnerable to harm. That idea, however, requires understanding public legal orders as having a distinctive type of moral significance of their own.

As I sought to explain in chapter 2, and in more detail in *Force and Freedom*, public officials stand in status-like relations rather than proprietary or contractual ones to those over whom they exercise power. Most of my emphasis in those discussions was on the duties owed by the officials to those people. Relationships of status have another dimension as well: the beneficiary not only has a claim to the loyalty of the official or fiduciary; the beneficiary also has a claim as against third parties that they not interfere with the fiduciary relationship. If a public official accepts a bribe, the official wrongs those over whom the official's power is exercised. The person paying the bribe also commits a wrong against those people. So, too, if someone forcibly removes a public official from office; that person wrongs those who are entitled to the services of that official. These examples illustrate the sense in which relationships of status are, as Kant introduces them, "a right to a person akin to a right to a thing." The wrong is different from the wrong of inducing the breach of a contract, which can only be committed through the participation of the person in breach; someone who attacks the contracting party wrongs the person attacked but not those to whom that person owes contractual (that is, *in personam*) obligations. By contrast, depriving citizens of the services of a public official who owes them a duty to govern on their behalf, either by inducing a breach of those duties or by preventing the official from carrying them out is a wrong against both the official and those citizens. That is why I seem to endorse both collectivist and individualistic accounts, and claim that an invasion, even a bloodless one, is both a wrong against the state and a wrong against the citizens, the two possibilities that Stilz regards as mutually exclusive. They are not mutually exclusive, however, because the wrong against the citizens is wrong against a right that each of them enjoys as a member of the public, rather than the violation of a right that they could have in a state of nature.

Stilz also criticizes the collectivist position as I seem to articulate it. Following Kant, I argue that even despotic states—everything short of barbarism—enjoy the right to political independence and territorial integrity. Stilz expresses concern about just how inclusive this account is:

> So long as a state does not commit slavery or genocide, thereby degenerating into "barbarism," it has a right against interference by other states, even if it commits other quite grievous wrongs—including torture; rape; famine; denial of rights to free speech, association, and political representation; or mass repression. (Stilz, p. 56)

I am not certain that everything on this list avoids barbarism, but I am confident that states that commit some of the wrongs that she correctly identifies as grievous—especially the denials of civil and political rights such as free speech, association, and political representation—are still entitled to political independence and territorial integrity as against other states. Indeed, I believe that even those nations—some of which provide robust civil and political rights to most of their own citizens—that are reported to have committed officially and publicly endorsed torture in recent decades have not lost their right to political independence and territorial integrity. Other nations are not entitled to invade them, even if doing so had a realistic prospect of preventing future torture.

Stilz goes on to point out that the Kantian view, as I articulated it, protects the political independence and territorial integrity of nations that have acquired territory through force, provided that that acquisition has later been secured through a comprehensive peace treaty. The list of nations that fit this description is rather longer, since almost all of the world's nations have historically acquired at least some of their territory through force. I did not claim, and indeed expressly denied, that they are entitled to claim new territory through force. Instead I followed Kant's claim that, having entered into a comprehensive peace treaty, their claim to their territory is good, even though (unlike other enforceable contracts) the peace treaty is typically the product of duress. The right to political independence and territorial integrity is not a right to be free of criticism, or free of political and diplomatic pressure from others or even, in certain circumstances, economic pressure from others. It is only the right to be free of invasion.

In order to back up these striking claims, I want to say little bit more about each. First, I want to explain why I think that the right against invasion is both an individual right and a collective right. In order to do so, I want to draw a distinction between the idea that the Kantian state is justified in terms of its function and the idea that its claims are merely functional and therefore fungible with anything else that performs the same function. In introducing her example of a bloodless US invasion and annexation of Canada, Stilz assumes that the interrupted rightful condition will immediately be replaced with a new one. This seems to me to be the wrong way to think about individual rights in this case and more generally. The difficulty is not that any imagined transition is likely to be bumpy, but rather that the immediacy of the replacement does not change the nature of the juridical disruption, even if it makes the factual disruption less disruptive in terms of the settled expectations of individual human beings. The wrong of invasion is distinctively public, but the more general structure of wrongdoing applies more broadly: the fact that the wrongdoer is able to provide a substitute for the object of the right does not make the wrongdoing permissible. If I wrongfully damage your property, I am liable to you in damages; if I wrongfully take your property, you are entitled to reclaim it or to receive an equivalent from me. If I use your property without your permission, I must surrender my gains. In none of these cases, however, does the availability of a remedy, or even the guaranteed future provision of one, make my wrong against you less wrongful. I am not allowed to take your property and replace it with more than enough cash to purchase a replacement. I am not even allowed to do so when the object of your right is fully fungible, such as, for example, if I embezzle money from your account and repay it with more interest than you could have earned. I am not allowed to do so even if you would have sold me your property for less than I am prepared to pay. Nor can I enlist you as my silent partner, using your property and giving you a share of the profits without consulting you. The underlying idea here is that I cannot take it upon myself to answer these questions for you, even if there is undisputed or even incontrovertible evidence that the answer that I give is the one that you would have given. Private right governs the way in which things stand as between us; in the case of property, it governs the ways in which things stand between us with respect to things, rather than with respect to their use, the value, or the usefulness of those things. The fact that the objects of these rights will often be exchangeable does not mean that people can force transfers on each other, taking it upon themselves to compel others to engage in involuntary transactions on the grounds that they would have been legitimate if voluntary. For you to have something as your own as against me—to the exclusion of all others—makes a central reference to the thing, rather than to the particularity of the use to which you put it or even the possible uses to which you might put it. I do not get to decide that the substitute I give you is just as good, or even better, than the thing I take from you; that is what it is for that thing to be yours.

Stilz might respond that the analogy with property is misleading, because the argument for setting up a civil condition is cast in terms of the specific functions that it performs. As a result, she might suggest that the members of a political society do not have political independence or territorial integrity subject to their choice in the way in which you have your property subject to your choice, such that you alone are entitled to determine the purposes for which it will be used. A rightful condition cannot be used in this way, except in the vacuous sense of using it to be the particular rightful condition that it is.

The point of the analogy is different: the point is that others may not take it upon themselves to decide questions for a person. Although the need for a rightful condition can be described in functional terms, once it is in place, its members have rights as against outsiders to remain in that legal condition. Although it is unlike a property right in other respects, the general form of wrongdoing is that same: subordinating one (natural or artificial) person's choice to that of another. It is not up to those outsiders to decide for them whether they will remain in that particular rightful condition. That is the point of the private law analogies: the wrong of theft or converting property is not that it causes a net loss, but rather that the other person wrongfully takes control of something that is properly subject to the owner's choice. That is the sense in which interfering with another person's property is a form of domination; that is also why it is a wrong that can only be analyzed relationally. It is what the wrongdoer (the thief) does to the rightholder (the owner). The same point applies to the nation: each nation is entitled to be politically independent because no other nation is its master. That is why it is wrong for another nation to make itself its master through violence, even if it suffers no loss of function. It is no less wrong to do it through the threat of violence or through deceit or corruption. If the invading force bribes public officials to stand down, the invasion may go smoothly, but the invaded country is wronged by having its public officials induced to act for private purposes of enriching themselves. The apparently seamless nature of the transaction does not make it any less wrongful, because the members of the rightful condition were entitled, as against outsiders, to live under it. Those outsiders therefore commit a formal wrong when they displace law. In so doing they replace law with force, even if their objective of subsequently replacing it with new law is entirely successful.

Stilz argues that her view also sets a higher threshold on state legitimacy.

> On my Kantian reductivist view, then, independence has two dimensions: a legitimate state must not only "rule on behalf" of its inhabitants (by protecting their private independence), it must also reflect the shared will of most cooperators on its

> territory to establish justice through a particular institutional configuration (thus protecting their political independence). (Stilz, p. 57)

I agree with Stilz about the significance of the shared will of those present on the territory, but I understand its significance differently. As I explain in my response to Forst, the Kantian theory of right requires a distinction between the boundary conditions of something being a rightful condition, that is, a condition of law rather than force alone, and the ideal case of it. The relation between the two sides of this distinction is familiar in Kantian philosophy, as it is the distinction between something falling under an ideal concept and the ideal concept itself. I had earlier put this in a somewhat misleading way, borrowing broadly Aristotelian or Rawlsian vocabulary to say that the regulative principle for a thing depends upon the nature of that thing. This way of putting it was misleading insofar as it gave the impression that the nature of the thing could be identified apart from its regulative principle. When Kant says of the act that forms a state that "the original contract is only the idea of this act, in terms of which alone we can think of the legitimacy of a State,"[52] he uses the ideal in order to characterize something as a defective version of it. The state is under a duty to bring itself more fully into conformity with its own idea "the only form which makes freedom the principle and indeed the condition for any exercise of coercion, as is required by a rightful constitution of a state in the strict sense of the word. Only it will finally lead to what is literally a state."[53] My claim is that, as against other states, anything that minimally satisfies the postulate of public right is entitled to political independence and territorial integrity. As I note in my response to Pinheiro Walla, to think of a people as occupying some region of the earth's surface is already to think in terms of it acting collectively and so under the concept of public right. As such, it has a right against invasion. Just as the question of acquisition must bring rational concepts to bear on empirical particulars, so, too, must the idea of a shared will. The things we do together as a people is share a rightful condition. Our collective agency consists in each of us living with each other under laws, conforming to those laws at various levels—refraining from committing crimes, paying our taxes, driving within the speed limits, and, most generally, living together in a way in which a plurality of separate persons can interact with their conduct governed by words rather than force, without giving up either our separateness or plurality. I do not mean to deny that there are more and less adequate versions of a shared popular will. My claim is only that any empirical instantiation of it will be defective in relation to its ideal. Returning to the issue of national defense against aggression, the wrong of aggression is always the wrong of substituting force for law. Adequate representation of a shared will is not a precondition of law that can be wrongfully violated in this way. To deprive someone of their participation in a legal order is a wrong both against the legal order and against that person, because it takes that person out of a condition of peace and puts them in a condition in which force decides. It is the same wrong even if the legal order from which they are removed was defective, and it is the same wrong even if the invader has a realistic prospect of success at bringing in a superior legal order.

[52] 6:315.
[53] 6:341.

An analogy with individual self-defense against non-lethal attack brings this relational structure more fully into view. Stilz appears to understand states' entitlement to defensive force on something like an analogy with a right to defend yourself against attack based on your autonomy as a self-governing being. The claim would then be that a nation is entitled to defend itself if another nation interferes with its autonomy, which is in turn analyzed in terms of an interest that its individual members have participating in its autonomous political structures. While acknowledging both the value such participation provides to individuals, and the way in which representative republican government is more fully in conformity with the idea of the original contract than despotic government is, I do not believe that the right to national defense is conditional or dependent on it. Indeed, the analogy with a non-relational conception of individual autonomy points to the difficulty of this conditional view of defensive rights. If someone is not living an autonomous life, instead being drawn in by this or that short-term temptation, responding to social pressure or cultural or religious indoctrination, or always deferring to others, that person still has a right against forcible interference by others, including well-meaning others who hope or realistically expect to interfere in a way that will leave that person with at least as much (non-relational) autonomy as before.

The same point applies to what Kant describes as the right of human beings, that is, the right to be in a rightful condition. If you are in a rightful condition, others wrong you by taking you out of it, even if they will quickly put you in the new one. They need to respect your rightful condition because of the right of human beings, not because of the harm that you suffer from being taken out of it. The idea that wrongdoing is to be traced to material loss is the mistake of all empiricist accounts of right.

Once more, the issue does not depend on welfare but on choice: the outsider does not have the political standing to move those people from this rightful condition to a superior one even if it can be done at no cost. The nature of the wrong against which national defense is a right cannot be formulated or expressed except in terms of the concepts of right. The inhabitants are not deprived of something the value of which can be characterized without reference to how things stand between legal orders. In Kant's time, the nations of Europe with which he was familiar were all despotic. Even those that allowed ordinary citizens to vote at all limited voting rights to a small subset of the adult male population. It was nonetheless wrongful for one of these nations to attack another so as to resolve some dispute between them. The nature of the wrong is in the subordination of legal order to violence.

In my discussion of colonialism, I had meant to emphasize the way in which the citizens as a collective body are passive with respect to some aspects of their affairs. In order to make that claim, then, I must provide some formulation of the sense in which the citizens are a collective body. My one hesitation about accepting Stilz's proposal is that there are places in which it seems to me to set the bar for active citizenship too high. Many examples might be used to make this point, but the boundaries between the new states formed during the period of decolonization following World War II provide a clear case. Those boundaries often reflected boundaries of colonial conquest, rather than anything central to the lived experience of the inhabitants of those areas, and sometimes the boundaries were drawn in ways that cut across cultural and economic connections between the inhabitants. For example, the division of the

Middle East through the Sykes-Picot Agreement, and the nation-states that came into being as a result, reflected the perceived interests of the British and French governments after the breakup of the Ottoman Empire. The creation of post-colonial states in Africa reflected the quasi-proprietary lines that emerged from the rush by European powers to acquire territory through the illicit means of private trading companies. Belgian law declared that Leopold, acting exclusively in a private capacity owned the company that owned the Congo.[54] Those ill-gotten borders later became the boundaries of colonies, and later of independent nations. To think of these states as entitled to independence, there is no reason to insist that, as an empirical matter, the inhabitants of each of the states that were thereby demarcated shared a united will with their co-nationals and with nobody else.[55] Some of the rulers and administrators of those states were despotic and corrupt. These are reasons for improving those conditions, and some of them might be grounds for eventual redrawing of borders, in accordance with Kant's remarks about the second preliminary article of perpetual peace, which grants permission to "postpone these laws into effect, without however losing sight of the end . . . permitted only to delay doing so, lest implementing the law prematurely counteract its very purpose."[56] But none of them are grounds for any country to invade another; each nation is entitled to be self-determining as against the others, even if it is not doing an especially good job of being internally self-determining under a clearly developed united will.

I now return to Stilz's observation that I endorse both individual and collective versions of the claim that states are entitled to political independence and territorial integrity. Like Renzo, she supposes—by considering first individual and then collective versions of explaining the wrong of violating political independence—that these two versions are separate. But on the Kantian view, they cannot be distinguished: although the state matters because the individuals do, it has rights against other states in its own right. It is the way in which those individuals enjoy their freedom together, rather than a tool or instrument that could be replaced by a different equally effective one. Violation of political independence is a wrong against both the state, that is, the citizens considered as a collective body, and also against each of them, considered separately, even though the wrong against them severally is mediated through the wrong against the collective. Rather than saying that it could only be a wrong against the collective if it was a wrong against the individuals making it up, considered apart from their membership in the collective, or that it is a wrong against the individuals only indirectly because the collective is the basic bearer of rights, the Kantian view insists that it is a wrong against the members because they have united themselves into a collective, which they must do in order to enjoy their rights. Unlike private associations, the state is a mandatory form of cooperation. The structure creates a duty to enter a civil condition while, at the same time, creating a right to be in that civil condition. The members of the civil condition are wronged by being deprived of their membership, not only by being deprived of benefits that membership brings. Invaders wrong

[54] See Stephen Press, *Rouge Empires* (Cambridge, MA: Harvard University Press, 2017).

[55] See Katrin Flikshuh's illuminating discussion in *What is Orientation in Global Thinking?* (Cambridge: Cambridge University Press, 2017).

[56] 8:347.

them, *omni et singli*, by subjecting them to force, and by making their legal order subject to force or the threat of force. An invasion makes whether they are in a rightful condition depend on the choice of outsiders. By framing the distinctive wrong in relational terms rather than in terms of causal contribution to benefits such as individual independence, autonomy, or well-being, which can be understood in the absence of legal order, the Kantian view explains the wrongfulness of invasion in terms of what one collective does to another, rather than in terms of what happens to members of one collective as a result of what another collective does.

1.E. The Peremptory Status of Cosmopolitan Right (Reply to Peter Niesen)

Peter Niesen's chapter raises a different question about the status of various claims that I defend on Kant's behalf. Niesen proposes to contextualize those claims in relation to a broader reading of the nature and ambitions of the Kantian idea of public right, and cosmopolitan right in particular. Building on his pioneering work on the positive role of cosmopolitan right—the part of public right governing relations between an individual human being and the legal orders of which that person is not a member—Niesen identifies three aspects of cosmopolitan right, which he characterizes as humanitarian, spatiotemporal, and communicative. He suggests that my account overemphasizes the spatiotemporal aspect and underemphasizes its humanitarian and communicative aspect.

In addition to this three-part articulation of cosmopolitan right, Niesen also suggests that public right always comes in two forms: provisional and peremptory. The provisional/peremptory distinction is familiar elsewhere in the *Doctrine of Right*: in the absence of authoritative and binding legal institutions, acquired private rights remain provisional. In the case of property, provisional ownership entitles you to hold something of which you are in current physical possession, in anticipation of entering a rightful condition, but, it is merely provisional rather than peremptory, because when you are physically separate from the object of your right, others are under no obligation to leave the object untouched.[57] Kant gives a clear specification of the solution to this problem in the case of private right: we must enter a public legal order, in which each is secure in what they have, making those acquired rights peremptory, that is, binding even when the owner is separated from the property.

Niesen suggests that a parallel two-stage sequence applies in the first two of the three forms of public right that Kant articulates. In the first, the case of the right of the state, every public legal order is provisionally rightful, but only becomes fully rightful if it becomes a republic. In the second, the case of the right of nations, nations must join the league of states capable of resolving their disputes peacefully, "as if before a court." Niesen reads these as transitions from an incomplete to a complete version of the relevant structure. On this reading, then, the provisional/peremptory distinction provides an explanation of the sense in which the first two "definitive"

[57] 6:256.

articles of Perpetual Peace are definitive: they mark the completion of the right of the state and the right of nations, and also specify the point at which they are fully binding. No such transition is specified in the case of cosmopolitan right; rather than talking about what an already existing thing must become, Kant specifies it negatively. In *Towards Perpetual Peace*, he writes that cosmopolitan right must be limited to "the right to visit." In the *Doctrine of Right* the same limit is expressed in terms of a prohibition on settlement of already occupied regions, leaving space only for offering oneself for commerce with others in distant lands, an offer those others are free to accept or reject. Niesen regards the absence of a characterization of the peremptory form of cosmopolitan right not as a failure on Kant's part to fully spell out the view, but as his response to the need for further development of legal and political institutions:

> Of course, foreigners will have access to domestic courts to contest any apparent violation. But the domestic jurisdiction, however well-disposed and law-abiding it may be, will still judge along the lines of what seems right and good to it. It therefore appears likely that, since Kant fails to discuss a mechanism for adjudicating conflicts under cosmopolitan right, what he is outlining in both *Toward Perpetual Peace* and the *Doctrine of Right* must be provisional cosmopolitan law, liable to institutional consolidation and transformation. (Niesen, p. 67)

The problem Niesen articulates is thus that the question of who is to decide in a conflict about cosmopolitan right, together with the question of what the relevant norm of decision will be, seems to admit of only two answers, neither of which gets entirely beyond merely provisional right. On the one hand, if the visitor makes a claim under cosmopolitan right and the receiving country adjudicates that claim, the country seems to be using its own legal norms, and so to be doing what seems good and right to it; if, on the other hand, the visitor gets to adjudicate, then the visitor seems to be able to import another legal order into the receiving country, once again, one party to an interaction gets to do what seems good and right to it. Although a public legal order is a party to either of these defective modes of adjudication, as I understand it, Niesen's suggestion is that its ability to impose resolution on the dispute is entirely unilateral; the fact that the choice is between one or another unilateral imposition does not repair that defect. The problem of public right is to find a common standpoint from which disputes can be resolved and enforcement authorized where necessary, exactly what Niesen suggests is missing in either of these arrangements. In the individual case, the standpoint is provided by a third (moral) person, entitled to judge and act on behalf of everyone, consistent with the right of each to independence (as individual human beings) from the others. In the international case, too, the standpoint is supposed to be provided by public international legal institutions charged with acting on behalf of the nations of the world considered as a collective body, once more, consistent with their independence (as public legal orders) of each other. Solving the problem of right in these cases requires ascending to a higher level that governs relations between juridical equals. The challenge, in the case of cosmopolitan right, is to specify a comparable standpoint governing relations between an individual human being and a state of which that human being is not a member.

These two strands in Niesen's argument come together in his claim that my account does not adequately distinguish between the right to travel and the right to visit, where the right to visit involves interacting with those at one's destination. Niesen plausibly proposes that enabling the right to visit might sometimes mean limiting the right to travel, and suggests that a mature peremptory law of cosmopolitan right requires political and institutional development.

I did not mean to deny that cosmopolitan right requires institutional development, or to claim that that it was complete in the absence of global institutions. I want to resist, however, making either of these claims by applying the provisional/peremptory distinction to it. That is a distinction that applies to private rights, which are merely provisional in the absence of a public legal order. Kant does not characterize the transition from provisional to peremptory private rights as one that changes the matter of those rights; instead, he argues that public right "contains no further or other duties of human beings among themselves than can be conceived in the former state; the matter of private right is the same in both."[58] The "laws of the condition of public right, accordingly, have to do only with the rightful form of their association (constitution), in view of which these laws must necessarily be conceived as public."[59] As Kant deploys it, then, the provisional/peremptory distinction does not change the subject matter of right, only the form of its lawgiving. That change in form makes everything that was merely provisional now be peremptory, that is, binding.

It is difficult to see how the provisional/peremptory contrast as Kant applies it would apply to any of the three forms of public right. The transition from a state of nature to a civil condition makes private rights binding; what further form of bindingness would be provided by introducing republican government? Republican government is superior to despotism from the standpoint of the Idea of the Original Contract, but the transition does not make some aspect of public right that was previously provisional now be peremptory, that is, binding. Kant is adamant that even despotic governments have a right to rule, and that "a people has a duty to put up with even what is held to be an unbearable abuse of supreme authority."[60] Their exercises of authority are defective, but not provisional in the way in which private rights in a state of nature are provisional.

The same point applies to the right of nations: prior to the establishment of a league of nations, each nation was entitled to be secure, that is, it had already a right to political independence and territorial integrity. Other nations were bound to respect that right, even though in the absence of a league of nations the situation is unstable. It is therefore difficult to see what is made peremptory by the transition. The same point applies to cosmopolitan right. If some way could be found to make it apply, it could not change the matter of cosmopolitan right in the way in which Niesen contends it changes.

In order to make private rights peremptory rather than provisional, a public rightful condition does not need to have a republican government; anything that minimally satisfies the postulate of public right generates rights that are peremptory rather than

58 6:306.
59 6:307.
60 6:320.

provisional. Despotism is a defective form of a rightful condition, not a provisional version of it. War, including Grotian war, that is, war as procedure, is a wrongful version of an international order. Both share with provisional right the fact that their effects (e.g., statutes enacted by a despotic government, lands acquired through patrimonial acquisition) will sometimes require reversal by their later rightful form. But the sense in which these must be tolerated, via a permissive law, is temporary, and so unlike the provisional claims of private right. With respect to cosmopolitan right, the same structure applies. Colonization is a violation of cosmopolitan right, prior to the creation of a "peaceful, even if not friendly, thoroughgoing community of all nations on the earth that can come into relations affecting one another."[61] The right of the state, the right of nations, and cosmopolitan right together form a system of peace in which the principle of freedom under law is everywhere realized. None of them can be fully realized; nothing in experience can be adequate to the idea of public right in any of its three forms. The provisional/peremptory distinction thus does not apply; as Kant points out, even the idea of republican government is an idea of reason, to which nothing encountered in experience can be adequate.[62] It is not from experience alone that we learn that every existing state is at least partially despotic, every league of nations operates partly through the private choice of its members, and any realization of cosmopolitan right must be incomplete, because the law of the place visited cannot be fully public in relation to the visitor.

On my reading, then, cosmopolitan right is the third form of public right because of the structure of public right itself. It is required for completeness, that is, to bring all interactions under law. As a result, the law under which the visitor is judged must be properly public in relation to that visitor. Sometimes that will mean that, if the claim concerns some acquired right of the visitor's, the receiving country must determine whether something falls within the scope of that right by consulting the visitor's own legal system, that is, looking to the visitor's law to determine whether the right was acquired. In so doing, the claim of private acquired right is brought under law by the receiving country's courts. That is the juridical structure of what Niesen calls the humanitarian aspect of cosmopolitan right. If you are to be able to visit without being met with hostility, the receiving country must take notice of your claims of right. That does not entitle visitors to import their own legal order into the receiving country's legal system, as in the case of the visitor who attempts to brings slaves to another country. The right to visit is just that; the person who shows up must not be met with hostility, and must be understood as a bearer of rights, both innate and acquired, and of what Kant calls "the right of human beings as such," that is, the right to be a member of a rightful condition. The visitor must therefore not be a treated as outside of all legal order. Instead, the procedures for processing the claim must treat the visitor as a temporary visitor from some other legal order, and so entitled to the security of the person and any acquired right, including property, contract, and family rights,

[61] 6:352

[62] "Every actual deed (fact) is an object in appearance (to the senses). On the other hand, what can be represented only by pure reason and must be counted among ideas, to which no object given in experience can be adequate—and a perfectly rightful constitution among human beings is of this sort—is the thing in itself." 6:372.

and to have such claims addressed through the norms of what lawyers call private international law (but which Niesen correctly describes as "international private law"). The visitor's legal claim is a claim to be treated as a member of some legal order with the concomitant protections of it. This claim is always already peremptory. It holds as against every legal order; having exited the state of nature, each legal order is already in a rightful condition, and cannot act as if the visitor remains in a state of nature in relation to it. The visitor's cosmopolitan right to visit is therefore always already peremptory as against the legal order that is visited.

Understood in this way, I now see that Niesen is correct that I added an improper condition to my characterization of cosmopolitan right when I suggested that "other legal orders do not need to permit private transactions between [foreign visitors] and their inhabitants." I misread Kant's approval of China and Japan's treatment of foreigners. That approval was directed at their response to the conduct of Europeans who had come to conquer under the cover of trade. It was not an illustration of a structuring feature of cosmopolitan right.

The peremptory nature of cosmopolitan right understood as the right to visit does not mean that further global institutions are unnecessary to properly give effect to it. Just as the right of nations is incompletely specified in terms of their independence, and the transitions to a "peaceful, if not friendly" mode of coordination, so, too, many pressing questions of international law are left unanswered by the right of nations. The right of nations does not say anything about whether there should be an international criminal court, created by the nations of the world, whether aggression should be among the crimes it punishes, or whether the punishment of war crimes should be left in the hands of the victim, or be treated is something for which states enjoy universal jurisdiction. Nor does it provide the matter of a solution to problems of climate change. The right of nations tells us only that certain institutions must be set up, and that such institutions can create no further or other duties between nations than can be conceived in their absence, and that the appropriate principle for their operation is peace, understood as "the principle of freedom under law." So, too, with additional cosmopolitan institutions, designed to secure peace.

Niesen also argues that I have understated the importance of Kant's conception of commerce in my analysis of the right to visit; I presented it as though it was simply a matter of the right to travel. I concede that he is correct about my understatement and offer two thoughts in response: first, I should have been explicit that the right to travel is predicated on the possibility of interaction; those occupying the spaces in between do not get to determine whether two people in different locations can enter into such arrangements as they see fit. The resulting norms make it possible for someone to travel with no intention to interact, but the right to travel gets its point from the right to visit. Second, wherever individuals are, whether at a final destination, or during a stopover, they can propose interactions with those they encounter. As Niesen's powerful pre-COVID-19 examples of responses to tourism in the world of cheap air travel and Airbnb illustrate, the right to visit and propose terms of interaction are not always beneficial to those who find themselves receiving visitors. If Barcelona or Venice (or the governments of Spain or Italy) wish to restrict short-term rentals to maintain housing affordability, they are entitled to do so in accordance with their more general mandate of securing a rightful condition of which all of their inhabitants are members. This is just a more specific instance of the general power of states to regulate their internal affairs and therefore attach conditions to visitors.

Niesen expresses the concern that I characterize cosmopolitan right too narrowly; conversely, I worry that his conception of it is problematically narrow in a different way. The complaint that it is violated when Venice or Barcelona are overrun with tourists who do not wish to interact with residents, or that some of the streets in European cities to which tourists flock have no local businesses, must suppose that economic transactions do not count as interaction. A Kantian formal view must have a maximally permissive conception of interaction; free beings are entitled to interact on the basis of crass commercial interests or lowbrow tastes. If you and I want to travel to some faraway place to meet to discuss Kant, but we both only want to do so with a good supply of familiar foods and proper espresso, we reveal ourselves to be narrowly parochial in our tastes. The aggregate effects of people like us and local residents who see opportunity in our tastes might undermine the ability of the local economy to maintain its distinctiveness. Our parochial tastes are a problem of virtue; their contribution to the aggregate displacement of local residents and businesses is a problem of right that the place that we propose to visit is entitled to address. But neither of these flaws is a refusal on our part to engage in *commercium* with our local hosts. As individuals, our hosts are entitled to agree or refuse to interact on those terms, although it might be imprudent for them to do so if we are rich and they are not. So, too, their governments can restrict the terms on which visitors can interact; such restriction presupposes interaction, rather than its absence.

Niesen also suggests that I overstate the spatial aspect of cosmopolitan right. He proposes a non-spatial interpretation of cosmopolitan right when he attributes a version of Grotius's *Mare Liberum* view to Kant. Rather than seeing the high seas or uninhabited deserts as nodes in the system of public right, required in order to enable the right to exclude, Niesen proposes that "For Kant, engaging with one another via ships and camels across the seas or across deserts uses an affirmatively common resource. It does not just take place under the idea of original possession in common, but makes actual use of a joint resource that mankind share." (Niesen, p. 74) This characterization seems to me to be both interpretively and systematically problematic. The idea that the high seas are the common heritage of human beings is ambiguous between a proprietary idea, which, like the Grotian idea of possession in common from which it is derived, is subject to Kant's critique in his discussion of it in the *Doctrine of Right*, and a very different idea that sees the high seas as Kant describes them, that is, as analogous to public roads on which every human being, as a member of a global public, has a right to pass and re-pass without obstruction by others. The difficulty with the Grotian reading is that the idea of original common ownership on which it depends cannot be made out; it seems to suppose that there is a group that already stands in an authority relation to individual human beings. That is, as Kant makes clear, the idea of common ownership presupposes the very concept of ownership, understood as an authority relation, which it is supposed to explain. Grotian contractors coming together and deciding to privatize property could only be entitled to do so if they already had the right kind of standing as against their current and future members.[63]

[63] 6:251. I discuss this in more detail in my response to Pinheiro Walla.

On the contrasting Kantian view, as the third form of public right, cosmopolitan right must treat the high seas as public rights of way. The natural global infrastructure is protected from unilateral acquisition because it is the condition of cosmopolitan right. I can see how my wording led Niesen to conclude that I treated public rights of way as remedial, as I articulated the point in terms of the problem of wrongful domination that would arise if all land were privatized. But it is not a remedy in the sense in which remedies ordinarily are cognized in terms of corrective justice, a way of giving someone something back that, as Kant puts it, would "only preserve what is mine undiminished."[64] It is not that anyone had (or for that matter lacked) a right to travel and communicate in the absence of public legal institutions; at most some people had the factual capacity to do so. The existence of public legal order makes the right to travel and communicate into a juridical, rather than a merely factual issue, because the legal order must find some way in which to guarantee that the exercise of private rights does not enable some to landlock others. Because a state of nature is a condition in which might makes right, its replacement by a system of freedom under law must be complete; it doesn't leave anything outside of law in a non-rightful condition of wild lawless freedom.

Niesen also raises questions about the ability of the spatial characterization of cosmopolitan right to explain its humanitarian aspect. Kant's remarks about vulnerability in the case of the person who would perish if turned away has two components. First, in the case of someone like a shipwrecked sailor, to send that person out into the high seas without a secure vessel would be to remove a human being from safety to a position of danger. That is inconsistent with the sailor's innate right, but no complex juridical analysis is required. Just as the transition from a state of nature to a rightful condition does not introduce further or other duties among human beings than those that can be conceived in the state of nature, so, too, setting up public legal institutions does not entitle one person to force another into a position in which the latter might perish, because nobody ever had a right to do that. Entry into a rightful condition gives its officials powers that no private person could enjoy, but it could not give anyone an authorization to move someone other than an attacking aggressor from a place of safety to one of danger. The shipwrecked sailor also has the right to be beyond reproach, to be presumed to have done no wrong, and so to not merit hostility upon arrival. This is not a humanitarian condition, at least if "humanitarian" is taken to refer to some Grotian idea of need. It "is not a philanthropic (ethical) principle but a principle having to do with rights,"[65] that is, a matter of the relationship between the choice of the sailor and the nation of which the sailor is not a member on the shores of which the sailor reaches a place of safety. This then raises the question of what happens when the sailor sets foot on the land, and cannot be sent away because of the peril entailed by doing so. Here the sailor is not yet a member of the rightful condition, but nonetheless subject to its laws. This is just the juridical structure within which Niesen raises his other challenge, the question of how to manage this relationship consistent with the independence of the state as a legal order and the independence of the visitor as an individual human being who is not a member of that legal order. The solution

[64] 6:271.
[65] 6:352.

is for those laws to be public in relation to that sailor, which means that the laws must be consistent with the sailor's "rights as a human being," even if the sailor is not to be made a citizen, or remains only a passive one.

The situation of the refugee is different from the sailor in one way: in the version of the manuscript that Niesen quotes I misstated this point, treating repudiation or attack by the refugee's own nation as a juridical analog of death. I now see that this fundamentally mischaracterizes the situation, representing the peril faced by the refugee as if juridically perishing is analogous to being cast out to sea. Thanks to Niesen's generous prodding, I see that the parallel must be understood differently: the difficulty faced by the refugee is that, if the refugee's state has made war upon him or her, or stripped the refugee of citizenship, then the host state needs to take the refugee in because otherwise the refugee would be in a state of nature in relation to it. As Kant explains in a footnote to *Perpetual Peace*, "For if only one of these were in a relation of physically affecting another and were yet in a state of nature, the condition of war would be bound up with this, and the aim here is just to be freed from it."[66] Cosmopolitan right requires that all interactions be brought under law, which means that in the case of the visitor seeking to interact with residents, it must satisfy the conditions outlined before. The situation of the person who has nowhere else to go juridically is perilous because, as a citizen of no nation, the refugee is in a state of nature. The peril is importantly different from the natural peril faced by the shipwrecked sailor, but the solution is the same: the host country must take the refugee in, treating the refugee as it would treat one of its own citizens. That requires that the refugee receives the full protection of the laws. The refugee is entitled to reside, perhaps only on a temporary basis, until situations and circumstances improve in the place the refugee fled (though perhaps not). While there, however, the refugee is fully subject to the host country's laws, and so gets the protections that the shipwrecked sailor would get. Although this response is sufficient to address the humanitarian concerns Niesen articulates, it is not grounded in humanitarianism. Like the case of the shipwrecked sailor, the protection that must be provided to the refugee is a matter of right, not philanthropy.

Niesen's characterization of cosmopolitan right in terms of the provisional/peremptory distinction opens the possibility that a peremptory division of property and territorial jurisdiction would differ from the current provisional one, and thereby "compensate for the vulnerabilities" of those who may have "lost out in the global race for appropriation." (Niesen p. 75) Much could be said in favor of this proposal, but I do not think that it can be brought within the reach of a Kantian conception of cosmopolitan right. Kant nowhere suggests that borders and boundaries are provisional; he says instead that the reversal of wrongful acquisitions—his examples are the acquisition of a state by gift, purchase or inheritance—is required, but can be postponed. Kant's wording emphasizes that this directive is addressed to the ruler; he does not characterize it as a task for the kind of global institutions he describes in the second definitive article of Perpetual Peace, or for any future institutions of cosmopolitan right.

[66] 8:349.

This does not rule out the possibility of global institutions seeking to address the situation of those who have lost out in these non-rightful transactions. As I explain in my response to Olsthoorn, it does, however, preclude any possibility of those who lost out using force to claim what they take to be their global fair shares. Both states and global institutions are under duties of right—which no one is in a position to enforce against them—to repair not only past effects of wrongful colonial conquest, but also to address dire circumstances even if they arose without any wrongdoing, such as the dependency on rich by poor people or countries. This is not, however, because, on the Grotian model that Niesen entertains, everything is really held in common and so the division must not, as Grotius puts it, "depart from equity" more than is necessary. Instead, it is because for human beings to live together under laws, all interactions must be brought into ever greater conformity with the requirements of freedom under law.

1.F. The Peremptory Status of Territorial Rights (Reply to Alice Pinheiro Walla)

Building on her important earlier work about territorial rights, Alice Pinheiro Walla criticizes a different aspect of my account of territory, my claim that separate legal orders are already *sui iuris* as against others prior to the creation of a systematically organized global legal order. She objects on two distinct grounds: first, she argues that the Kantian argument presupposes territorial rights that are not based on any concept of specifically legal ordering; and second, that the existence of a legal order generates claims only against other states, and not against individual human beings. She therefore concludes that my argument that each legal order is *sui iuris* fails to engage with one of states' most salient features, their claim to control their borders, not only as against other states, but also as against individual human beings. In place of the account she criticizes, she proposes an alternative model of an international rule of law, founded in common possession of the earth's surface, which would make bounded territories legally irrelevant. Pinheiro Walla concludes that bounded territories are a temporary expedient in the contemporary world because human institutions have not progressed to the point of such an international order; she characterizes the hard borders between states as we now find them as a sign that states are still in a state of nature in relation to each other. Like Niesen, she supposes that a truly rightful international condition would fundamentally change familiar norms of the global order.

Before taking up the specifics of her argument, I should clarify something about the ambitions of mine. I agree that there is a duty incumbent on states to implement an international rule of law, and to secure their rights in a way that ensures that the rights of every human being are secure as against everyone else. My main concern in *Kant and the Law of War* was with how things stand between a plurality of separate states when that form of order is not in place. Following through on Kant's analogy between the state of nature between individual human beings and between nations, I suppose that how things stand as between nations conceived as separate legal orders is not changed by the creation of a global legal order. Each is entitled, as they were prior to its creation, to be independent of the others. I had not intended to suggest that international law does not, much less that it should not, limit the ways in which states can act, internally or externally. My main focus was on the way in which global legal institutions give effect to requirements of peace that are already in place. The Kantian account of international legal order proceeds in two steps. The first of these requires creating an analog of what Kant characterizes as a society "compatible with rights," that is, bringing the international state of nature up to the level of a condition of anarchy, in which there is freedom with neither force nor law, which is, as Kant explains in the *Anthropology*, a state of nature as an idea of reason.[67] That is the point of the preliminary articles of perpetual peace. The second preliminary article, which Pinheiro Walla quotes, emphasizes that the possibility of peace depends on the state's territory not standing to it as an external object of choice. As I understand it, the point of that prohibition is to remove one of principal grounds of war, that is, as a way of resolving legal disputes between sovereigns by depriving them of their subject matter. If sovereigns have private, patrimonial rights with respect to the states and territories over which they rule, then they can litigate such things as purchases, gifts, and inheritances. For the regular war writers such as Grotius, Pufendorf, and Vattel, war just was this type of litigation; unsurprisingly they included these titles of private right as possible grounds of war. The second preliminary article forbids these means of acquisition and so forbids litigating them through what Pufendorf calls the "arbitration of Mars."[68] The third preliminary article removes the means through which wars are conducted by eliminating standing armies; the fourth removes the incentive to war by precluding the possibility of investing in a way that war is required to turn a profit. The first, fifth, and sixth preliminary articles are fundamentally different: they constitute a condition in which nations are independent of each other by having binding bilateral peace treaties, prohibiting interference in each other's internal affairs, and prohibiting ways of conducting a war that make peaceful interaction impossible. Taken together, these preliminary articles are insufficient to create an international legal order; they describe a realm in which each legal order is independent of the others, but with no alternative to war as a mode of dispute resolution.

The second stage in the Kantian view involves the creation of an international order in which conflict can be resolved peacefully. As I will explain in more detail in my response to Flikschuh, the definitive articles of perpetual peace correspond to the three forms of public right that Kant identifies in the opening section of Public Right in the *Doctrine of Right*. They characterize what it would be for all human interactions to be fully governed by the principle of freedom under law. The establishment of a global just legal order leaves the *inter se* obligations of individual states as against each other in place.

Pinheiro Walla is correct that this model of the right of nations is silent on many important questions about the powers of states, most notably their ability to engage in international trade or control immigration. Those are questions for cosmopolitan right, the third title of public right, which gives some explanation of how, when, and

[67] Anth, 7:331.

[68] Samuel Freher von Pufendorf, *Two Books of the Elements of Universal Jurisprudence*, Translated by William Abbott Oldfather, 1931, Revised by Thomas Behme (Indianapolis: Liberty Fund, 2009), p. 57.

why states are entitled to control their borders and when they are not permitted to do so, as well as explaining why part of what they must work toward is a rules-based international order, as well as a rules-based international migration regime. But their entitlement to have trade policies, as well as their entitlement to, within limits, control their borders, does not await the full realization of a system of cosmopolitan right (which is, like the idea of republican government, an idea of reason "to which no object given in experience can be adequate"[69]). As I understand it, the point of each of the three forms of public right is to bring human interactions under law. Kant's argument is that right must be present in relations between individual human beings and the public authority of which they are members, between separate states understood as public authorities, and between individual human beings and public authorities of which they are not members. This third form—cosmopolitan right—requires that any enforcement in relation to a visitor be brought under law, by making it properly public in relation to the visitor. When the visitor is a member of some rightful condition, publicity requires acknowledging the visitor's right to propose terms of interaction, which necessarily includes security of person, and of acquired rights, so that those rights are secure for the visitor. Although, as I suggest in my response to Niesen, the idea of *commercium* includes more than economic transactions, it necessarily includes them also. So the visitor cannot lose title to goods merely by visiting. Provided that the right to visit is public in relation to the visitor in this way, the visitor can have nothing more than the mere right to visit, with no right to settle. So long as the host country and the visitor stand in the right kind of relation mediated by public law, any enforcement of rights is properly public, and so subordinates force to freedom under law. The visitor can therefore not stand on any right in claiming to be entitled to settle, for example, based on some idea that (as suggested by the Salamanca scholastics) the earth should be divided up in a way that will enable each person to satisfy the divinely ordained obligation to preserve themselves.

With that preliminary framing of the ambitions of my discussion in place, I now turn to her more specific arguments.

Group Acquisition. The first of these suggests that a legal order is not required for territorial rights; rejecting my claim that European explorers encountering the indigenous inhabitants of regions that they visited were under an obligation to treat them as already in some kind of civil condition, Pinheiro Walla suggests that "the first acquisition of a *group* is itself *sufficient* for imposing a duty on others to respect the occupation of land." (Pinheiro Walla p. 83)

I reject the claim that first acquisition by a group is sufficient to impose a duty on others because of the difficulty with the idea of acquisition by a group. For Kant, both the concept of acquisition and the concept of a group as an agent are rational practical concepts, not empirical ones, even though they must be brought to bear on things encountered in space and time. As I explain in my response to Forst, for Kant, a state is an idea of reason to which no empirical thing can be adequate. Instead, Kant considers certain structural features—in particular, acting under authority—to determine what counts as the act of a group. To borrow Hobbes's language, a group acts if

[69] 6:370.

someone "personates" the group, such that that person's acts are the acts of the group. That is why Kant insists that a revolutionary cannot speak and act on behalf of "the people" of a group, because the group in question exists only through its public legal institutions. The revolutionary cannot be the voice of a public institution, and in opposing and seeking to overthrow public institutions, the revolutionary repudiates the only possible form of a public institution. Within the Kantian setup, then, there is a problem with talking about a group having acquired land without it being a civil condition of some sort. Taking physical possession of something only counts as acquiring it in a way that has consequences for rights if it is brought under an omnilateral general will, that is, understood through the juridical idea of a people acting together. In *Perpetual Peace* Kant describes the constitution as "the act of the general will by which a multitude becomes a people,"[70] suggesting that a people can only be thought of as such through its legal institutions. In the *Doctrine of Right*, he writes that "The act by which a people forms itself into a state is the *original contract*."[71] On either account, the key issue is whether, in the absence of something like a constitution, a people can be thought to have arranged uses of land in a way that binds its members or outsiders.

The difficulty, then, is how to articulate an idea of group acquisition that is not the act of a juridically constituted people. The passage Pinheiro Walla quotes refers to interaction with "a people" (*Volk*)[72] as do others, using the same term that Kant uses to describe "a multitude of human beings." Merely considered as a multitude, however, it is difficult to see how they could acquire anything; at most they could acquire things severally, but, so understood, their acquisition would be only provisional, even *inter se*. When Kant considers collective acquisition, discussing the example of Mongolia where "all the land belongs to the people, the use of it belongs to each individual, so that anyone can leave his pack lying on it or recover possession of his horse if it runs away, since it is his,"[73] he represents collective acquisition as marking the transition from provisional to peremptory property rights in movable things.

Kant's other remarks about peoples who have not organized themselves into recognizably legal orders might suggest that he needs some other account of collective acquisition to explain the wrong of colonialism. As I understand Kant's general mode of argument, the question does not concern an objective factual (or third-person normative) state of affairs. The interpersonal morality that is the subject of the *Doctrine of Right* is just that—interpersonal. Its directives do not tell you how to orient yourself toward states of affairs, or toward yourself, as some interpretations of Kant's ethics seem to contend. It is concerned with how you orient yourself toward others, both in particular, (the subject matter of, among other things, the norms of private right), and with respect to the possibility of future orientation (the subject matter of, among other things, the *in bello* norms). The only concepts available to the visitor interacting with a multitude of human beings encountered in "distant parts of the world"[74] are these same interpersonal concepts, which are necessarily mediated by the fact that

[70] 8:352, 8:371.
[71] 6:315.
[72] 6:266, 6:353.
[73] 6:265.
[74] 8:858.

both the visitor and those visited are members of their respective public legal orders. The visitor's interactions must be governed by the possibility of legal relations with the inhabitants; the visitor cannot exit legality and enter a state of nature for the purposes of the visit. Such visitors purport to bring with them rights secured by their own legal order, and to be entitled to its protection. So the question of how they interact with the people who are already there must be governed by the fact that they are arriving as members of an established legal order; they cannot suppose that they have arrived in inhabited places "belonging to no one." But the only way to avoid counting "the inhabitants as nothing"[75] is to allow them to remain in the condition they are in, and to treat their occupation of the territory as rightful, even though the visitors might have no way of conceiving what that rightful relation is, except by some kind of analogy. That is why Kant says that "it follows that all nations stand originally in a community of land, though not of rightful community of possession (*communio*) and so of use of it, or of property in it."[76] That community is not a matter of shared governance but rather of possible physical interaction—the same concept used in explaining community of substances in the *Critique of Pure Reason*. The visitor must conceive of the "people that has already settled in the region," under the category of substance, that is, as independent. Nothing more (or less) could delimit the rights of the visitor; all that the visitor can do is propose interactions, even if the visitor has no straightforward way of determining who is in charge.

The question of how travelers arriving on their own initiative on a populated region of the Earth's surface should treat its inhabitants is therefore a question of how things stand between those inhabitants and the traveler's home legal order. The question is not asked from a third-person perspective as to whether those inhabitants are in a state of nature or rightful condition (how things stand between those human beings and their own institutions and practices) but rather whether the traveler's home legal order must interact with those human beings considered as a collective body (a people, which Kant defines as a multitude of human beings). The question of whether they meet relevant criteria cannot be a question for the visitor's legal order, because no legal order is entitled to stand in a supervisory relation to another.

Kant makes the formal version of the same point in his discussion of how things stand between a person and the legal order in which that person lives in his response to Bouterwek. Bouterwek challenged Kant's claim that there can be no right to revolution, questioning how this could be established a priori without investigating the specific credentials of the legal order in which one finds oneself. Kant's reiterates his argument that a lawful order is an idea of reason, to which nothing given in experience can be adequate. In order to satisfy yourself that you are under no obligation, "If then a people united by laws under an authority exists, it is given as an object of experience in conformity with the idea of the unity of a people as such under a powerful supreme will, though it is indeed given only in appearance, that is, a rightful constitution in the general sense of the term exists."[77] That is, the fact that something can be understood as the people acting together is sufficient to put its members under an

[75] 8:358.
[76] 6:352.
[77] 6:372.

obligation of right. In your own case, you are not entitled to conclude from the defects in the legal order in which you find yourself that you are in a state of nature and thus entitled to disregard it.

In order to move from the relation between a legal order and its members to the case of someone encountering a social formation, the necessary intermediate step is to think about how things stand between separate legal orders. Each is required to presume the other to be a legal order. As I noted briefly in my response to Forst, the situation of individual human beings provides a parallel: right demands that you interact with others in certain ways, and limits the use of force to those uses consistent with the status of other human beings as rightholders. This might be thought to raise a question of the criteria another being must satisfy in order to place you under an obligation. In order to be a rightholder as against you, that being does not need to first prove to you that it is transcendentally free, as judged from a third-person or impersonal point of view. Nor does that being face the burden of proving to you that you would somehow be contradicting yourself if you mistreated it. The Kantian theory must repudiate any such line of inquiry, because nothing encountered in experience could show someone else to be a free being. The unconditional duty to respect the rights of others could not be conditional on the accumulation of sufficient empirical evidence. The point is not merely that we do not have enough evidence, or that people are likely to answer such questions in self-serving ways; the point is rather that questions of juridical standing are not empirical. Nor is the Kantian claim that each of us extrapolates from our own case to the case of others. Each us has no empirical access to our own freedom either. Instead, we are under an obligation to treat human beings as free, and the class of cases to which that obligation applies is just the class of beings with human form.

The situation of travelers arriving in distant lands is parallel: the fact that there is a social formation is a law for their legal order and therefore for them; they cannot uproot or destroy it because it falls under the principle of right, even if they do not judge its juridical capacities to be properly developed. The Kantian argument that in the absence of a shared public lawful condition, individual human beings "each has its own right to do what seems right and good to it and not to be dependent upon another's opinion about this,"[78] does not give each license to deal with others contrary to the principles of right. Although neither individuals nor states need to defer to the judgments of others about what has transpired between them, that does not entitle them to refuse to accept that another human being has rights, or that another social order is a rightful condition. Kant's scathing remark about the "veil of injustice" in scholastic apologetics for colonial conquest contemplate a situation in which human beings "hold out no prospect of a civil union"[79] noting that even in such a condition, concepts of right preclude wrongful means.

The adequacy of the way in which the people has organized itself is therefore not a question for the visitor or the visitor's legal order; as against them, that mode of organization must be regarded as rightful. The outsider lacks standing to replace law, or what it must treat as law, with force. The visitor is not entitled to conclude that the

[78] 6:312.
[79] 6:266.

distant land on which another people "has already settled"[80] is in a state of nature, and so is not entitled to force its inhabitants into any form of civil union (let alone a defective one such as a colony) or to treat their claims with respect to land or movable objects as anything other than peremptory.

Legal Order without Territory. Pinheiro Walla introduces the intriguing possibility of a public legal order that is not territorially bounded. This possibility can be understood in two very different ways. She gives the example of German reunification; on that model, previously separate legal orders have merged into one. I certainly agree that territorial differentiation of legal orders could be consistent with a system in which there is full mobility of people and goods. That possibility does not, however, show that such full mobility is required by cosmopolitan right, or that the limitation of cosmopolitan right to the right to visit is an incomplete or provisional version of it.

Rather than supposing that all three forms of public right must finally merge into a single unitary state with open borders, Kant supposes that the principle of freedom under law can be realized in three different ways. His claim that if "the principle of freedom under law is lacking in any one of these three possible forms of rightful condition, the framework of all the others is unavoidably undermined and must finally collapse"[81] does not say that rights are merely provisional until a unitary cosmopolitan order is achieved, but rather that anything that compromises any one of them—through despotism, war, or claiming more than or denying the right to visit—compromises the others. Each of these is a way in which interactions between human beings can fail to be properly public; the lack of publicity in any of these three forms undermines it in the others. The visitor who claims more than the right to visit treats the legal order that is visited as counting for nothing; the legal order that denies the right to visit treats the stranger as in a state of nature in relation to it. When Kant writes that "acquisition will always remain only provisional unless this contract extends to the entire human race,"[82] he is not claiming that everyone must stand in an unmediated relation to a unitary global legal order, but rather that all human interactions must be mediated by the form of public law appropriate to the relations they instantiate.

Pinheiro Walla fills out the idea of non-territorial legality by imagining a division of the world into a system of merely personal rather than territorial jurisdiction: what you have, and what rights you have depends on whether you were in the red, blue, or green group, which has no systematic non-accidental relation to where you are located. In such a world, the rules that govern you are a function of your group membership; in such a system "The relations between citizens of different public legal orders are regulated by international rules and principles specifying which jurisdiction has the right to coerce whom and in which case" (Pinheiro Walla p. 85). Membership in one of the colors restricts not just how members of that group can treat the members, but also the ways in which others can. The red norms govern the ways in which blues and greens can treat reds. Suppose, for example, that red has a rule that protects its members against a certain level of noise. If a red-member lives near greens and

[80] 6:353.
[81] 6:311.
[82] 6:266.

blues, red-member can red-restrict the amount of the noise that they make. Again, if a green-member purchases land from a member of blue, green law now governs that land. Green law thereby governs any nearby reds and blues. Further rules govern potential conflicts, just as choice-of-law rules determine which law is applicable when, for example, a resident of one legal order negligently injures a member of another on the territory of a third.

I have two concerns about this proposal. First, I am not sure that it preserves the idea of a plurality of separate legal orders, rather than simply merging into an order in which everyone stands in an unmediated relation, and only preserving some factual characteristic that will in certain cases lead to differential treatment. In order for the rules and principles that determine who is entitled to coerce on which occasions to be consistent with the freedom of everyone, they would have to create courts of general, rather than membership-based jurisdiction, binding on everyone, and before which any human being can bring any other to resolve an alleged dispute. The source of those rules and the courts that enforce them would only be consistent with freedom under law if they stand in an unmediated relation to every human being on earth. This would then leave membership in red, blue, or green only a vestigial role, which would, as Kant notes, "contradict the presupposition"[83] of the right of nations; they would not be separate or independent legal orders, but dependent and subordinate ones.

Second, even if the first problem could be solved, it would not show that territorial rights are only an expedient, that is, "required when there is *lawlessness* between states, i.e., in the absence of an international legal order that regulates the relations between two previously isolated legal orders," or that "the *rationale* for territorial rights is not the protection of a public legal order but the fact that possession of land is still unregulated and therefore can be *contested* in inter-state relations" (Pinheiro Walla p. 86). Instead, the primary individuation of legal orders is territorial, because they govern outer freedom, the interactions of spatially embodied beings who always interact somewhere. Like Niesen's suggestion that cosmopolitan right has different provisional and peremptory forms, Pinheiro Walla's conception of the final form of the right of nations supposes that territorial rights are provisional not in the sense of not being fully binding absent proper international law, but in the stronger sense that they would have a different *content* in a peremptory system. As I explain in responding to Niesen, this is a fundamentally different distinction than the one that Kant draws in the case of private rights. Private rights become peremptory when public legal institutions guarantee noumenal as well as empirical possession; you do not have a property right in the absence of public legal institutions. That is why human beings in a condition of externally lawless freedom "do one another no wrong at all when they feud among themselves" in the absence of public law.[84]

Further, it is difficult to know what could correspond to merely empirical possession in the case of the right of nations. Each nation, understood as a legal order, is not in factual possession of its territory; again, a legal order exists as an idea of reason, applied to an empirically identified particular. As such, rightful possession explains the attribution of factual possession: a nation possesses territory by exercising authority

[83] 8:354.
[84] 6:307.

over it. Once again, non-state peoples must be understood to be acting collectively and so as exercising authority, even if that exercise lacks republican form.[85]

Temporal Asymmetry. Pinheiro Walla shares the concern raised by Forst and Niesen that the Kantian view, as I articulate it, attaches too much weight to the results of past wrongful transactions. Although she does not suppose that force should be used to reverse them, she proposes a different rationale, writing

> For pragmatic reasons, we must assume that a territory was acquired in good faith. Even if territory was acquired by violent means, its legitimacy must not be questioned if appropriation of the territory was ultimately *successful*, lest we allow instability to threaten an emerging and fragile rule of law. However, *future* acquisition must be rightful and certain ways of acquiring territory must be ruled out in the future. This principle should not be applied backwards since it could endanger the establishment of a rule of law. (Pinheiro Walla p. 87).

As I explain in my response to Forst, I agree that there is a temporal asymmetry. Pinheiro Walla sees this as an empirical requirement. I think that Kant understands it formally, just as he understands the wrong of revolution formally. In his discussion of revolution, he argues that how things stand between a legal order and its members does not depend on it having come about in the right way. The same point applies to its external relations, that is, how things stand as against other legal orders. International legal ordering must begin with peace, and peace is still peace even when it is made by enemies (as it normally is); claims as between nations are legitimate and binding because peace is the precondition of peaceful, that is, procedural, negotiation. When Kant remarks that there can be no procedure in a state of nature,[86] he is not saying that war is therefore acceptable, but rather that the possibility of peace requires accepting that the past is settled, that is, that no past acquisition provides a ground for war. He makes the same point in *Towards Perpetual Peace* when he remarks "For the prohibition here concerns only the way of acquiring, which from now on shall not hold, but not the status of possession."[87]

Possession in common. Pinheiro Walla argues that

> imposing an obligation on all others requires assuming *that we are already in common possession of that thing with all others.* (6:352) Unless we assume that we *originally* possess all land in common (*communio fundi originaria*), we cannot conceive the idea of someone being *wronged* by others who interfered with one's possession of land.... As Kant rightly recognised, we must already stand in a relevant legal relation with all others in order to be able to speak about rights to external things, and consequently about the violation of these rights (*Läsion*). We must therefore postulate

[85] The identification of collective as acting presupposes some differentiation between its executive and legislative functions, because the execution must be in accordance with a rule that is in principle specifiable apart from acting on it. That is why Kant remarks in *Perpetual Peace* that Athenian-style democracy with plenary powers on all matters is "strictly speaking without form." 8:352

[86] *Naturrecht Feyerabend* 27:1377.

[87] 8:347.

an original juridical community in regard to all external things that can become the matter of acquisition as a normative starting point. (Pinheiro Walla p. 91).

Kant's discusses common possession in relation to the original acquisition of land, and again with respect to cosmopolitan right. I had read the *communio fundi originaria* passage as treating possession in common as a potential obstacle to acquisition rather than as its basis, an obstacle that can only be overcome through entry into a civil condition, because acquisition is a unilateral act that puts others under a new obligation with respect to the object acquired. Prior to acquisition, individuals were at liberty to use or interfere with the object; Kant writes that "Original possession in common is, rather, a practical rational concept which contains a priori the principle in accordance with which alone people can use a place on the earth in accordance with principles of right."[88] That is, prior to acquisition, no one does wrong with respect to land unless they displace someone from where that person already is. Acquisition changes that, and so must go beyond common possession, rather than being grounded in it.[89] Indeed, the original community of land enters as the basis not of the rational title but of the empirical title of acquisition (what Kant refers to as *quid facti* rather than *quid juris* in the *Critique of Pure Reason*). The common possession mentioned in Kant's discussion of cosmopolitan right raises a different question, not about individuals but rather concerning the relation between nations that "nature has enclosed ... within determinate limits." The first form of community, then, is relevant to the way in which one private person can acquire a right to exclude as against others, and is answered in terms of entry into a civil condition with those others; the second to the way in which relations between nations, and their disjunctive (that is, exclusive and exhaustive) possession of the habitable portions of the earth, bears on the situation of those who leave their own rightful condition and find themselves elsewhere. As an element of public right, this second form of community is not presupposed by original acquisition. Instead, a nation, understood as a legal order, comes into existence by exercising jurisdiction somewhere, that is, on some portion of the earth's surface, over the human beings who are present there. To acquire any particular piece of land requires setting up a legal order on the region of the earth's surface that include that piece of land. The manner in which nations possess the earth in common "is not a rightful community of possession and so use of it or property in it; instead they stand in the community of possible physical interactions (*commercium*)."[90] The possibility of interaction requires conceiving them as members of a whole, each member of which is free to offer to engage with others. This does not require an antecedent overarching legal order. Instead, a legal order brings law to some region of the earth's surface, and cosmopolitan right governs the relation between that legal order and visitors from other legal orders.

Involuntary Migrants. I certainly agree with Pinheiro Walla that involuntary interactions are different from voluntary ones, that, as she puts it, "While host countries maintain the right to refuse *voluntary* interactions under the condition that they do

[88] 6:262.
[89] 6:267.
[90] 6:352.

not treat peaceful visitors with hostility, they have no right to refuse those whose interaction is *involuntary*" (Pinheiro Walla section V). We disagree, however, about the grounds for this conclusion and its implications. Cosmopolitan right emphasizes the right to visit, where visiting and offering to engage in *commercium* is paradigmatically voluntary. The kind of community in which nations stand does not explain why involuntary migrants must be protected. Those who would otherwise perish—shipwrecked sailors or today's climate refugees—are a case of involuntary interactions, but the way to explain their claims is to focus on the fact that they are exercising something different from the right to visit. If they have a right to settle, it is not because they are claiming their fair share of a common possession. It is instead, as I said in response to Niesen, because no one can ever rightfully be removed from a place of safety to a place of danger. I think that political refugees are in yet another category; they are not demanding their fair share of the divided earth, but rather to be treated as members of some rightful condition. If they have been deprived of their citizenship or face perils if they return, the claim they make as citizens of the world is emphatically not a claim to continue being a citizen of the world, but rather to become a citizen of someplace in particular. Their demand is that the place of safety that they reach take them in not as someone in a state of nature, but as a member of some rightful condition.

More generally, the fundamental difficulty with attempts to ground the rights of migrants in common ownership of the earth is that it leaves out the fundamental public-legal aspect of citizenship. The person who is entitled to be taken in is not exercising a right that can even be thought in a state of nature; it is a cosmopolitan right, a right that all human interactions be brought under public law. So it seems to me that the right way to understand their claim to be admitted is precisely by understanding it as different from those of voluntary interactions, a case in which the normal right of a nation to control its borders cannot be made to apply, rather than as supposing that the case of the entitlement to control borders is just a defective and provisional form of legal ordering, which never really applied anyway. The Kantian picture captures the moral and legal structure of the situation. The person who claims to be entitled to ignore borders, or those who say that the world should have entirely open borders because we possess it in common, represents legal and political institutions as having no moral significance of their own. The Kantian picture is different. When you really have no place else to go, you can stand on your right as a citizen of the world, your right to have your interactions be brought under the principle of ordered freedom limited by law.

1.G. Prevention and Remedy (Response to Alon Harel)

Where Benbaji, Renzo, and Stilz all seek to define the grounds of political independence and territorial integrity through other accounts of legitimacy, Alon Harel's chapter is prepared to accept the Kantian conception of the nature and basis of that right, but questions my claim that it leads to a limitation of the grounds of war to national defense. Building on his important earlier work on why law matters morally, Harel argues that the same grounds that generate a right of national defense also give rise to a right to engage in remedial wars. As he puts it, "there is no difference between the case in which the right of a nation to independence is being presently compromised by aggression and the case in which it is being presently compromised by failing to undo the products of past aggression" (Harel p. 100).

Harel's argument is the mirror image of one made centuries ago by the Jesuit theologian Francisco Suárez. Suárez, too, saw a symmetry between defensive and remedial wars, supposing that both were designed to uphold rights. They differ in one important respect: Suárez thought that remedial wars were the basic case, and regarded defensive wars as preemptive versions of remedial ones; Harel moves in the opposite direction, trying to assimilate remedial wars to defensive ones.

Before saying more about the difference between protection and remedy, let me first say something about the relevant conception of an independent legal order. Harel understands the relations between political independence and territorial integrity in what seems to be a broadly proprietary sense. Successful aggression violates the victim's independence by depriving it of something that it already had, that is, its political independence or territorial integrity. Harel reasons that the victim's entitlement to defend the territory against aggression rests on the importance of that territory to its independence. He therefore argues that the theory of national defense must address the question of whether a nation may engage in a "war of self-defense only when the anticipated aggression endangers its independence" (Harel p. 99). On this understanding, a nation would not be entitled to defend itself if the territory in dispute was not important to its function, such as, for example, the islands between Denmark and Sweden that were the subject matter of a series of wars in the eighteenth century. The Kantian view does not conceptualize territory in terms of its usefulness to the nation's ability to function. Instead, it views political independence and territorial integrity entirely relationally; they come as a package because the nation's territory is where it exercises legal authority. Harel, by contrast, conceptualizes territorial integrity non-relationally; on his understanding, a nation's territory is something in which the state has an interest important enough to justify it in using force to protect it. According to this view, if a nation will remain independent once the territory is seized (as is sometimes said about the situation of Finland in its Winter War against the Soviet Union, in which the Soviet Union attacked only after Finland declined its offer of an exchange of comparable area of largely unoccupied and undeveloped territory) then defensive force would not be warranted. Conversely, if a state needs the territory, its need for it will not diminish once it is lost; if its need to take it back is pressing enough, it can use force to reclaim it. On Harel's non-relational account, then, the symmetry between remedy and defense is direct: losing the territory would not (at least in the short to medium term) diminish the need the state has for its territory, so would persist as an interest in reclaiming it.

This argument to establish symmetry between defending and reclaiming territory has additional consequences, some of which are troubling. Rather than focusing on how things stand between aggressor and defender, it makes a version of Grotius's claim that the right to self-defense applies whenever self-preservation is at issue, apart from any wrong by the aggressor. Although Harel's example concerns taking back wrongfully taken territory, its focus on the state's non-relational need for it makes no essential reference to the fact that it was taken wrongfully. Instead, the account he proposes focuses only on how badly the state needs the territory. This way of setting up the issue has the surprising consequence that not only is there a symmetry between defending and reclaiming, but, cast in the same usefulness-based terms, there appears to be a symmetry between both of these and acquiring territory that a nation has never had before, as well as things other than territory. It thus extends not just to remedial wars but also to what Olsthoorn, in his chapter, identifies as resource wars. If the interest that one state has in having something that another state currently has is sufficient grounds for going to war, how that other state came to have that thing is irrelevant. On Harel's non-relational account, then, the weight of the nation's interest appears to be sufficient to determine whether it can demand or seize another nation's territory, for example, to serve as a buffer against an anticipated invasion, or indeed to use as a staging ground to invade another nation against which it has a grievance. That is, it would be entitled to attack a third party in order to position itself it to defend against an anticipated attack, or to acquire a staging ground from which to correct a past wrong. So, for example, on this view, if the Soviet Union's factual claims were correct, on Harel's reasoning it would have been entitled to seize the Finnish territory it required to defend itself against the anticipated German attack. The basic difficulty, then, is that the non-relational account appears to render the concept of interstate wrongdoing irrelevant. Indeed, it also raises the possibility that two nations—Finland and the Soviet Union in the present example—could both need the territory, and so both be entitled to use force, one to capture and the other to defend, or, later, one to reclaim and the other to defend.

The Kantian relational conception rejects such grounds of war because it rejects the idea that a state's right to territorial integrity depends on the non-relational interest the state has in its territory. Instead, it understands the morality of war exclusively in terms of the difference between law and mere force. The distinctive wrong of aggression is not that something bad happens or that the state suffers a significant setback to an important interest in preserving itself, governing itself or providing for its inhabitants. Instead, the distinctive wrong of war is that law is substituted for force. On this relational conception, then, the existing legal order, consolidated and effective within the territory, is the holder of a right against invasion by any other legal orders. Once a war has ended and a peace treaty concluded, each of the former belligerents is entitled to political independence, which is inseparable from an entitlement to territorial integrity with respect to its new borders.

Kant was vividly aware of the cycle of violence generated by the claims of remedial war, but his objection was not exhausted by the familiar thought that, as a matter of fact, claims of private right with respect to states, such as those that he rejects in the second preliminary article of perpetual peace—purchase, gift, inheritance—tend to lead to disputes that will never be resolved to the satisfaction of both parties. Instead, his argument was that it is a mistake to look to whether a peace treaty led to the correct result. That a legal order is in place is sufficient for others to have an obligation of right against interfering with it. The consequence of this is to make the past irrelevant. Harel concedes that once things have changed enough it may be unjust to enter into a remedial war, but the changes in circumstances are not the kind of factual changes he seems to have in mind. The juridical change of a new legal order now ruling the territory is different from the kinds of factual changes and expectations on which he focuses. A legal order that has consolidated its position cannot be displaced by force, because law cannot rightfully be displaced by force.

Again, that does not mean that it is always clear whether this has happened, only that the point at which it happens is the point at which war ceases to be defensive and becomes remedial, and, as such, is prohibited. Several of Harel's examples—his description of the situation between France and Germany, and Egypt and Israel—turn on the unclarity of the situation. The Kantian account does not resolve questions about particulars; its aim is rather to show that everything turns on how things stand between the parties, and that nobody could ever have standing to displace a legal order from the territory over which it exercised jurisdiction. One consequence of this is that there is sometimes unavoidable indeterminacy to the question of the individuation of wars and whether a war has ended (the claim by some US government lawyers that the 2003 Iraq war was the continuation of the 1990–91 Gulf War reveals the limits of appeals to this kind of ambiguity). There may also be ambiguity as to whose legal order is in place in the period immediately following conquest or even the end of the war. My Kantian claim is that those ambiguities are different than the question of whether there is an authorization, whether a legal order's entitlement to exercise jurisdiction over a territory survives its replacement with the exercise of legal authority by a different legal order.

I should have been clearer about the requirement for public legal institutions to authorize remedial force. Harel rightly points out that questions of ongoing wrongdoing are no less open to disagreement than are questions about past wrongdoing. Indeed, in the interpersonal case, the past will sometimes be less controversial than the current situation. But my claim was not that people are prohibited from enforcing controversial judgments; it was that the use of punitive, remedial, or administrative force presupposes the existence of a common public institutional standpoint. As such, it requires public lawmaking and public adjudicative institutions governing any use of force. Even in the limited instances in which the legal order permits self-help—taking back my stolen bicycle when I see it leaned against the wall somewhere—the permissibility of my doing so is subject to law and can be tried and tested in court. Within a public legal order, defensive force is also subject to public lawmaking and adjudication. In the absence of a public legal order, however, it can be self-authorized because it is consistent with the independence of the separate parties in a way in which none of punitive, remedial, or administrative use of force can be. In resisting aggression, you are not presuming another to be a wrongdoer; instead, any use of force by another can be taken as unauthorized, and inconsistent with your right to be beyond reproach, that is, your right that force not be used against you except subject to public legal proceeding that can credibly purport to judge and act from a standpoint that represents you.

This difference between a public-legal-order analysis of remedial force and a state-of-nature analysis of defensive force provides a response to Harel's treatment of the example of the Six Day War between Israel and its neighbors. As he describes it, the case is one in which it was not clear whether Egypt was planning aggression or acting defensively. Thus the question of whether Egypt was an aggressor is controversial, and, depending on specifics of the case that are beyond the scope of the present discussion, it may be that each side in the conflict judged itself to be under threat. On the Kantian analysis, this would, if the factual assessment is correct, lead to the conclusion that each was defending itself against the other, because neither is under an obligation to defer to the other's judgment that it is wrongfully attacking. This is a perhaps unusual but not conceptually impossible situation. One of the defects of an international order without fully effective public adjudicative institutions is that in such a situation each nation can only do what seems good and right to it. The subject matter with respect to which things seem good and right is the existence of a threat.

Harel has an additional argument, which, mirroring Suárez, points to the relevance of wrongdoing. If wrongdoing merits defensive force while in progress, he reasons that it must also merit remedial force after the fact.

As I try to explain in chapter 3 of *Kant and the Law of War*, this type of argument faces two different obstacles, because any use of force consistent with freedom must be consistent with the freedom of everyone in three ways. First, it must fit into a system of reciprocal limits on freedom which are, at the same time reciprocal limits on the use of force. Wrongdoing is therefore a necessary but not sufficient condition for defensive or remedial force. The second requirement is that the parties using the force be entitled to do so. The third condition is that the force be appropriate to its rationale. In the case of defensive force, both of these conditions can be satisfied even in a state of nature: force may be used to prevent a wrong or stop a wrong that is in progress; in resisting aggression the defender is standing on its right to be free of force unless a wrong against it has been established by a body entitled to hold it accountable. The second and third conditions apply differently to remedial force. In a public legal order, self-help is generally ruled out, except in narrow cases and even then, it is subject to legal oversight. Instead, you can bring another before a court, to demand a remedy for a wrong that is personal to you. One of the effects of being in a state of nature, a condition in which public legal institutions are absent, is that your private rights do not fully survive their own violation, because others may rightfully resist your attempts to exact a remedy for a wrong done to you. If you attempt to reclaim what you contend is yours, the person from whom you reclaim it does not need to defer to your claim, and so may resist your efforts to remedy (what you take to be) a wrong.

But the international case prohibits remedial war for an additional reason: war hands not only that particular transaction but everything over to savage violence, that is, it not only decides a contentious claim through force rather than words and procedures, but creates a more general situation in which everything is subject to force. One does not need to go as far as Grotius's view that war eventually escalates into a situation in which the dispute is not about something in particular but only about who is in charge to see this point.[91] As soon as peace is abandoned, and force permitted to decide, the question of whether to accept the current results of force is itself a matter of the belligerents' assessment of their prospects of future success. Remedial war hands everything over to savage violence in this way, because the party against whom the remedy is sought might not accept the results of the initial contest of force. Defensive

[91] Book III of *The Law of War and Peace* includes chapters on such topics as "Of the right of killing enemies in formal War, and of other violence against the person"; "Of Ravaging and Pillaging Property"; "Of the right of acquiring things captured in War"; "Of the right over Prisoners of War"; and "Of Lordship over the Conquered."

war does not create a situation of general violence in this way, because the initial aggression has already introduced a condition of force. The defender is entitled to resist that force in order to restore legality.

That means that a peace treaty is importantly different from other treaties or private contracts, none of which are binding if coerced. A peace treaty is normally, perhaps inevitably, the result of force or its prospect. It is binding because it is the way out of a condition in which interaction is governed by force; it creates the context in which disputes can be resolved by something other than force.

As I explain in my responses to Flikschuh and Forst, the idea that a peace treaty is binding even if it is unjust and even if it consolidates the results of past wrongdoing is inseparable from the thought that the creation of a legal order is fundamentally an extralegal act. In the absence of public legal procedures, no transition is governed by law. That does not mean that it cannot be judged to be good or bad or just or unjust. It means only that the transition from a condition where each can only do what seems good and right in its own eyes to one in which law creates a shared standpoint for the governance of human affairs and the resolution of conflict is not itself governed by any already existing framework. If it needed to be, the creation of such a framework would require that it already be in place. That is why Kant writes that it is wrong to look into the founding of a rightful condition "with practical intent," that is, with the intention of using force based on what one finds, and also why it is a mistake to ask "whether power came first and law arrived only afterwards, or even whether they should have followed in this order."[92] Kant makes this point in the context of a discussion of revolution, and so his immediate focus is the transition from a state of nature to a rightful legal order. But the same thing is true of a peace treaty. It is a way of partially exiting a state of nature; as such, there is no public procedural standpoint from which to judge it. As I explain in chapter 2, this does not lead to the conclusion that a peace treaty can have any terms whatsoever, but to the opposite conclusion that it cannot have terms inconsistent with independent legal orders. A putative peace treaty that violated basic *ius cogens* norms, such as the prohibition of slavery and genocide, would not be a peace treaty at all, because it would be inconsistent with peace between independent nations.

Harel offers yet another justification for his conclusion when he writes that "a state has a duty to undo wrongs that it has committed in the past" (Harel p. 97). As I explain in my response to Forst, the existence and importance of a state's duty to address historical wrongs does not entitle any other state to use force to make it comply with it. As is always the case, the question of whether a duty exists does not resolve the question of who, if anyone, has juridical standing to enforce that duty. As Olsthoorn notes in his chapter, some contemporary revisionist writers endorse redistributive wars as remedial, contending that the affluent have a duty to provide the poor with adequate distributive shares, and concluding from this that the poor must also have an executive right to enforce those duties in case of noncompliance. In the absence of a court, the idea that there is an ongoing authorization to enforce the duty to undo past wrongs is even less plausible. Harel acknowledges that this picture is wrong in the case of

[92] 6:318.

punishment, but it is no less wrong in the case of remedies. In both, public legal institutions have powers that no private person could rightfully exercise in their absence.

The right to remedial wars is therefore not "an inevitable by-product of the principle that nations are independent of each other" (Harel p. 100). To suppose that it is rests on the thought that a nation's territory is like its property, and that property wrongs can be remedied at the sole initiative of the party who claims to have been wronged. However, in the relevant sense, a nation's territory it is not like property: its territory is where it exercises jurisdiction. During the course of the war, territory may switch back and forth and so jurisdiction may, at times, be indeterminate. But once jurisdiction is settled after a war, it is now the new nation's territory.

The underlying thought is that whoever exercises jurisdiction on a territory has better title than anyone who does not. Where a defensive war protects the exercise of jurisdiction, a remedial war displaces it. That is the sense in which, although nations are entitled to be independent of each other, their independence presupposes that each nation is entitled to be the nation that it currently is, and so to continue to exercise jurisdiction over the territory that it currently does. It thus imposes a sharp asymmetry between past and future: taking territory through force is prohibited. That prohibition requires that territory that has successfully been taken through force in the past—where the measure of success is the consolidation of legal rule and order—cannot be taken. Those who rule have a better claim to rule than those who do not rule a given territory. Otherwise, as I will explain in my response to Flikschuh, peace would be impossible because it would always need to precede itself.

1.H. Wrongdoing and Questions of Degree (Reply to Massimo Renzo)

Massimo Renzo's chapter raises a challenge to the Kantian view of national defense, arguing that it is unable to explain why defending one's legal order is sufficiently important to merit the use of lethal defensive force. Building on his important contributions to revisionist theory, Renzo takes up two aspects of this question. The first concerns the distinctive kind of wrong that aggressive invasion constitutes; reading the Kantian account of the state in broadly functional terms, he suggests that there could be cases in which the changing regime would lead to no loss of function, and so the Kantian view, on this reading, fails to justify anyone doing anything to stop such an invasion. Renzo puts his main challenge this way:

> But it is not clear what it is for a legal order to be wronged. If by "legal order" we mean a set of institutions and norms, then the claim that a legal order can be wronged sounds puzzling. The claim is more plausible if it's understood as being about the members of the legal order, as some other passages in Ripstein's book seem to suggest. But although this claim is more plausible, it raises the question of why the members of V would be wronged by A. Why should it matter to them whether the civic condition required to protect their independence is provided by a particular legal system rather than another? (Renzo p. 108)

The first worry, then, is that any wrong against a legal order must be a wrong against its members, but those members could only have an interest in having some legal order, rather than the one that currently rules over them. Renzo continues, "If it's in virtue of the vitally important functions performed by V that the right to independence of V's members is not threatened by their subjection to V's political authority, why should it be threatened by their subjection to A's authority, once we assume that A would be performing the same functions?" (Renzo p. 108)

The second concerns the weight to be attached to protecting an existing legal order, given the regrettable aspects of all uses of force, including defensive ones. Renzo concludes with a proposal to understand national defense differently, rendering it consistent with the individualism of much recent revisionist just war writing. As he articulates his alternative, resort to defensive war can be justified because of the great value for individuals of participation in collective self-determination.

In responding to these objections, I will first take up the broad question of the justification of defensive force. Renzo frames the issue in terms familiar from the recent revisionist just war literature, asking whether the loss of national self-determination could be bad enough to offset the harm of killing done to prevent it, let alone killing on a large scale. After articulating a response to this way of framing the issue, I explain why Renzo's alternative account is vulnerable to his own objection. I then respond to his charge that the focus of the Kantian account of the state on its function in solving a problem about right makes states fungible, introducing a theme I already took up in my response to Stilz. I conclude by suggesting that his way of setting up the problem, in terms of a comparison of costs and benefits broadly construed so as to include moral wrongs as costs, makes it effectively impossible for his own account to ever justify the use of defensive force.

When Renzo asks how a set of institutions can be wronged, he expresses puzzlement, and moves immediately to the question of how the individual human beings living under those institutions can be wronged, assuming that any wrongs against them must be specifiable without reference to those institutions. That is, he assumes the form of reductive individualism that I discussed in my response to Stilz. Thus he stipulates that the victorious conqueror's regime does at least as good a job of protecting the innate, acquired, and perhaps also civil rights of the inhabitants. In such a situation, he concludes that the conqueror does no wrong to the inhabitants, and so concludes further that the conqueror does no wrong at all.

That seems to me the wrong way to frame the wrong. Rather than moving directly from the question of whether interests are set back to the question of how much they are set back and so how much the other people's interests can be set back to prevent that from occurring, the identification of the wrong and the question of responsive conduct must be separate stages in a properly structured inquiry. The question of responsive conduct is secondary to the question of whether there is a wrong. This thought is familiar in other contexts: rather than asking (as, perhaps, a utilitarian might) whether punishment or other responsive conduct is justified in a particular case by weighing the interests on each side, ordinary moral thought supposes that a wrong must first be established, and, having been established, then goes on to ask whether punishment is warranted. The Kantian approach is more structured still, asking not only about whether responsive conduct is justified, but also about who is entitled to make the determination that a wrong took place, and who, if anyone, is entitled to carry out responsive conduct if warranted.

Renzo treats the problem of justifying responsive conduct as unstructured. He sets up the problem by drawing on contemporary just war literature, choosing one of two approaches to proportionality that figure prominently in it. For the path that Renzo does not take—exemplified by Thomas Hurka's writings on proportionality—questions of proportionality are downstream of questions of just cause. According to this approach, there can be just cause for going to war which no one is allowed to act on because the consequences of doing so would be too horrific. Hurka's example is the Soviet invasion of Czechoslovakia in 1968, which, he suggests, would have given NATO countries just cause for going to war against the Soviet Union. Nonetheless, it would have been wrong for them to do so because the moral costs of doing so—worldwide war—would have been catastrophically high. By attaching priority to the question of just cause, Hurka's approach, like that of the ancient and medieval founders of the just war tradition, treats proportionality as a restriction on otherwise justified conduct. Further, Hurka argues that there are qualitative restrictions on just causes; only the right kind of cause can be just, regardless of its degree, because questions of degree are secondary. Although ancillary benefits to fighting the war might outweigh some of its costs, none of them can provide a positive justification for going to war.

Rather than following Hurka in treating questions of proportionality as limiting the occasions on which ways in which a just cause can be pursued, Renzo follows David Rodin in treating proportionality as dispositive of whether the cause is just. Rodin describes national defense as a "myth" because the harm it prevents is generally not important enough to justify the loss of human life. On this analysis, there are no qualitative questions about whether something is a just cause, only quantitative ones about its degree. So where Hurka would judge NATO intervention in Czechoslovakia in 1968 to be a just cause but disproportionate because of its catastrophic consequences, Rodin would deny that there was a just cause at all. Instead, on this view, which Renzo adopts, the weightiness of the just cause must be weighed against the negative consequences of fighting for it.

This approach therefore turns the determination of whether there is a just cause into an essentially comparative exercise. The comparative nature of the proposed exercise is the source of the difficulties that Renzo identifies with the Kantian account of defensive war, and underlies his objections to that account. It is also the wrong way to set up the problem; it not only creates challenges for the Kantian approach, but basically ensures that nothing can qualify as a just cause. That is why, as I will explain later, his proposed alternative to the Kantian approach cannot address the problem as he frames it.

The source of the difficulty, however, is with the framing of grounds of war in comparative terms; it creates two related problems, both of which are intractable. First, a thoroughgoing comparative analysis does not have a sufficiently robust concept of wrongdoing. Second, the lack of a robust concept of wrongdoing prevents it from sustaining a systematic distinction between aggression and defense. On the comparative analysis than Renzo assumes, there is nothing distinctive about the use of force; instead, the comparative analysis must either criticize most uses of force on the basis of their bad consequences or, alternatively, treat all uses of force as presumptively objectionable, and so submit aggressive and defensive uses of force to the same form of moral scrutiny.

The Kantian view is fundamentally different: it takes as its starting point a principled distinction between aggressive and defensive force. In the case of war, aggression is wrong both as against the state at which it is directed, and wrong "in the highest degree" because it "hands everything over to savage violence," that is, it initiates a condition in which force decides. At the level of the interpersonal wrong between states, if one threatens the other with violence, responsive violence on the part of the defender is not the initiation of a condition of violence, but rather nothing more than resistance to it. If one person grabs another, and the second pushes the first away, the second is not initiating a condition of violence. Nor is the second initiating a condition of violence if the first threatens the second with violence and the second resists. If a mugger says "give me your wallet or I will stab you," the victim who knocks the knife out of the mugger's hand is not the one who initiated the condition of violence. This point is easy to overlook if, following many contemporary just war writers, you suppose that proportionality operates as an open-ended restriction on enforcement of rights, or, as many of them do, an open-ended permission to do otherwise forbidden things provided that the offsetting benefits are good.[93] At the root of all of these proposals is the thought that it is wrong to cause harm unless doing so will produce some greater good. That is why contemporary just war theorists have such a difficult time with national defense: it will cause harm to the aggressor's fighters, as well as foreseeable harm to the defender's combatants and to civilians on both sides. If the issue is framed in these accountancy terms, the harms caused by defense against aggression can only be outweighed by the prevention of still greater harms. And, so the argument goes, the loss of political independence or territorial integrity (as such) is usually not serious enough to justify a defensive response. In some cases, an armed response may be legitimate, but if it is, it is not because of the threat to political independence and territorial integrity, but rather the threat to the interests of the individuals in the nation that is under attack.

This way of thinking about proportionality has become prominent in the philosophical literature, but it can only make its way into discussions of self-defense if the problem is set up in what, from the Kantian perspective, is the wrong way. The setup preferred by contemporary just war writers begins with the thought that the basic focus of moral prohibition is causing serious harm. From this idea, a category of wrongdoing is built up, and, correlative with it, some idea that people have rights against (certain types of) harm. Beginning with this idea, then, any kind of response to any kind of wrongdoing would need to be sensitive to the type, degree, and extent of harm. Unsurprisingly, many contemporary writers introduce a foundational idea of proportionality to suggest that the amount of harm done in response to ongoing or threatened harm must be limited by the amount of harm that it itself causes. Thus the entitlement to defend against an aggressor is, on this model, restricted by some kind of balance between the badness of the harm caused by the aggression and the badness

[93] E.g., Jeff McMahan, "Proportionality and Necessity in Jus in Bello," *The Oxford Handbook of Ethics of War*, Edited by Seth Lazar and Helen Frowe (Oxford: Oxford University Press, 2016).

of the harm caused by the response. This model provides an explanation of why you cannot kill someone rather than give them your pen even if killing them is the only way to stop them from taking it; the harm done (the killing of the mugger) is disproportionate to the harm prevented or good achieved (the securing of the pen). This approach ultimately rests on some idea about the appropriate distribution of harm. For example, Jeff McMahan explicitly says, "the determination of liability to defensive harm is a matter of justice in the *ex ante* distribution of unavoidable harm."[94] Using excessive force creates an overall increase in the amount of harm, and so cannot be justified as the just distribution of the smaller antecedent harm.

Any such approach faces a significant challenge, because it needs some principled way of determining the scope of the distributive exercise. Yet distributive justice only operates on or within groups of people who can be in some way characterized in advance of the distributive exercise. If you are giving grades to a group of students, your guiding aim must be to assign them on the basis of the quality of the academic work submitted by each of them: that is your mandate. This narrow mandate means that you are supposed to only attend to the students whose work you are grading, and, further, that you are only to attend to their work: their other merits and vices should not enter into your assessment. In this case, you have scope restrictions both with respect to who you grade and with respect to grounds for the grade. So, too, if you are charged with allocating some scarce resource, or, with a task that some philosophers take to be fundamental, seeing to the overall distribution of resources or welfare in your society. Different philosophers have different views about this, with some advocating distribution tied ideas of desert, others to ideas of responsibility understood in some other way, and still others to well-being. Each of these is a much broader scope restriction than in the grading example, but each is, nonetheless, a scope restriction.

In a strictly comparative analysis, it is not just that the nation is not entitled to defend itself when the harms caused by a defensive war are higher than the comparatively small harm of the loss of political independence or territorial integrity; the very same comparative structure could, in principle, also license an invader to invade (or better, threaten to invade) another nation whenever the anticipated benefits of doing so are greater than the cost. The structure emerges because the party under attack is not permitted to use defensive force in conditions in which capitulation is a better outcome than defense. The same result obtains even if the probability of successful defense works to discount the value of defense, so that if defensive force is unlikely to be successful, capitulation does even better in comparative terms.

The source of this result is already contained in the setup: the entire exercise is comparative rather than rights-based and transactional. Rather than asking whether the aggression is wrongful, the inquiry subordinates questions of wrongfulness to those of degree, and treats the question of whether something is even presumptively wrong as the conclusion of the comparison of various states of affairs, the value of which does not depend on how they come about.

This may seem a surprising conclusion, given the ways in which so much of the contemporary just war literature works from individual examples about which

[94] "Self-Defense against Justified Threateners," in *How We Fight: Ethics in War*, eds. Helen Frowe and Gerald Lang (Oxford: Oxford University Press, 2014), p. 117.

readers are expected to have firm intuitions, such as a paid hitman attacking someone, or runaway trolleys being turned. The difficulty comes with the explanation of those firm intuitions in terms of an essentially comparative analysis. If the analysis is comparative, then the causing of harm and the prevention of harm are weighed on a single scale, even if certain harms (for example, directly caused ones, intended ones) are weighted more heavily, on the grounds that it is worse to cause harm than permit it, or to intend harm than merely foresee it, or worse to harm opportunistically (by using a person's injury or death as a way of achieving something else, for example, by dropping someone in the path of a runaway trolley) rather than eliminatively (by getting rid of someone to stop them from being an obstacle or posing a threat). All of these characterizations, however, are themselves comparative, rather than cast in terms of any idea of wrongfulness, because the comparative approach has no conceptual space for any idea of a principled prohibition. The range of cases in which the use or threat of force is permissible is considerable, since the same pairwise comparison must apply to aggressive as to defensive force.

Any such comparative exercise needs some kind of currency in which to conduct the comparisons. Cast in these terms, however, ideas of political independence and territorial integrity do not look important enough to justify foreseen loss of life, let alone to justify intentionally killing the aggressor's combatants. Because the exercise is fundamentally comparative, however, it is not just that defensive force is not warranted to protect them; their existence also appears to count for nothing in the aggressor's determination of whether to attack. If the defender is not entitled to defend political independence or territorial integrity because they are not important enough, then the aggressor presumably does not need to attach any weight to them either. If the invasion will bring about some good—access to resources or opportunities, or, a buffer zone to defend against unanticipated attack, or a shorter route to carry out a justified attack—the fact that it threatens the political independence and territorial integrity of a nation not only forbids that nation to defend itself, but should be of no concern whatsoever to another nation contemplating aggression.

This difficulty manifests itself in Renzo's proposed alternative to the Kantian view. Renzo recommends focusing not on the negative idea of independence but the positive idea of autonomy. On this view, it matters to each individual's autonomy that groups of which that individual is a member be able to govern themselves through processes that are appropriately sensitive to that individual's participation in them.

> The point here is that in cases of collective action, what matters to me, at least insofar as autonomy goes, is not that the group acts as I wish, but rather that it acts in ways that allow its members (including me) to have an adequate input in a process of collective deliberation, so that they can see whatever the group does as an expression of their collective agency. (Renzo p. 110)

Yet this proposal faces all of the difficulties that Renzo presses against the Kantian view, in particular its inability to explain why defensive force could ever be justified, as well as additional difficulties of its own. As long as the idea is that responsive conduct must be keyed to the interests of the parties, assessed not merely in their fact but in their degree—as introduced by his example of the limit on the amount of force that can be used to stop someone from stopping you from voting—the same question can be asked about just how important it is to participate in these practices of collective agency. You cannot kill someone because they are stopping you from voting, even if voting is an expression of collective agency. Although Renzo writes that "After all, it seems clear that I may resort to defensive force to protect my right to act together with the members of certain groups, such as my family" (Renzo p. 112). because his entire apparatus turns on questions of degree, however clear this might seem, whether defensive force can be used in any particular case will depend entirely on how bad it will be for others to have that force used, and what aspect of acting with family members will be prevented. If we are holding a family discussion to decide some important matter, such as arrangements for an elderly relative, we are not allowed to use defensive force to silence the person making loud and distracting noises nearby, even if this really is the only opportunity we will have to address this pressing issue. The family analogy faces the further difficulty that the wrong of interfering with the collective action of the family is actually many different wrongs, including violations of the private rights of the members. Your entitlement to use force to stop someone from separating you from your children does not depend on the idea that you have a right as against those people to act together with family members. Even if the family case was straightforward, Renzo has not explained why other, less intimate forms of collective agency such as the state (or the academic department of which he is a member) would be important enough to justify lethal force.

I assume that the reason that Renzo begins with the small-scale example is not only because readers can be expected to agree that family members are entitled to protect each other and the family unit, but also because he is committed to denying the Kantian claim that the state is *sui generis* as a form of collective agency. The cost of doing so is that he makes it even more difficult to show that being deprived of participation in it is a serious loss for the individuals that make it up. In many democracies, large numbers of people are cynical or apathetic about politics, and many do not bother to vote. It is not apparent whether Renzo thinks that those people would be wronged by an invasion. The smaller-scale examples are no less puzzling. Academic restructuring raises many difficult issues, but, when, as many universities have done in recent decades, separate departments of botany and zoology are replaced by departments of ecology/evolutionary biology and cell/systems biology, the fact that everyone is now in a different self-governing collective does not seem to be a deep problem at all. Few would think that botanists should head to the barricades over such a matter, even though departmental membership is a fundamental aspect of the identity of many scholars and scientists, a not insignificant number of whom have greater attachment to their department than to the nation in which they happen to have found employment.

Turning to the case of the state, although Renzo suggests that a single instance of voting is not important enough to justify lethal force, he offers no explanation of how multiple instances of voting might add up to something significant enough. His method of aggregation, whatever it is, would appear to make it more seriously wrong to interfere with large states than small ones, or those with a higher life expectancy, since they would be deprived of more votes. None of these factors seem relevant to the wrong of invasion, yet they must figure in his model of how to scale up from the wrongfulness of using force to prevent a single vote being blocked to the legitimacy of force used to resist annexation.

Because Renzo frames the issue in terms of proportionate harm, his alternative proposal faces an additional difficulty: it appears to authorize taking things from other people so as to enable valuable forms of collective agency. Whether it is permissible for a nation to invade its neighbor would depend on whether doing so would assist the invader's exercise of collective agency more than it would compromise that of the invaded country. Further, the invasion itself might be an expression of collective agency. Suppose a democracy decides through its collective processes to invade a non-democratic country and overthrow its government, installing one that is friendlier to the invader's own collective agency and ambitions. Based on the assessment of proportionality that Renzo uses to level his charge against Kantianism, it seems that he would have to reserve judgment on such an example, depending on how much collective agency there was on each side.

Having criticized Renzo's emphasis on collective agency, I should now note that on the Kantian view, the state is in important ways a collective agent, and an invasion does indeed violate the collective agency of its citizens. As I explain in my response to Stilz, those rights are not pre-institutional, however; they exist only because of and through institutions. But they are rights to belong to the particular institutions to which those people belong.

Although a legal order does solve problems that would make the absence of a legal order a condition of violence, it does not follow from this that legal orders must be interchangeable. That would only follow if the "function" of entering a rightful condition was to improve a state of affairs, along some comparative dimension of assessment. Renzo seems to have something like this in mind when he asks "whether it should matter to them" which government people have. The thought seems to be that the point of a rightful condition is to create a superior state of affairs, not measured in terms of welfare or happiness, but instead in terms of people enjoying their rights. This then sets up Renzo's puzzle: If the invasion leads to a state of affairs that is no worse than the current state of affairs, why would anybody mind?

But the nature of the wrong is not to be found in the reduced value (along any dimension) of the *ex ante* and *ex post* states of affairs. The distinctive wrong is the wrong of aggression, which is wrongful because it both subordinates one legal order to another and creates a condition in which force alone decides. Whether it does or should matter to them which government they have, no other government is entitled to take it upon itself to decide that question for them, or to cash in on their actual or rational indifference.

Renzo frames the problem as one about outcomes; the organizing idea underlying the *Doctrine of Right* as a whole is that interpersonal morality is about relations of standing and authority: the question of right is always the question about who gets to decide about some matter. Your right to your own person precludes others taking it upon themselves to determine the purposes you will pursue, or to interfere with your capacity to do so. If you have a property right, it means that others are not entitled to determine the purposes for which that property will be used, or its condition. Questions of public right concern the public standpoint with respect to interactions between human beings. If human beings have united themselves into a rightful condition, however inadequate, they are entitled, as against each other, to have its public institutions govern their interactions. Unless there is some kind of affirmative act through which some other natural or artificial person acquires the power to disrupt that rightful condition, that other natural or artificial person is not entitled to interfere with or disrupt it. That is, what Kant elsewhere calls the *quid iuris* question—by what right—applies to any claim by an outside party to answer such a question for those who are members of the legal order, questions to which that legal order already contains institutions for answering. The ideal case of those institutions would be, as Stilz helpfully puts it, one in which citizens were makers as well as takers of the laws, active in relation to the laws rather than merely passive recipients of them. That ideal case would be also a strong form of collective action in the sense to which Renzo appeals. Any such defects do not, however, give anyone else license to take it upon themselves to replace those defective institutions with institutions of the outsiders' choice. It only looks otherwise if it is supposed that there is always a kind of implicit universal jurisdiction for a person or agent to take it upon itself to answer a question for someone else. The whole point of Kantian right is that if we think of people as entitled to be independent, any such claim to authority needs to be justified by showing it to be consistent with the universal principle of right, both in terms of the rights of individual human beings and the right of human beings as such, that is, their membership in a rightful condition. Kant has specific arguments to show that a public legal authority can have that kind of power over its members, and that a precondition of having that authority is being in a position to (more or less effectively) exercise it. That argument is not consistent with and so does not permit the claim that anyone with a significant prospect of successfully fulfilling that function thereby has the authorization to usurp the existing legal order. If they lack the authorization to do so, they can rightfully be prevented from doing so. Nations are entitled to be independent of each other.

Renzo frames things differently, and, by focusing on the welfare effect on individuals of such a transition, identifies possible cases in which the net cost of the invasion could be neutral. So, too, with cases in which one nation annexes a currently unoccupied region of another's territory. Indeed, such invasions could, assessed only in terms of their welfare effects, mark improvements, perhaps resulting in a better means of dispute resolution or granting the inhabitants of the territory new opportunities to participate in self-governing institutions.

The substitution is nonetheless still wrongful, not as an interference with actual or even possible uses the inhabitants might make of their legal order, but rather as an interference with their being in this legal order. The problem, then, is not that there is no possible substitute; the wrong lies in the substitution even if there were to be no flaw in the substitute, or it was in some way functionally superior. As in the case of private rights, the key issue concerns whether someone has the authority to change (even improve) an existing rightful condition. Whether the members of that rightful condition are adequately represented or provided for is a question that an outsider cannot take it upon itself to determine for them. If it could, they would be no better than a colony. Indeed, an international order in which powerful states were entitled to take it upon themselves to repair what they saw as the defects of weaker ones would just be a new form of colonialism, because it would always be up to the Metropole to decide whether its subjects are functioning adequately. Even if they chose not to invade, it would be entirely up to them.

2. Part II

2.A. Closure and Determination: Kant's Philosophical Ambitions (Reply to Katrin Flikschuh)

Katrin Flikschuh's chapter addresses the question of whether right imposes closure in order to engage a deeper question about the philosophical ambitions of the *Doctrine of Right.* In *What is Orientation in Global Thinking?* Flikschuh offers powerful arguments to show that Kant's focus on property and the state are only one possible determination of the requirements of right; her chapter extends that line of thought to the forms of international relations. In *Kant and the Law of War*, I argued that the purpose of law, and entry into the legal order, is to provide closure with respect to disputes between individual human beings. Although she concedes that positive law requires closure, she argues that the requirement is pragmatic, that there is no metaphysical requirement of closure and, further, that the structure of the metaphysics of right is not itself closed. Contrasting my position on Kant's argument about a league of states with one defended by Joachim Hruschka, Flikschuh argues that the metaphysics of morals may provide no answer to such a question. She does not mean to dispute the importance of closure at the level of public legal institutions: in order to determine "with mathematical exactitude" what belongs to each, a legal order must contain either an answer or a procedure for determining how to answer who is to answer the question, and the factors the decision maker is to consider. Only then can each person subject to that legal order securely have what is their own. Flikschuh's challenges the claim that the doctrine of right as a whole is itself closed, that is, that in each of the three forms of public right—the right of the state, the right of nations. and cosmopolitan right—there must be a specific structure that is appropriate to the relevant form of relation.

In order to respond, I should begin by clarifying what is and is not at issue in our disagreement. Both of us accept that positive law must impose closure on disputes between those living under it, and neither of us supposes that all of the determinate answers provided by positive law are themselves already contained in or fully specified by the metaphysics of morals. Neither of us supposes that the metaphysics of right is more precise than positive law.

We disagree, however, on the question of why positive law must be exact. Flikschuh's case for exactitude appears to be empirical and contingent:

> The law as idea promises to give us standards that are in an important sense free of the frailties of human judgement. And of course, so far as possible, actual law has to have these qualities if those subject to it are to be able to trust that law both secures them from arbitrary state power and defends the equal rights of each from incursion by others. These are good reasons for expecting positive law to be applied with precision and impartiality. However, they are ultimately instrumental reasons that do not

> really help to explain the current philosophical fascination with the very idea of law's precision and exactitude. (Flikschuh p. 131)

I do not think that the need for closure in positive law is instrumental in this way. Nor do I think that its main aim is to secure trust empirically. More significantly, I do not think that positive law provides the degree of precision that Flikschuh suggests. It provides closure by determining who will address the question and the relevant grounds on which it is to be addressed. Rather than enabling people to flawlessly predict the legal consequences of various courses of action, positive law makes questions of interpersonal right determinate, typically before the fact but sometimes only after the fact. On this understanding, the need for "mathematical exactitude" has its origin in the idea that things must stand a particular way between any pair of independent beings if neither is to be subject to the choice of the other. I do not, however, think that this can be achieved in a way that is in any sense free of the frailties of human judgment: the person who decides what the law demands in a particular case is called a "judge" because that person occupies a public role the point of which is to exercise judgment; in some disputes the person appointed by the parties to decide it is called an "arbitrator" because they are charged with deciding rather than judging. Public legal institutions are required in order to provide a common standpoint from which authoritative judgments can be made on behalf of everyone. If two private persons have a disagreement in the absence of a legal order, they can perhaps appoint a third person and agree to accept that person's decision as binding. If they do not agree, their dispute is unresolved, or perhaps resolved by questions apart from its merits. Right requires more than the mere resolution of conflicts; it requires a court of general jurisdiction before which parties can bring each other to receive an answer on behalf of everyone. The idea that the court in particular, and a legal order more generally, must constitute a standpoint capable of acting, judging, and enforcing on behalf of everyone generates the problem of public right and with it each of the three forms that public right must take. The need for a public standpoint is a requirement of the metaphysics of morals, but does not answer specific questions of the kind that are essential to the possibility of a shared public determination of what belongs to each. Instead, the metaphysics of morals explains why public institutions are required, and the distinctive sense in which public institutions must be legal: positive law is "what is laid down as right"[95] because it is more determinate than right alone.

Although the metaphysics of morals demands that closure be imposed on particular disputes, that is, that they be publicly determinable, it imposes closure in a different sense than positive law does. Kant's argument that there must be exactly three forms of public right—the right of the state, the right of nations, and cosmopolitan right—is structured a priori by the three categories of relation identified in the *Critique of Pure Reason* in strict parallel with the relation between the three titles of private right. Just as the difference between property, contract, and status relations is not a matter of positive law, or the product of any kind of indeterminacy in the Kantian framework, but rather brings the concept of right to bear on three forms of interaction, so too, the

[95] 6:311.

division of the public right is part of the metaphysics of morals. Kant is explicit about this relation in the case of the three titles of private right: he expressly identifies them with the three categories of relation, and notes that there are exactly three because if the distinction between persons and things is applied to itself, it yields four possibilities, the fourth of which, a right against a thing, is impossible, because a thing cannot be a bearer of obligations. (This is the very argument that Kant makes in *Theory and Practice* and reiterates in the *Doctrine of Right* to demonstrate the moral impossibility of a slave contract.) This formal-categorial structure of private right does not generate its entire doctrinal structure; positive legal doctrine and officials charged with developing and applying it are necessary to provide closure in relation to concrete disputes.

The three forms of public right are generated when the distinction between a state and its members is applied to itself. Once the two distinct categories—of persons and of states—is accepted, they stand in four possible relations: the relation of a state to its members, the relation of one state to another state, and the relation of a state to someone who is not a member of it. The fourth possibility would be a state considered as a member of another state—Kant's characterization of colonialism in his argument for its wrongfulness. The three possibilities Kant considers therefore correspond to the three categories of relation articulated in the *Critique of Pure Reason*: substance, causality, and community.

If the right of nations is understood through the categories of relation in this way, then it must have a form that is consistent with it. It is not optional to choose either a binding, enforceable federation which states can be compelled to join or a voluntary league. Instead, the voluntary league fits within the structure of public right that Kant has articulated, because it characterizes how relations can be rightful between nations that are conceived as independent of each other. This does not establish that a plurality of independent nations is a necessary condition of right; it specifies instead what it would be for there to be rightful relations between a plurality of nations, such that none is subordinated to another, but also such that each is not somehow subordinated to all the rest in a global despotism.

Reading the right of nations in light of Kant's architectonic concerns reveals the sense in which closure is required for a metaphysics of morals. It is different from the closure that a legal order provides, and rests instead on the postulate of public right, which Kant characterizes before introducing the "definitive" articles of Perpetual Peace as "the postulate on which all the following [i.e. definitive] articles are based is that all men who can mutually affect one another must belong to some civil constitution."[96] The categories of public right are closed, not in the sense of resolving every dispute, but rather because the principle of freedom under law must be applicable to every form of interaction; Kant makes this structural point in both *Perpetual Peace* and the *Doctrine of Right*. In the former he writes that

> every Constitution must be
>
> (i) one in accord with the right of citizens of a state, of individuals within a people (*ius civitatis*),

[96] 8:349; see also 6:307.

(2) one in accord with the right of nations, of states in relation to one another (*ius gentium*),
(3) one in accord with the right of citizens of the world, insofar as individuals and states, standing in the relation of externally affecting one another, are to be regarded as citizens of a universal state of mankind (*ius cosmopoliticum*). This division is not made at will but is necessary with reference to the idea of perpetual peace. For if only one of these were in a relation of physically affecting another and were yet in a state of nature, the condition of war would be bound up with this, and the aim here is just to be freed from it.[97]

The Doctrine of Right provides the same tripartite division:

> Since the earth's surface is not unlimited but closed, the concepts of the right of a state and of a right of nations lead inevitably to the idea of a right for a state of nations (*ius gentium*) or cosmopolitan right (*ius cosmopoliticum*). So if the principle of outer freedom limited by law is lacking in any one of these three possible forms of rightful condition, the framework of all the others is unavoidably undermined and must finally collapse.[98]

These parallel passages also provide my response to Flikschuh's question of whether Kant is inconsistent across multiple writings. *Towards Perpetual Peace* draws a fundamental division between what Kant calls "preliminary" and "definitive" articles of perpetual peace. The preliminary articles mark the transition from a condition of barbarism in which force always decides, to one of anarchy, in which a plurality of separate states interact in a way that is consistent with their respective independence. The basic rule for such an order—the preliminary rule of peace, that is, the precondition of peace ever being achieved or maintained—is that law can never be displaced by force. That is, the preliminary articles are ways of achieving a condition in which violence is not in fact effective. The definitive articles specify what counts as peace in its various relations. It can never be fully realized. The three definitive articles of Perpetual Peace are definitive, rather than preliminary because they constitute (define) what it is to be at peace, that is, for human interactions governed by law rather than force. They also correspond to the three forms of public right—the right of the state, the right of nations, and cosmopolitan right—in that republican government, the league of states, and a right to visit are the pure forms of the three forms of public right. The two accounts are therefore consistent, each focused on the principle of right appropriate to the relevant relation between legal orders and individual human beings living under laws.

In describing this as an organizing structure, I do not mean to claim that its structure fully determines the content of positive law, only that positive law must be structured in a particular way if the principle of freedom under law is to hold in every relation between any two juridical persons; for every possible form of relation between human beings, there is a corresponding form of public right.

[97] 8:350.
[98] 6:311.

This formal argument does not guarantee that anything satisfies any of the three forms of public right, let alone that all of them are satisfied. If, for example, as a result of whatever geographic, environmental, economic, political, agricultural, and military history the world happened to have, there turned out to be only one state in the entire world, there would then, by default, be a world state. This result would not be inconsistent with the right of nations. Kant says that the right of nations must govern interactions between independent states. Since none may set itself up as the master of another, none can coerce another into joining a world state. So, too, with cosmopolitan right: if there was only one state in the world, there would be no cosmopolitan right. But if there are a plurality of states, cosmopolitan right must govern relations between human beings and states of which they are not members. The full development of the right to visit is partly indeterminate and requires development through positive legal doctrine. In other writing, Hruschka represents cosmopolitan right as some version of the WTO. I think that is much too specific, but agree that cosmopolitan right does require a rules-based international trade regime, while being open-ended with respect to what those rules should be.

Flikschuh writes, "Assuming that no one can compel himself, the demand that states comply with the demands of international right falls considerably short of coercible rights obligations" (Flikschuh p. 127). She later fills out this thought by describing what she sees as a paradox,

> Significantly, Kant ends up where he does not as a result of any faulty reasoning on his part. To the contrary, his reasoning is rigorous: the unenforceability of rights relations between states follows from the enforceability of rights relations between persons. It is, as it were, because rights relations between persons are coercible that rights relations between states cannot be coercible. And yet this yields the paradoxical conclusion that rights relations both are and are not coercively enforceable. (Flikschuh p. 128)

I do not see this as paradoxical. It is an instance of a perfectly general structure rather than any sort of problem: the possibility of enforcement must, on pain of regress, depend upon an enforcer which is not itself subject to further enforcement. That is why Kant both insists, against Hobbes, that the state has duties to its citizens and denies that the citizens have coercible rights against it. The possibility of peace depends on all uses of force being subject to legal oversight, and, however complex the system of oversight of the overseers might be, ultimately turns on enforcement that is "irresistible."[99] But, on pain of regress, the enforcer could not be subject to further enforcement. That does not mean that in the absence of effective public international legal institutions that a nation that goes to aggressive war is not subject to enforcement by other nations. They are authorized to use force to stop the aggression. It is just that there is no further enforcer charged with making sure enforcement is carried out. So it is not just that relations between states can never be coercively enforced; they can

[99] 6:316.

be in this limited sense. What cannot be coerced is a requirement that a state join in a dispute between other states.

This brings me to a more general point about the connection between right and coercion. Kant does not claim every right can in fact be coerced; instead, the requirement of right reflects the fact that claims of right always concern external freedom, and so they are enforceable in their subject matter. All enforcement must be brought under law. When Kant puts the point this way, particularly but not exclusively in his anti-revolutionary writings, he is not denying that there could be a system of American-style "checks and balances," so as to somehow distribute the factual power of enforcement among multiple agents so that each oversees the other. To suppose that he rules out such a possibility is to imagine him to hold the Hobbesian view that the sovereign, at the apex of the legal order, is not subject to law at all, a position he expressly repudiates in "Theory and Practice." Instead, his point is that enforcement requires an enforcer who, on pain of regress, must have at least some powers that are subject to oversight, but are not subject to enforcement. Both the concept of right and the structure of republican government demands that uses of force be subject to law. The only way to achieve that is to have an enforcer who is bound by law, and so is not entitled to use force at its own initiative. Just as the enforcer (executive) does not make the laws that apply to it, so too, the legislature does not enforce the laws that it enacts. To allow either of these things would be to collapse legislation and execution of the laws into despotism. Therefore, the enforcer cannot be subject to enforcement.

Nor is there any paradox in the Kantian idea "that, necessarily, justice—rightful law—emerges from lawlessness" (Flikschuh p. 129). Interpersonal morality requires the establishment of institutions so that interactions among human beings can be brought under law. There are better and worse, and wise and foolish ways of setting up those institutions and so of bringing human conduct under law. But there are no legal ways. The creation of a legal order is always an extralegal event. Precisely because foundings are beyond law, Kant remarks that one must not look into historical origins with practical intent. It is not only that we are likely to find ugly things there. Instead, the broader point is that the use of force is barbaric because it hands everything over to savage violence, making it impossible to resolve any question on its merits. Legal ordering is a condition of peace rather than war precisely because it is the condition in which disputes can be resolved on their merits. As I note in *Kant and the Law of War* and my response to Forst, the implication of this adamantine distinction between peace and war is that the merits of a transition to peace cannot be a precondition of peace. If they had to be, peace would need to always precede itself and so be impossible.

Flikschuh also questions my rejection of political teleology. I have systematic Kantian reasons for setting aside the political teleology: Kant's political teleology gets thinner and thinner throughout his writings, until, by the end of the *Doctrine of Right*, it requires only that we be able to satisfy ourselves that the idea of a future peace is not conceptually incoherent. Some of his earlier and more robust formulations of it join an eighteenth-century debate about teleology, theodicy, and the philosophy of history. Kant's aim in his interventions in that debate is to see if any sense can be made of what he concedes is the "nonsensical course of human affairs." I do not think that the doctrine of right requires any teleological scaffolding. I read it instead as the culmination of Kant's critical project. Some might complain that architectonic concerns overwhelm Kant's argument, but I see things differently. If anything, architectonic concerns make the argument more orderly and systematic, in ways that may leave some readers disappointed at Kant's seeming indifference to the messiness of human affairs.

Bringing human interaction under law is the only way in which it can be governed by words rather than force. The only way to do that is to bring all uses of force under law. That is the task of the law, and that is why peace is the final end of the doctrine of right, why practical reason's "irresistible veto" is that there shall be no war, neither between you and me in the state of nature nor between nations, and, fast forwarding to our century, nor shall there be war between a nation and those who are not members of it but are subject to its power.

2.B. Redistributive Wars and Other Private Wars (Reply to Johan Olsthoorn)

Johan Olsthoorn situates questions of *ius ad bellum* in the context of broader debates about global justice, considering in particular the way in which those who are severely resource-deprived in the current global order might have rights to use violence in order to acquire means of subsistence. Building on his past work on both Grotius and contemporary just war theory, Olsthoorn considers two different versions of this argument, one grounded in the contemporary human rights debate and the contemporary just war literature; and the other in an older, Grotian idea of a right of necessity. Although he thinks that the Grotian view provides for "more capacious grounds" for such a war, through his distinction between the existence conditions and the conditions of lawful exercise of a right, he suggests that both arguments are of more theoretical than practical interest, since various other general conditions for initiating war, such as necessity, and, even more significantly in the case of subsistence wars, reasonable prospect of success, will probably never be satisfied. Given the vast defensive arsenals assembled by affluent nations around the world to protect their institutions, citizens, and supply chains, there is no realistic prospect of a successful subsistence war giving the global poor the resources that they need. Further, the cost of waging such a war against well-equipped militaries with sophisticated weapons would be so high that assembling weapons with any prospect of success would require resources that could go some distance toward alleviating the problems that are the grounds of war.

Olsthoorn's interest is in the theoretical question, and in particular the narrow question of whether rights to subsistence could authorize a private war, fought not by an organized state but rather by individuals, each acting in their own personal capacity rather than on behalf of any larger public legal institution. The idea of a private war figures prominently in Grotius's writing, and had been largely neglected in recent centuries, but has received renewed attention under the influence of reductive individualist just war theory. Reductive individualists do not think that there are any novel principles governing war that do not apply to individuals interacting with each other outside of it; as Jeff McMahan puts the point, war differs from other violent conflicts only in its scale and complexity.[100] On this analysis, then, such traditional just war requirements as proper authority and that individuals act on behalf of a nation, and any distinctions between soldiers and civilians or mercenaries are at best pragmatic compromises with morality, parts of what Christian Wolff called "the voluntary law of nations."

Before turning to the specifics of his argument, it is worth noting that Olsthoorn may overstate the practical obstacles (at least within the revisionist framework he is discussing) to initiating such a war if it is just. A powerful country might decide for strategic reasons to help poor people fight against other affluent powerful countries. If the poor have a right to engage in private wars of subsistence, others may rightfully come to their aid in pursuing those wars. Perhaps those others could produce better results by meeting the needs of the poor directly, rather than assisting them in their effort to extract them from other responsible affluent people. Olsthoorn emphasizes other moral factors, such as prospect of success or necessity, that might make it impermissible to exercise a right, but any such factors are contextual. The entitlement to enforce a right is limited by the rightholder's actual situation and viable options, even if multiple other parties could and should do various things that would make resort to violence unnecessary. If there really is nothing distinctive about war other than scale and complexity, then general principles of assistance would seem to apply. Others may assist a rightholder upholding basic rights, even if those others could improve the situation in other ways. Those in extreme need are presumably entitled to act from the situation in which they find themselves. Cécile Fabre, the contemporary just war writer whose work Olsthoorn discusses in most detail, argues that past wrongdoing and indeed even past contribution to a wrong in question does not prevent someone from being permitted to use force to prevent the continuation of that wrong.[101] At least within the revisionist just war framework, then, the question is not purely theoretical.

Olsthoorn points to a fundamental difference between the Grotian and contemporary approaches to private wars of subsistence: for contemporary just war writers, the right to kill in war (or on any other occasion) always depends on what contemporary writers describe as the "liability" of the person being killed. The near-consensus view in this literature is that liability to being killed is established through the person's actions, for which the person must be responsible. The relevant conception of responsibility is weaker than culpability. According to standard treatments in the revisionist just war literature, someone who is not culpable, because they act under a blameless mistake of fact, or even under duress, can still be liable to being killed if they are non-culpably responsible for the creation of danger. On this view, someone can be responsible for a risk if they could have avoided creating it.

Olsthoorn brings out an important tension running through such views; although the relation between responsibility and liability is sometimes presented as though it is a matter of conceptual entailment, the concept of responsibility is scalar, that is, admits of degrees, while liability is Boolean—a person either is or is not liable. Although the aggressor and the defender might have both prevented the conflict, the aggressor

[100] Jeff McMahan "Laws of War," in The Philosophy of International Law, eds. John Tasioulas and Samantha Besson (Oxford: Oxford University Press, 2010), p. 505

[101] Fabre, *Cosmopolitan War* (Oxford: Oxford University Press, 2013), p. 113–114.

by avoiding the use of force, the defender by staying home, the aggressor is more responsible. Olsthoorn argues that the scalar nature of responsibility leads to a situation in which the right to wage a subsistence war is already indeterminate, prior to any questions about the prospect of success, necessity, or proportionality. The requirement that the target of the force be the responsible party fails to fit any kind of systematic division between the affluent and the poor. Different affluent people are responsible to differing degrees for the plight of the global poor. Writers defending the idea that the affluent are responsible, and so, in principle, subject to being killed to improve the situation of the global poor, can only reach this conclusion by having an expansive conception of responsibility that includes receiving benefits from the currently structured global order. Such an account seems overbroad, as many or even most people in affluent countries are not in a position to avoid participating in the economies of the societies in which they live, and so in the global order.

I want to suggest a reply that Fabre and contemporary just war writers could give to Olsthoorn's argument, before proposing a way of extending that argument. The reply would point to the complicity of pretty much everyone in affluent countries in their economic order, showing that they all participate in an order that they know, or should know, causes great suffering to the global poor. The argument might continue by emphasizing that such complicity is sufficient to make every one of them liable to some degree. On views of liability that figure in the contemporary revisionist just war literature, assessments of liability are always fundamentally comparative. On this view, force can be used whenever transferring the risk of death from a person with a lesser degree of responsibility to another with a greater degree will be effective in saving the life of the less responsible one. Thus, for example, Jeff McMahan suggests that a driver who, despite driving carefully, loses control of a vehicle is liable to defensive force by the pedestrian toward whom the out-of-control car is headed, because the driver was in a position to foresee that even careful drivers of properly maintained cars sometimes lose control.[102] Applied to the case of subsistence wars, if individual members of the global poor are less responsible for the risk of death that they face than affluent people are, then any (non-responsible) poor person may permissibly select any (more responsible) affluent person from whom to exact resources, and use force against that affluent person whenever doing so would improve the overall distribution of harm, leading more responsible people to die instead of less responsible ones. That is, the reductive individualist picture, at least when it frames the issue of liability in terms of a proper micro-distribution of risk of death from less-responsible to more-responsible people, does not require shifting the burden to the *most* responsible person; if someone is more responsible for a risk to another than that other person is, that person is liable to force, even if there is some third person who is even more responsible. Fabre makes this point explicitly, contending that responsible people can be killed opportunistically, not only to stop them from making a direct and immediate contribution to wrongful harm, but also to break the spirit of other contributors to that harm.[103]

[102] Jeff McMahan "The Ethics of Killing in War," *Ethics* 114 (July 2004): 693–733 at 729. The same example occurs in McMahan, *Killing in War* (Oxford: Oxford University Press, 2009), p. 165.

[103] Fabre, *Cosmopolitan War*, 122.

Therefore those waging such wars could choose their targets strategically, killing those whose deaths are most likely to be effective, either directly or indirectly, in alleviating the threat.[104] Indeed, even in cases of individual self-defense or war between nations, the reductive individualist model does not restrict uses of force to those who pose the greatest or most direct threat; instead, the person who is unjustly endangered can kill any responsible person if the killing would be effective in protecting an innocent, non-responsible person. Contemporary just war writers could therefore argue that private subsistence wars could be justified based only on the greater responsibility of affluent people. The other practical difficulties perhaps still apply. There may be no circumstances in which any particular act of killing will be effective in saving even one life, and so in practice no private war could ever be justified. Then again, the effect of this reframing of the issue is to treat all minimally responsible affluent people as targets at all times, so that killing large numbers of inhabitants of affluent countries in order to pressure governments to change the terms of trade might fall within this account of just war. Low-cost terrorist acts, targeting large numbers of affluent people to sow a climate of fear and so lead the nations in which they live to transfer resources or revise terms of trade might face a realistic prospect of success.

Fabre hopes to resist such apparent implication of the analysis she proposes by emphasizing proportionality,[105] suggesting that most affluent people do not contribute enough to make them liable to being killed, but it is difficult to see how that claim can be maintained given the comparative nature of the determination of liability. On most revisionist views, proportionality attends to the seriousness of the harm being prevented, balancing it against the degree of force used or degree of collateral damage. Killing someone to stop them from taking your pen is disproportionate because of the relative weight of the interests in life and property in a pen; this conception of proportionality is supposed to apply in abstraction from any questions about culpability, leading to the conclusion that even the most malicious pen thief cannot be killed. By contrast, it is a commonplace of the revisionist just war literature that someone who is non-culpably but somehow responsibly attacking can be the subject of lethal force where necessary. Although the revisionist view could perhaps be modified to set a minimal threshold for responsibility, that would not be enough to limit the operation of proportionality, which, (as I note in my response to Renzo) many revisionists treat it as a continuous function of degree: the greater the good to be achieved, the more harm can be caused in its pursuit. The only way to bring such a conception of proportionality to bear on some idea of degrees of responsibility is to openly concede that the use of force has a punitive dimension, introducing focused proportionality requirements with respect to the culpability or responsibility of the offender. Many revisionist writers have sought to distance themselves from such ideas of culpability, proposing instead to, for example, authorize the use of force on the grounds that doing so will shift the unavoidable risk of death (which will fall somewhere) from a non-responsible person to a somewhat responsible one.[106] Given that affluent people are in a position to educate themselves about the needs of others, they seem to be in at least

[104] Fabre, *Cosmopolitan War*, 122.
[105] Fabre, *Cosmopolitan War*, 117.
[106] McMahan, "The Ethics of Killing in War," 729.

as bad a position as the careful driver. The careful driver is a philosopher's example, and the pedestrian has the opportunity to save herself because she is equipped with a ray gun that will vaporize both car and driver. The prospects of success are lower in the subsistence war, but, so too, the technologies might be less elaborate, including, as I already pointed out, opportunistically killing large numbers of minimally responsible civilians so as to generate pressure to change government policies or terms of trade or lead to an immediate transfer of resources. Such acts of violence are an instance of the form of individualism on which reductive individualists insist: the right is in principle enforceable, against all duty-bearers, and may even be enforceable in practice. The argument is therefore of more than analytical interest and concern.

My aim in identifying realistic occasions on which the revisionist account could justify subsistence wars is not to show that they are in fact defensible. They never are. The point instead is to bring out the presuppositions of the revisionist account and thus the stakes of the debate more sharply. Extraordinary circumstances might sometimes prevent someone from enforcing a genuine right, for example, as when a creditor cannot recover from a debtor because of the latter's insolvency, or the expiration of the statute of limitations, or when the only way to protect yourself from some wrongful but minor injury would kill large numbers of bystanders. The supposed right to engage in a subsistence war is unenforceable in a different, non-contingent way, because the alleged right is supposed to hold without being a right against anyone in particular. Here, the advantages of the Kantian view are clear: rights are always relational rather than comparative. They take as their subject matter how things stand between one person's power of choice and that of another, rather that the comparative weight of competing interests. Private persons are only permitted to use defensive force; they are not permitted to engage in vigilante enforcement of distributive justice. Fabre, like other revisionists, works instead with a non-relational interest theory of rights, supposing that someone could have a right to a certain state of affairs or option set whenever some urgent interest of the rightholder is important enough to put others under a duty. The picture of a right to subsistence thus entails a duty on the part of everyone to see to it that a global institutional context be set up in such a way that that interest is satisfied. The well-being-focused individualism at the heart of the revisionist program commits it to the conclusion that the duty cannot be borne by a collective or an institution, but must instead be borne by each individual, with collectives and institutions at most instruments for the joint discharge of the duty because of practical empirical obstacles to almost any individual discharging the duty on their own. It is hardly surprising that the remedial response to their failure to do what is required of them is radically indeterminate, because the requirement that they face is indeterminate in just the same way. The root of the difficulty is the assumption that someone can have a right to be in a certain condition, rather than that every right is a constraint on the action of some other natural or artificial person. An enforceable right to be in a certain condition inevitably leads to open-ended and indeterminate enforcement of it. The indeterminacy that Olsthoorn identifies in the enforcement of the right arises because it is already contained in the supposed right itself and its correlative duty.

Olsthoorn contrasts the contemporary just war account with the earlier Grotian idea of a right of necessity. He points out that the Grotian idea of a right of necessity is not indeterminate in the same way, because it is a right *in rem* to whatever specific object those in need require to preserve themselves. Unlike the right to subsistence in the contemporary literature, this right, too, will probably never justify an actual war. But it does not involve the same indeterminacy, precisely because the rightholder can simply take the object in question; the claim to make use of it does not depend on any past wrongdoing, but on need alone. As such, it is claimable against whoever has the thing that the rightholder needs. On this model, there is no problem of identifying the correlative duty-bearer: it is any person who owns or controls the object that the person in need has selected to satisfy the relevant need; among these people, the right of necessity permits the rightholder to pick the duty-bearer. That feature might seem to make the Grotian view more attractive; many critics are appalled by the fact that common law countries do not impose tort liability on people who fail to make easy rescues. Grotius limits the right of necessity so one person in need cannot simply transfer their need to another, placing other people in the condition from which they have just escaped. Applied to current global conditions, however, there is no shortage of people who would not be placed in need by being compelled to transfer a significant proportion of their resources. In the world in which we find ourselves, then, the Grotian right of necessity would entitle those in need to demand rescues that are not easy by any measure, and to kill those who failed to provide them. Yet the idea that someone can be killed for failing to make a difficult rescue is likely to attract fewer supporters.

Olsthoorn puts the point in terms of those who prevent those in need from satisfying their needs, but if there really is a right of necessity, it is not only a right to unobstructed access; it is a right to provision. The Grotian picture does not fully express this thought, because, as Olsthoorn emphasizes, in his picture, the right of necessity is derived from the compact that sets up a system of private property in the first place. Any such arrangement must be consistent with everyone's right to self-preservation. As such, the Grotian account insists that the operation of property rights is inherently limited to cases in which their exercise is consistent with every person's right of self-preservation. As Olsthoorn notes, Grotius treats that right as arising "directly and immediately from the Care of our Preservation, which Nature recommends to every one."[107]

This proposed right to self-preservation can be understood in two different ways, either relationally, as a constraint on the conduct of others; or, alternatively, non-relationally. On the relational interpretation, its Hohfeldian correlate is a duty to rescue on the part of others. Alternatively, it can be understood non-relationally, but then it becomes more difficult to represent it as right at all. If it avoids the indeterminacy of the just war approach, is does so only because the rightholder is able to select the duty holder by selecting an object that will satisfy his or her need. If there are multiple options—multiple sources of food, shelter, or access to opportunities—the Grotian right of necessity binds the owner of whichever one the person in need selects. Perhaps it would be better to describe it as a power of necessity rather than as a right, given that it is the entitlement or capacity to place another under an obligation. On the non-relational reading, then, the person in

[107] Grotius, *The Law of War and Peace* 1625: II.1.3, also III.1.2.1.

need has a right that does not have a correlative duty on anyone's part, only a conditional liability to being selected as the duty-bearer. By selecting you, the person in need thereby puts you under a new obligation. Indeed, the person in need can choose a rescuer who is not responsible for at all, even if more responsible parties are nearby. If the Grotian account is interpreted in this non-relational way, then it is more similar to the contemporary just war account than Olsthoorn suggests. Both accounts permit anyone who could have done something to alleviate the needs of the particular attacker to be selected as a target of redistributive force. By failing to give the thing up in anticipation of being selected, the person makes themselves a target of force where that force would be effective. Both are doctrines of private war; McMahan's remark that war differs from other violent conflicts only in its scale and complexity masks the extent of the private violence that revisionist just war theory embraces.[108]

Thinking about subsistence wars shows a fundamental problem for both the idea of private wars and the idea that rights are based on self-preservation, rather than on the relation between the choice of one person and another. The Kantian relational account rejects both private wars and wars of subsistence, because it recognizes that peace can only be achieved through public institutions, not through unlimited license to engage in unilateral action.

2.C. Human Rights in Conditions of Barbarism (Reply to Aravind Ganesh)

Aravind Ganesh's chapter brings Kantian theory to bear on the applicability of human rights norms in conditions of war and of barbarism. Standard human rights instruments, such as treaties and other international legal documents, represent those rights as holding against states; on some readings, they are understood in terms of addressing problems inherent in the state system. Ganesh argues that they must be of broader application, despite the fact that they must still be understood in terms of legal and political institutions. Taking up examples of human rights violations by illegal occupying powers, he develops the idea of *in bello* rules in a strikingly original way. On his model, human rights do not hold simply between every individual human being; they are institutional in a fundamental sense. Nonetheless, even institutions that cannot claim any modicum of legitimacy are bound by them. Ganesh shows that representing human rights in this way shows that these norms are not instruments in the service of a morality that is not structured by them, but are fundamental principles governing human interaction. Illustrating this point with the example of a case in which the applicability of human rights jurisdiction depended on whether

[108] McMahan writes that someone can be attacked if "they have become an obstacle to the prevention or correction of a serious wrong, or in which they pose an unjust threat to others. In these circumstances, they will have no justified complaint if they are subject to proportionate harm in the service of the just cause, or in order to avert an unjust threat for which they are responsible." McMahan, "The Morality of War and the Law of War," in *Just and Unjust Warriors*, eds. David Rodin and Henry Shue (Oxford: Oxford University Press, 2008), p. 22.

the occupying forces had taken custody of civilians before shooting them, he shows that only a non-instrumental Kantian account can explain the scope of human rights obligations.

Ganesh points to a general principle of pleadings to make his point: you cannot answer a charge of wrongdoing by asserting that you were committing a different wrong. This way of putting things has many advantages, to which I want to first draw attention, before explaining my reservations about it. First, a word about its scope: the principle itself applies only with respect to the elements of the wrong of which you were accused. If you are accused of murder, you can provide an alibi in the form of security footage of you committing a bank robbery in another city at the time of the crime; it only precludes your exculpating yourself by showing that you did indeed carry out the conduct of which were accused, but insisting that it qualified as a different crime. The obstacle to any such exculpation is that it does not exculpate at all with respect to the commission of the act of which you have been accused. The obvious advantage of appealing to this principle is that if the *in bello* rules are identified as limits on the means of defense, and the distinction between those who are and those who are not part of the war is articulated in terms of who is and who is not part of an aggressor's armed forces, Ganesh's proposal provides an elegant account of what goes wrong when the aggressor says, "but I was not fighting a defensive war; I was fighting a remedial one, and so it was appropriate for me to target those who were wrongfully retaining things that I believe are properly mine." The difficulty with the assertion of any such "justification" is that the principle of distinction precludes attacking those who are not part of the war, and so put limits on the ways in which even a defensive war is conducted. As such, the principle of distinction restricts the means of protection available in a defensive war, but the restriction does not follow from the fact that it is defensive; it follows from the fact that it is war. The same restriction therefore applies to wrongs committed by aggressors and those purporting to fight remedial wars. It does not restrict the means of achieving whatever purpose you might have to those that are inherently connected to that purpose.

Applied to the case of human rights, Ganesh explains how an occupying military exercises powers and is thus bound by the norms that apply *in subsidium* of the exercise of those powers. Thus it is not permitted to avoid accountability by claiming that it was not exercising those powers legitimately. Indeed, the example of belligerent occupation through which Ganesh introduces his analysis makes this clear not only with respect to human rights. A legitimate belligerent occupying power (that is, one that comes into occupation of another nation's territory in the course of, and for the duration of, a defensive war) is prohibited from attaching too much weight to its own military needs in proportion to the carrying out of its responsibilities, and prohibited outright from moving its civilian population into the occupied territory, or the residents out of it. The same norms apply to an aggressor that comes into occupation of the territory; it is also prohibited from having a disproportionate impact on the civilian population, or moving its civilian population into the occupied territories. Nor can it escape these strictures by insisting that it was not fighting a defensive war, or that its occupation of the territory was not legitimate.

Despite the advantages of this approach, my one hesitation concerns Ganesh's formulation of it as a principle about what someone accused of a wrong "may be heard to say." A court of law plainly operates with such a principle governing pleadings, and there is an interesting question about how it relates to other principles governing what can be said during the course of the operation of courts. Estoppel provides an interesting counterpoint; in some cases, it prevents a litigant from arguing something even though that thing is true and relevant to the matter currently before the court. The person who has falsely claimed that his property belonged to somebody else, which is then seized by that other person's creditor, is not allowed to assert his property right in an action for conversion.[109] The question is whether this is a principle for courts—like Kant's example of the rule of market overt (*vindicatio*) according to which a court must recognize as valid the purchase of a stolen object provided that it was made for consideration in a public market regulated by police ordinances—or whether instead it is a first-order principle of right to which a court must give effect. Kant's "four cases"—a contract to make a gift, a gratuitous loan, market overt, and acquiring guarantees through oaths—are all cases in which the principle for a court leads to the result opposite to what natural law would seem to demand.

Describing the principle to which Ganesh appeals in terms of what a person may be heard to say makes it seem like a principle for courts. Yet in other cases in which courts talk about what can be said—for example, in claiming that it "it does not lie in the mouth of the fraudulent person to say that they [damages] could not reasonably have been foreseen,"[110]—the court appears to be articulating a norm of conduct. The fraudster is blocked from saying something plainly false, because the one thing that the fraudster can foresee is that the dupe will believe and act on the false claim. The fraudster's entire enterprise is built on exactly that assumption. In this case, then, talk about what can be said is just a way of expressing an underlying norm of conduct; the rule the court upholds is the rule to which ordinary people are supposed to conform.

Turning to Ganesh's example of the shepherds, the relevant legal norm appears to govern the occupying power directly. It is not allowed to assert that it is engaged in illegal activity not because of the court's unwillingness to accept such a claim in justification, but rather, because the assertion is irrelevant. Human rights norms apply to those exercising political power, whether legal or illegal; they are presupposed by the very structure of exercising power.

One way to understand the relation between the two claims Ganesh considers would focus on the fact that the occupying power is seeking to avoid accountability for one wrong by asserting that it has committed a different and indeed more serious wrong. Ganesh suggests that if the occupying power claims not to have been exercising authority, it effectively confesses to having committed a war crime. It is easy to see why an occupying power would hesitate to implicate itself in this way. Nonetheless, I hesitate to characterize this as a situation in which the occupying power faces a choice between two options, each of which makes it a wrongdoer. Although an occupying power would not want to concede it had committed a more serious wrong, the problem is not that it ends up being wrongdoer either way. Instead, it is that choosing the other option does not relieve it of answerability under the human rights norm it has violated. That norm prohibits using positions of power and authority in ways

[109] *Horn v. Cole*, 51 N.H. 287 (1868).
[110] *Doyle v Olby (Ironmongers) Ltd* [1969] 2 QB 158.

inconsistent with the rights of those subject to that power or authority. It applies to the exercise of power: the fact that a particular exercise was illegitimate changes nothing; any claim that its exercise of power was illegitimate shows that it committed both wrongs, not that it committed a war crime rather than a human rights violation. Even defective exercises of political power are subject to the norms that govern its exercise. In this, the political entity that seeks to escape responsibility for human rights violations by denying its own legitimacy is like the driver who seeks to contest a speeding ticket by insisting that his license had expired, or the kidnappers who seek to answer the charge of child neglect by insisting that their custody of the child was illegitimate. Violating one prohibition does not exempt a person from responsibility for violating others, even if the same act violates both.

Ganesh makes a further point: even if the postulate of public right is not satisfied, and so the exercise of power is not legitimate, some restrictions on exercises of power that follow from ruling over people still apply. This is an important insight, which raises a further question about how to understand the scope of the application of the postulate of public right. Even people who are in what is, in significant respects, a state of nature, are subject to these obligations so long as they are exercising some form of political authority. Ganesh limits his discussion to cases of occupying powers, which exercise political authority in a way that, for example, a warlord does not. This suggests that the distinctive wrong can be committed only by something that, at least at some other time and place, satisfies the postulate of public right. A warlord is the kingpin of a criminal gang, an organization that commits crimes but does not rule. An occupying power is different, because it at least purports to rule.

This issue requires more attention than I can give it here, but may have the potential to explain the way in which even an aggressor state whose combatants commit war crimes differs from "a common robber band." In the Nuremberg tribunals, the chief British prosecutor, Sir Hartley Shawcross, contended that the Nazis were no different from a common robber band; the same claim was made at the Tokyo tribunals by Joseph Berry Keenan. They were making two interrelated points: first, they were rejecting the idea that war is a kind of gentleman's agreement in which opposed armies resolve conflict through violence—roughly the regular war picture defended by Grotius, Pufendorf, Wolff, and Vattel—and with it the thought that being a member of the military conferred special legal status permitting otherwise forbidden acts. As I explain in *Kant and the Law of War*, Shawcross, Keenan, and the contemporary revisionists go wrong in supposing that the *in bello* rules of war provide special permissions, rather than introducing additional prohibitions that are specific to war. With Ganesh's help, I now see that, if anything, I understated the case. An aggressor's commission of war crimes is worse than a common robber band. As Ganesh observes, we can understand the condition of war under the idea of peace, and if we do, we not only identify the distinctive wrong committed by those who conduct war in ways inconsistent with the future peace; we find the moral and conceptual basis for understanding the distinctive obligations that even such an aggressor owes to those over whom it exercises power, and the distinctive and compound wrongs that it commits when it abuses those powers. If that power were exercised peacefully and legitimately, those exercising it would owe obligations to those over whom it is exercised. The fact that the exercise is not legitimate does not generate any additional permissions or dispensations; as well as the occupation being wrongful, the acts carried out under it are also wrongful. Identifying the multiple wrongs is important to understanding the moral structure of the situation. The fact that the aggressor is engaged in activities that are already prohibited gives it no license to violate further prohibitions.

Let me conclude by applying this analysis to the *Al-Skeini* case that Ganesh raises, picking up on the example that he discusses,

> of two hypothetical scenarios, the first involving shepherds simply being shot by enemy soldiers, and the second involving those shepherds first being driven to a cave by those soldiers before being shot. Recall also that the court opined that human rights jurisdiction would not have arisen on the first set of facts, but that it might have on the second. An explanation, elusive on any of the mainstream approaches, can be provided on the Kantian account. In the second situation where the shepherds were ordered into a van and driven to a cave for a brief period, it can be said, however implausibly, that they were governed. (Ganesh, p. 165)

Ganesh's reasoning is that in such a case the only alternative to conceding the wrong would be the worse one of admitting to the commission of a war crime. Here is a different way of understanding the sense in which the shepherds were in fact governed. It is not merely incidental to the example that the shooting is done by enemy soldiers. They were not members of a roving robber band; as such, the soldiers must be understood not as private persons, who, were they to have shot the shepherds, would have been ordinary murderers. Instead, they are acting as minor public officials, and so, to use Thorburn's formulation, acting under authority. When a public official takes control of someone, they are acting in a public capacity, and so are subject to the norms governing the exercise of public power. Put differently, they cannot exempt themselves from their role as public officials, and so are necessarily subject to the norms that govern the occupants of their role. They must therefore act in conformity with human rights obligations. It is not that they could only claim to escape these obligations by acknowledging that they were doing something even worse, such as simply shooting unarmed civilians. It is that as soon as they exercise control over a human being, they are bound. In some cases they could commit both a human rights violation and a war crime, as an occupying power does when it engages in reprisals against the civilian population. To do so is a war crime, even though in committing it they are also violating their obligation to exercise power on behalf of those over whom it is exercised.

2.D. Public Offices and the Standing to Punish (Reply to Malcolm Thorburn)

Malcolm Thorburn takes up the issue of what is sometimes called the "combatant's privilege," the prohibition on punishing "war-making activities," in particular, the prohibition of punishing them for mere participation in an unjust war. This "equal criminal immunity of combatants" is, Thorburn notes, "probably the best example of a doctrine that appears to most moral philosophers to be either totally repugnant or else explicable only as a sort of depressing concession to reality" (Thorburn p. 172).

As Thorburn explains, the Kantian approach views this matter very differently. These differences have their root in the Kantian conception of the distinctive moral nature of legal obligation. The writers who Thorburn identifies as "legal moralists" suppose that morality is complete in absence of legal institutions, and that legal doctrines and concepts can be nothing other than directives, the point of which is to bring each person more fully into compliance with reasons that apply to them independently of those directives. As Thorburn describes the view, it sees the moral task as one of comparing "the substantive norms of a particular area of law with what the author takes to be the demands of morality and to use that comparison as the grounds for moral evaluation of the field" (Thorburn p. 171). As Thorburn explains, the Kantian view sets itself a very different question. All questions of legality and its morality turn on how things stand as between persons, with the basic problem as one of explaining how one (natural or artificial) person's choice can necessitate the choice of another, that is, put that other person under an obligation. The simplest case of this is in interpersonal interactions: your choice determines whether I may permissibly interact with your body and property in ways in which I otherwise could not. Public right governs "norms put in place as between some authority and its subjects" and so the question of justification concerns the conditions under which, and the ways in which, public legal institutions can impose obligations on those who are under their authority, that is, how one (artificial) person's choice can necessitate another. The Kantian solution focuses on explaining how such relations of authority can be possible between independent human beings, that is, human beings who are not subject to the choice of any other. The solution, as Kant develops it in the *Doctrine of Right*, and as I have explained it in various places, is to see that the ideas of independence and interpersonal authority are already mutually implicating: each person's entitlement to independence is part of a system in which everyone also has a certain kind of authority as against all others. The reconciliation of these independent claims to authority requires public right. A public authority is an artificial person—Kant's preferred eighteenth-century term is "moral person"—which can only act through natural persons, who in turn can only act for the artificial person by acting in official roles. Thorburn has done more than anyone to develop a systematic account of the public nature of official roles, emphasizing the ways in which public officials always act under authority, and how the permissible exercise of their public powers always depends on the mandates conferred by the authorities under which they act. The familiar legal and moral questions that shape so much of criminal law doctrine, particularly in the case of justification defenses, are, as he has demonstrated, fundamentally questions about whether someone acting in a public capacity does so within the scope of their authority.[111]

Thorburn ably explains how this account extends to the case of various types of public officials, and illustrates the way in which I sought to incorporate it in my treatment of the immunity of combatants from prosecution for mere participation in an unjust war. As Thorburn puts it, "To articulate the position, however, is not (yet) to provide an argument for it." He identifies several gaps in my account:

[111] Malcolm Thorburn, "Justifications, Powers, and Authority," *Yale Law Journal* 117 (2008): 1070–1130.

> First, he must show why we should ever think of soldiers as just the embodiment of their state, rather than as individuals doing criminal acts. And second, he must explain why states must defer to wildly implausible determinations by others states that they are under attack and need to take defensive action. Although Ripstein sets out all of the basic building blocks for the Kantian account of combatant immunity, he does not draw them together tightly in an explicit argument in chapter five of *Kant and the Law of War*. (Thorburn p. 179)

Thorburn illustrates his concern with the example of Nazi Germany's invasion of Poland in 1939. Imagining that a German soldier had been captured by Polish forces, he raises the question of whether such a person could be put on trial, had Polish institutions lasted long enough to do so. The puzzle, as he articulates it, is that Poland's right to political independence and territorial integrity was at issue in the invasion; Germany's was not. Further, Poland's entitlement to defend itself was based on a claim to wrongful invasion that was uncontroversially true. Why, then, should Polish legal institutions be bound by Germany's wildly implausible claim that the invasion was an act of self-defense, the only support for which consisted of photographs of concentration camp inmates who had been dressed in Polish uniforms and shot? In the unlikely event that the German soldier he imagines being captured could credibly claim to believe such claims, that would at most provide some sort of excuse, but excuses only arise after the wrong has been established. Nothing of the sort could provide a general exemption from even being put on trial. Yet the position of international law, and indeed Kant's position as I articulated it, preclude putting ordinary soldiers who do not violate the *in bello* rules on trial.

Thorburn proposes to fill in the missing pieces by offering an account of the public office of soldier, and what falls within its scope. From this he argues that punishing another state's soldiers falls outside a state's jurisdiction. I am grateful to Thorburn for providing a route to what seems to me to be the right result, and broadly sympathetic with his attempt to fill in the missing pieces of my account. I still have reservations about some of the details that he fills in. I will explain that account in more detail, before articulating my reservations and my slightly modified versions of his account.

Thorburn explains that the public office of soldier can only exist through a distinction between the soldier as a private person and the soldier as a public official. As a public official, it is not up to the soldier to decide whether to fight in some particular war. This part of his claim is exactly right; a nation can only defend its political independence and territorial integrity if its combatants act under authority, so that it (understood as a political unit) alone decides whether to defend itself, rather than each soldier making a private decision about whether to participate in responding to a perceived threat. The question of whether to act to prevent an attack on the nation—rather than on an individual human being—is a question that can only be answered by a properly constituted public authority. In this respect, national defense differs from individual cases of self-defense. As Thorburn has shown in his other work,[112] within a legal order, questions of the justification of defensive force are framed as they are because someone exercising force does so under authority, and each of the elements of the defense rest on this distinctively public authorization.

[112] Thorburn, "Justifications, Powers, and Authority."

In the case of a legal order under attack, any defense of it must be the act of the legal order as such. Although Thorburn does not take up this point, I take it that this thought applies even to a *levée en masse*, in which citizens take up arms to repel an invasion of their nation. Even in such a case, their action must be understood as something done by and in the name of the nation; although a *levée en masse* is often characterized as spontaneous, a spontaneous taking up of arms only qualifies if the enemy has not yet occupied the territory, the attack is imminent, and there is no opportunity to join organized military forces.[113] As in the domestic case, so in the law of war, self-help and self-enlistment is an extremely limited exception to the general requirements of authority. Thorburn's own work on justifications in the criminal law provides an illuminating explanation of the way in which even cases of self-help, such as self-defense or recovery of stolen property must be treated as acting under authority, in particular, as a form of self-deputization into a publicly limited public role.

The point of this brief digression into *levée en masse* is to underscore the importance of a requirement of proper authority in order for someone to qualify as a combatant, and so be entitled to prisoner of war status, and thereby exempt from punishment for participation. As Thorburn explains it, this requirement of acting under authority (even if, *in extremis*, one is a self-appointed official, one still acts subject to authority) is essential to immunity from punishment.

All of this seems right, and I am grateful to Thorburn for pressing me on this point. Still, I now need to say more about three issues. First, there must be some explanation of why acting under authority operates as a bar to prosecution in the case of war, rather than as an affirmative defense as it does in the domestic case. If a police officer shows that a search, arrest, or use of force was duly authorized, or if, *in extremis*, a private person offers a justification defense for making a citizen's arrest or using defensive force, if successful, it provides a complete defense. Such a justification defense, however, can only be brought forward after the elements of the offense have been established. Although it may make no sense to press charges in a case in which some justification defense is plainly available, it is not the same as the bar to prosecution the combatant enjoys. A prisoner of war does not need to offer a defense to a charge; the "combatant's privilege" blocks charges based on participation in the war.

Second, some explanation is required of why individual and state responsibility are mutually exclusive. If the state engages in a policy of perfidy or targeting civilians, that policy and the acts falling under it can be imputed to the state, as well as to the individuals who carry them out. Culpability for war crimes can work its way up the chain of command, just as the orders to engage in them work their way down the chain of command. By contrast, combatants who fight in conformity with the *in bello* rules are not responsible for the acts of the state that acts through them.

Third, and most fundamentally, a legal relationship between two legal persons—the public official and the legal order—does not normally change the moral or legal situation of anyone else, particularly not of persons who are not subject to that legal

[113] Article 2 of the 1907 Hague Regulations provides that [t]he inhabitants of a territory which has not been occupied, who, on the approach of the enemy, spontaneously take up arms to resist the invading troops, without having had time to organise themselves in accordance with Article 1, shall be regarded as belligerents if they carry arms openly and if they respect the laws and customs of war.

order.[114] Appointment as a tax collector entitles a person to take money from others, but does not really change the situation of those from whom the money is taken, because the state was already entitled to take it. The official's status changes the relation between the official and the state, and serves as a bar to punishment for acting within the scope of the office because the office commits no crime within its mandate. In the domestic case, a justification defense establishes that the defendant was acting under the state's authority; it is the relation between the action and the state that claims the right to punish wrongdoing that serves as a bar to punishment. Having authorized the official's action, the state could not be in a position to punish the official for acting within the mandate that it authorized.

The case of enemy combatants seems to be different, because the question is not whether the state that sends them to fight in an unjust war is entitled to punish them for doing so, (it is not). Rather, the question is whether the state against whom they unjustly wage war is entitled to punish them. Put differently, the challenge is to find an explanation of how the relation between unjust combatants and the state on whose behalf they act places any kind of limitation on the rights of the state they unjustly attack. More needs to be said about how this structure could bind a different state.

I will try to address the first and second challenges by addressing the third. Within a domestic legal order, the fact that someone is a public official authorizes that person to carry out a prescribed mandate. It does not change how things stand between police officer and criminal suspect, prosecutor and accused, prison guard and prisoner, or tax collector and taxpayer, because the legal order was already in the relevant relation to suspect, criminal defendant, prisoner, and taxpayer. Each official's acts are just the way in which the legal order exercises its authority in relation to particular people. That is why the arresting officer does not commit the tort of battery, the prosecutor does not defame the innocent criminal defendant, the guard does not commit the tort of false imprisonment, and the tax collector is neither thief nor embezzler. Conversely, those subject to these exercises of authority are not allowed to resist them. The prisoner is not allowed to escape or fight off the guard, and the taxpayer is not allowed to evade taxes or use force to prevent their collection. Within a domestic legal order, those who act properly under authority not only cannot be prosecuted: they cannot be rightfully resisted.

The situation of the aggressor's combatant looks importantly different: the aggressor's legal order does not stand in a relation of authority to the defender's combatants. The defender's combatants do not need to put up with the acts of the aggressor's officials; they are entitled to use force to resist them. That is, the defender is allowed to act through its combatants to resist the aggressor's acts. Although, as between the aggressor's political and military leadership and its individual combatants, the fact that the latter act under authority is relevant—the aggressor's own courts cannot hold them accountable for things done within the scope of that authority

114 Just as a contract between two persons cannot change the rights of a third. In other cases of special legal relations, the arrangement between two persons does not limit the rights of a third. A trustee or other fiduciary owes a duty of loyalty to the beneficiary of the trust, but that duty of loyalty does not empower the trustee to disregard the legal rights of others.

relation—the authority relation needs to be made relevant in some other way if it is to restrict the actions of the defender.[115]

The solution is to focus on how things stand between the aggressor's legal order and the defender's legal order. In the manuscript, I was not sufficiently clear about when this relation is and when it is not relevant. The state can only try to punish someone to whom it stands in the right relation, even if that person has committed a wrong that it would be entitled to punish had someone to whom it stands in the right relation done it, and even if another legal order would be entitled to punish the same act. In the case of the aggressor's combatant who kills on the defender's territory, ordinary criminal law principles would make the aggressor's combatants liable to prosecution and punishment by the defender's legal order; the wrongfulness of the invasion does not exempt the combatant from territory-based criminal law.

The relation between the defender's legal order and the individual combatant is not the problem; the relevant relation is between the legal orders of the defender and aggressor. A nation is entitled to defend itself against aggression—to stop the aggression—not to hold the aggressor accountable for it. As I argued in chapter 3 of *Kant and the Law of War*, remedial, punitive, and redistributive war are wrongful, and defensive war cannot be reframed as a version of any of these. The entitlement to stop aggression neither rests on nor generates any kind of forfeiture of the independence of the aggressor's legal order. The fact of victory, even victory over an aggressor, does not entitle the defender to demand compensation for the costs of war or to punish the aggressor for its wrong, because the aggressor's legal order remains independent, and so is not subject to the defender's authority or jurisdiction. As such, a victorious defender can neither impose new obligations on the aggressor's citizens nor suspend the power of the aggressor's claim, as against its officials, to determine for them when or whether they will fight. Like any such authorization, it applies even when the aggressor is mistaken or dishonest about the grounds for it. If the defender cannot obligate combatants to disregard their own legal order's judgment about the grounds of war, it cannot punish them when it is victorious.

The same point applies with respect to particular combatants who fall into its hands as prisoners of war: the fact of capture does not give the defender any new kind of authority over them. The Third Geneva Convention puts the point in terms of the prisoner: "the accused, not being a national of the Detaining Power, is not bound to it by any duty of allegiance."[116] This generates a bar to prosecution because the Detaining Power is not entitled to demand of them that they not carry out their official duties, even though by detaining them it can prevent them from doing so. That is not to say that they should carry out those duties, only that no other legal order can demand that they not. The result is a bar to prosecution rather than an affirmative defense.

The claim, then, is not that someone who acts under authority is not subject to punishment for anything they do, but rather that no other nation has authority over the

[115] Kelsen presents the rule that imputes the act to the state rather than the individual combatant as a positive rule of international law; although he elsewhere characterizes the principle of individual rather than collective responsibility as civilized, he does not explain the rules of imputation in those terms. Hans Kelsen, "Collective and Individual Responsibility in International Law with Particular Regard to the Punishment of War Criminals," *California Law Review* Vol. 31 (1943): 530–571.

[116] Art. 87.

actions of those who act within the scope of the legal order's authority. In another paper, Thorburn makes a similar point in terms of the idea of sovereign immunity, which he takes as a fixed point of the current international legal order: Sovereigns cannot be punished.[117] It is more than a fixed point of positive international law, however; it is rather, from the Kantian perspective, simply the form of independent legal orders. The structure manifests itself in other places also: if one legal order imprisons another country's national for a crime the national did not commit, that is ground for protest, but does not justify punishing the prison guard who detained the person should the guard fall into that country's power. In the same way, if a nation inappropriately withholds taxes on a foreign national's income, that person's home country is not entitled to charge the tax official with theft or wire fraud. The independence of the separate legal orders precludes putting these officials under an obligation to accept another legal order's judgment about whether the punishment or taxation is warranted. So rather than saying that the sovereign state cannot be punished for either of these things, it seems to me more perspicuous to say that the state's officials cannot be punished.

The point is not merely one about wording, however; sovereign immunity is a conclusion of my argument rather than a premise to it. The legal order cannot be punished for the acts of its officials acting within their proper mandates because those mandates are outside the authority of any other legal order. The account therefore does not rest on the idea that, having taken some form of an oath of office or allegiance when taking up the office, the officeholder is somehow bound to accept authority. The scope of promissory obligation is not so wide as to authorize wrongdoing by the promisor. The point instead is that punishing wrongdoing is an essential part of the state's mandate, and it is entitled to do so through its officials, and is not subject to punitive oversight by other nations. That does not mean that the officials have a moral obligation to do as they are told, only that no other nation can put them under an obligation to disregard what they have been ordered to do within the scope of their official roles.

A combatant who violates *in bello* rules—either by targeting civilians or engaging in some form of perfidy—can be prosecuted. That combatant's own legal order is already under an obligation to punish those acts, and international law must presume every legal order to have already prohibited violations of *ius cogens* norms. Indeed, the third Geneva Convention is structured by the supposition that the normal forum for punishment of war crimes is the combatant's own legal order, and the Statute of the International Criminal Court assigns it a residual role in cases in which the combatant's own legal order fails to prosecute. Further, combatants are required to disobey manifestly unlawful orders; a combatant who obeys such orders can be punished for doing so, as can the superior who issues it. The prohibited act is imputable to both by any and every legal system. The double imputation of the wrong, to both the superior(s) who order it and the combatant who carries out the violation, reveals the way in which a combatant who violates the *in bello* rules remains a combatant. That person does not become a civilian, but rather is better understood as a corrupt official, the corruption of whose act can go all the way up the chain of command.

117 "Soldiers as Public Officials: A Moral Justification for Combatant Immunity," *Ratio Juris*. Vol. 32, No. 4 (December 2019): 395–414.

The aggressor is also already prohibited from engaging in aggression, and its independence from other legal systems does not permit it to do so. But the point at which the defender is not allowed to direct the aggressor's legal system is not that the defender cannot prohibit the aggressor's already prohibited aggression. It is rather that the defender is not entitled to prohibit the aggressor's officials from accepting its determination.

This explanation of how things stand between the two orders then explains why the role of soldiers a public officials does not figure in a defense to a charge the elements of which have been established, but rather preempts the entire proceeding: in the domestic case, a justification defense goes to whether the defendant is acting under the authority that the state itself has conferred; in the prisoner of war case, the bar to prosecution reflects the fact that the detaining power did not have the requisite authority over the prisoner.

Thorburn is right that the role of soldiers as public officials figures in the bar to punishing them. In this response I have tried only to further fill out the presuppositions of the relevance of that role.

2.E. A Priori Concepts and New Institutional Forms I (Reply to Ester Herlin-Karnell)

The papers by Ester Herlin-Karnell and Bertjan Wolthuis and Luigi Corrias take up the question of the juridical status of the EU. Herlin-Karnell focuses on how things stand between it and other nations, and in particular, whether it is, juridically speaking, a form of a superstate that in some way limits the juridical independence of its members. Wolthuis and Corrias look at its relations to its members, asking whether it is a form of cosmopolitan right. Both papers are illuminating in the way in which they bring Kantian relational ideas of public right to bear on questions of novel jurisprudential forms. For Kant, there are exactly three forms of public right, corresponding to the three types of relations in which human beings and legal orders might possibly stand. The right of the state governs how things stand between a state and its members; the right of nations governs how things stand as between a plurality of independent states; and cosmopolitan right governs how things stand between a state and those who are not members of it. The questions these papers take up get their interest and importance from the fact that Kant's classification of possible relations presents itself as mutually exclusive and jointly exhaustive, but the EU seems to be a counterexample either to the claim of mutual exclusion or of conceptual exhaustion, as it does not fit neatly into any of them. Herlin-Karnell concludes that the EU is a new type of state; Wolthuis and Corrias that it is a distinctive and restricted type of cosmopolitan order. The nature of the EU is a topic of great interest and considerable practical importance, but not one to which I have devoted any attention. As such, what I say here will necessarily be very tentative.

Herlin-Karnell helpfully frames this question in terms of the idea of legal personality: the structure of the EU differs from both that of a unitary legal order and of a treaty between fully independent states. A unitary legal order has its own legal personality, and, since the Reparations Case, so too does the United Nations.[118] Yet the EU is structured differently from either of these cases. It is not a unitary state in any straightforward sense, since it is comprised of its members. But unlike the United Nations, it is not a world body that claims mandatory membership by other states. This then raises the question of whether it has independent legal personality in the international realm.

Herlin-Karnell connects this question to a more general discussion of defensive force and its relation to (and contrast with) preventive force. As she notes, the current international legal regime is broadly consonant with the Kantian approach: resort to war is only justified to preserve the peace, and so subject to a requirement that war is in response to an imminent threat. If the attack against which defensive force is to be used is in the future, then other means of preventing it might be available. As she also explains, the requirement of *ad bellum* proportionality also invites a Kantian interpretation: force must be narrowly addressed to the immediate threat. In this, then the current law repudiates the Grotian idea that the inner dynamic of law leads inevitably to escalation, so that every war is eventually about which sovereign is in charge of the other. Instead, the Kantian view is that all conflicts are to be minimized, and so any response to the use or threat of force must be narrowly tailored to the specifics of that use or threat. The court's choice of words in the Nicaragua case to which she refers is that force must be "necessary" and "proportionate," but these are separate requirements. Even necessary force can be disproportionate. The requirement of imminence and *ad bellum* proportionality jointly prohibit preventive war.

As Herlin-Karnell points out, many recent writers and world leaders have sought to blur these distinctions, some suggesting that those who are under pervasive threat of future attacks should be permitted to engage in preventive war, others arguing that the line between *ad bellum* and *in bello* rules should be relaxed. The only possible Kantian response to these suggestions is to insist that the point of the rules of war is not to create a level playing field or to ensure that those with a just cause always win. It is instead to protect the peace.

On this understanding, the imminence requirement for defensive war shares a structure with, but nonetheless operates differently than, the requirement that goes by the same name in the case of individual self-defense. Structurally, both individual and collective defense can only be used when no alternatives are available; imminence is the temporal manifestation of the lack of alternatives. In the individual case, the salient alternative is calling on the state for protection. If a police officer is nearby, it falls to the officer rather than the person under attack to prevent the attack; if the attack is already in progress, the option of calling on a public authority for protection is unavailable. The alternative of seeking help from a public authority is not available in the international case, but even it has, at least in principle if not in fact, an analog: force can be used to hold off an attack until the Security Council is able to act. The Security Council's ability (and frequently, its willingness) to act is lamentably limited. That does not undermine the structural parallel; individuals may find themselves under attack in a place where the police are absent or indifferent. The difference between

118 *Reparation for injuries suffered in the service of the United Nations*, Advisory Opinion [1949] ICJ Rep 174, April 11, 1949.

the interpersonal and international cases is that, as I explain in Chapter 3 of *Kant and the Law of War*, an aggressive attack in wartime is temporally extended. It is thus underway for much longer, raising the possibility of a more extended response to it. That in turn means that once the conflict has begun, defensive force could be used in a way that would seem preventive in the individual case, such as targeting someone who is not in the process of attacking, because they are a participant in an ongoing armed conflict. Herlin-Karnell gives the example of targeted killings, which aligns with the traditional rule that combatants can be targeted during wartime when they are not in the process of attacking. These examples are not exceptions to an imminence requirement, but rather implications of that requirement in cases in which the attack that provides the warrant for defensive force is itself temporally extended. For all of the same reasons, an international attack could, at least in principle, be imminent for an extended period.

Herlin-Karnell's chapter brings these more general questions about defensive force to bear on the question, both pressing and perplexing, of the distinctive legal character of the EU. As she points out, it has a distinct legal personality, apart from that of its member states. That means that acts can be imputed to it—it can be a wrongdoer, and where its public officials do wrong, it is answerable for that wrong. Conversely, a state, in particular, can be wronged, and that a wrong against it is distinct form a wrong against its individual members, just as a state is wronged by the mistreatment of its diplomats.[119] The EU's legal personality also means that it can impose distinctive claims on its member states, and potentially on the individual human beings residing within them. Yet it is not a state in the way in which any of its members is a state, or in which any non-member is. Nor is it an international entity in the sense in which the United Nations is, or merely a treaty lacking legal personality. Nonetheless, it is able to impose binding legal obligations between its members; among these are obligations of both solidarity—requiring member states to protect each other—and collective defense—requiring members to protect the whole. But in addition, precisely because it has an independent legal personality, the EU itself should be in principle entitled to defend itself, even though it lacks a military of its own, so that the only way it can do so is by calling on its members to come to its defense.

This structure raises an interesting conceptual question: what would it be for the EU to be under attack and so entitled to defend itself, other than for one or more of its members being under attack? Any defensive entitlement that the EU has as a separate legal person would have to be something over and above the several defensive entitlements of its members. A kinetic military attack on the EU would invariably also be an attack on one of its members, but, as I explain in my responses to Stilz and Renzo, a juridical person that is attacked is entitled to decide whether to use defensive force. So although a territorial attack on the EU would also attack at least one of its members, the question of whether to respond to that attack would be a question for it as well as for its attacked member. But it is also entitled to call on its members to defend it, and so to defend each other. As Herlin-Karnell emphasizes, this falls under the solidarity provisions of the EU *Charter*. The members each owe duties of solidarity to the others.

[119] For a general discussion, see Roland Portmann, *Legal Personality in International Law* (Cambridge: Cambridge University Press, 2010).

Herlin-Karnell emphasizes the solidaristic foundation of this duty of defense. Nations are entitled to come to each other's aid when they are under attack, and may undertake duties to do so through various mutual defense treaties. By focusing on solidarity as the basis, Herlin-Karnell raises an interesting question about the kind of legal entity that the EU ultimately is. If two nations enter into a mutual defense pact, each has a contractual duty under it to come to the aid of the other if attacked. Any such duty only holds as between the parties to the treaty. In the EU case, as Herlin-Karnell describes it, the duty only holds among member states, but it is different from any such pact, not only because it is more demanding, putting member states under an "obligation of aid and assistance by all the means in their power," as opposed to, for example, NATO's much weaker "assist the Party or Parties so attacked by taking forthwith, individually and in concert with the other Parties." More fundamentally, they differ because the basis of duties is different. Thus the NATO treaty binds its members to defend each other, but it does not bind them to defend NATO. The contrast, as Herlin-Karnell explains, is that each EU member's duty to protect each other is mediated by the duty to protect and preserve the EU as an "area of security, cooperation and justice."[120] An area of security could not exist unless secured; like a state, it has its juridical personality in a specific location on the earth's surface. As an area not only of cooperation and justice but of security, the plurality of states must be conceived as a unity without, however, abrogating the separateness of its members. Herlin-Karnell raises a series of issues about whether that separateness is actually preserved.

Herlin-Karnell addresses the issue of separateness by considering the role of judicial review. There are few decisions made by a state as consequential for its citizens than the decision about whether to go to war, but in practice it is among the least scrutinized of decisions. In the case of collective security, and the strong form of the obligation of collective defense, the challenge to judicial review is even greater. I share her concern; leaving the decision about whether to go to war in the hands of the executive branch, or the combined hands of the executive and legislature turns it into a political question in what might be regarded as the pejorative sense of that term. Leaders and elected representatives might have a short time horizon, looking to the next election, and might also overreact for narrowly partisan reasons, or fall under the sway of people with agendas of their own. At the same time, if the problem is simply one of poor quality decisions being made because of the incentives faced by various branches of government, is it is not clear that judicial review is a solution to it. Historically, when war is raging or looming, judges have not always done a very good job of curbing the excesses of state power, as the notorious *Korematsu* decision in the US illustrates.[121] Those frailties are compounded by the difficulty of representing the question of whether to go to war as a legal question of the kind that court is in a position to answer. The justification of defensive war is fact-sensitive, and the court is not in a position to operate as an independent finder of fact on such matters, and so will, perhaps inevitably, defer to the executive. In addition, any such process would have to take place *ex parte*, without the aggressor being entitled to claim that it is not attacking. That is, the court, in assessing the nature of the threat, can only ask whether

[120] Article 3(2) TEU.
[121] *Korematsu v. United States*, 323 U.S. 214 (1944).

proper procedures were followed in making the determination. This raises a further problem in the case of duties of solidarity and mutual defense: if one EU member state says it is under attack, the other members can perhaps investigate the truth of the claim, but as Herlin-Karnell describes the structure, they are obligated to come to its defense regardless of their own judgment about the matter. Terms of treaties are frequently litigated, but given the urgency of the occasions of legitimate defensive force, the suggestion that whether something falls within the terms of a mutual defense agreement could be subject to litigation seems different.

In addition to these empirical questions about judicial review, Herlin-Karnell's discussion also raises what must be, from the perspective of Kantian legal philosophy, a conceptual and normative question. She points out that judicial review is not inconsistent with democracy and is better understood as essential to it. Courts are charged with making sure that executive action is in accordance with law. That requirement is sometimes understood in a narrow, almost deflationary sense, that is, as looking only to whether, for example, an official acts within express or implied statutory powers. When it comes to international relations, different legal orders take up international law in different ways, but, at a minimum, treaties that have been expressly incorporated, as well as customary international law, are considered to be part of domestic law, and so relevant to the question of whether an official act is within statutory powers.

In addition to this judicial minimalism about whether a particular subject matter is within a particular official's mandate, there is a further question on the Kantian account of whether in a particular case, the official is acting for a properly public purpose. That is, the Kantian idea of the rule of law is inconsistent with the idea of prerogative power, and so opposed to the exercise of the power that is not subject to review with respect to the question of whether it was used for an improper purpose. Historically, that is, prior to the Kellogg-Briand pact and the UN Charter, the right to go to war was treated as a prerogative, not only internally but externally. In Kant's time, rulers claimed to be entitled to simply decide; later legal systems imposed procedural but not substantive restrictions upon the power to declare war. Kant advocates a much narrower account, writing that the people "must therefore give their free assent, through their representatives, not only to waging war in general but also to each particular declaration of war," and that "We shall therefore have to derive this right from the duty of the sovereign to the people (not the reverse); and for this to be the case the people will have to be regarded as having given its vote to go to war. In this capacity it is, although passive (letting itself be disposed of), also active and represents the sovereign itself."[122] This distinctive view of the direction of justification gives some explanation of why going to war cannot be a prerogative power—it must be done on behalf of the citizens. But if the member states of the EU are bound to come to each other's defense, then Herlin-Karnell must be correct that this shows the EU itself to be a state among states. Its members do not have the right against each other or the EU as a whole that Kant describes as "original" that is, to avoid getting involved in a state of actual war. Kant's own discussion concerns a world organization and the right to "avoid getting involved in a state of actual war among the other members

[122] 6:346.

(*foedus Amphiclyonum*),"[123] but the point carries over to this case. One way to determine whether something counts as an independent order is whether its members are entitled to refuse to get involved in each other's disputes. Herlin-Karnell's analysis shows that the EU's member states are not entitled to so refuse.

2.F. A Priori Concepts and New Institutional Forms II (Reply to Bertjan Wolthuis and Luigi Corrias)

Like the chapter by Ester Herlin-Karnell, the chapter by Bertjan Wolthuis and Luigi Corrias raises the question of how a Kantian approach to public right would analyze the legal structure of the EU. As I noted in my response to Herlin-Karnell, this is not a topic to which I have devoted any attention, and of course it is not one that Kant could even have imagined.

Wolthuis and Corrias focus on the kind of jurisprudential entity that the EU is, arguing that it cannot be understood using the concepts Kant articulates through either the right of the state or the right of nations. They therefore conclude, by process of elimination, that it must be understood as a distinctive type of cosmopolitan union. Wolthuis and Corrias are thus concerned with its internal structure: How do things stand between individual human beings residing within the EU, and the legal institutions that exercise power over them? Herlin-Karnell, by contrast, focused on its external structure, the way in which the area of peace, security, and cooperation governs the relation between the EU and nations and institutions outside of it.

Wolthuis and Corrias interpret the EU as a system of cosmopolitan law on the grounds of the rights that it grants to citizens of one member state in relation to the other member states, writing

> In the vocabulary of Kant's theory EU law has to be classified as an instance of cosmopolitan law, we argue, because this law grants the citizen of an EU member state rights in foreign states that are members of the EU. The citizen of an EU member state has the right to freely bring goods, services, and capital to foreign EU member states, for example. Trade tariffs or similar obstacles are prohibited. The citizen of an EU member state also has the right to travel, work, study and retire in the foreign state that is a member of the EU. (Wolthuis and Corrias p. 205)

The organizing thought is that EU law sets up a new set of relations between individual human beings and states of which those individuals are not members. The citizen of one EU member state is not (based on that alone) a member of another, but has considerably more than the right to visit to which Kant limits cosmopolitan right.

Wolthuis and Corrias understand the cosmopolitan nature of the EU narrowly, focusing in particular on the problem to which Kantian cosmopolitan right is the solution, that is, how things stand between a legal order and human beings who are not members of it. As such, their characterization is not limited to the narrow Kantian

[123] 6:344.

"right to visit," but, they argue, it is a different solution to the same structure of interaction between a state and a person who is not a member of it. So where Kant understands the form of interaction as governing its solution—there cannot be more than a right to visit because of the presuppositions of the problem, that is, because everyone must be entitled to interact with others and remain a bearer of rights wherever they are—as I understand their argument, Wolthuis and Corrias propose to distinguish between the problem and its possible solution by showing that more than one solution is possible. The problem concerns the form of relation, that is, how things stand between a state and human beings who are not members of it. They do not reject the Kantian idea that cosmopolitan right must be limited to the right to visit in the context in which he discusses it. They share the Kantian view, raised in opposition to those of the Salamanca Scholastics, that the refusal to permit settlement does not allow the nation whose settlers wish to move into another territory to use force to give effect to their demand. On their reading, the limitation to the right to visit is a possible negative solution: absent any other type of arrangement, every human being has the right to visit and, absent any further arrangement, can claim no further right.

As a negative solution it does not rule out the possibility of nations entering into arrangements that create a positive version of cosmopolitan right. As I understand it, they propose seeing the right to visit as a baseline, and then understanding a complex formation such as the EU as moving beyond that baseline, through an arrangement made among its member states. This then allows for a positive version of cosmopolitan right in addition to the negative one that Kant articulates.

Their presentation of the problem thus raises in a particularly striking way the question of what would qualify as a cosmopolitan set of institutions, as well as what, following Peter Niesen and David Owen, might be called the problem of "cosmopolitanism in one country."[124] The question of whether a limited cosmopolitanism is possible and, more to the point, what it would entail, is a deep and interesting question.

Before engaging with their analysis in detail, I should note one reservation about it: Corrias and Wolthuis follow Niesen in supposing that each of the forms of public right comes in both a natural and a properly juridical/institutional form. As I explain in my response to Niesen's chapter, the three forms of public right that Kant identifies in the doctrine of right correspond to the three definitive articles of perpetual peace. As forms of public right, they are, as Kant remarks, forms of a public rightful condition: "a system of laws for a people, that is the union of a multitude of human beings, or for a multitude of peoples, which, because they affect one another, need a rightful condition under a will uniting them."[125] This does not sound like an idea that can be thought of as applying in any way, even provisionally or defectively, in the absence of legal institutions, and so I am not convinced that there is anything that could qualify as a natural version of cosmopolitan right, as opposed to a public-juridical one. More significantly, as I point out in responding to Niesen, Kant's own treatment of the

[124] Peter Niesen and David Owen, "Cosmopolitanism in One Country," unpublished paper, 2012. In the 1920s, after the revolution failed to spread from the Soviet Union to the entire capitalist world (as Marx had predicted and on assumption of which the Bolshevik Revolution was seemingly predicated), there was a raging debate about the possibility of socialism in one country.

[125] 6:311.

difference between provisional and peremptory rights insist that there "are no further or other duties" in a condition of public right then can be conceived of as existing in a state of nature. Kant's setup of the problem, then, seems to preclude attaching any content to the provisional version of cosmopolitan right.

This may seem like a minor quibble, of no immediate relevance to understanding the distinctive juridical nature of the EU, because whatever exactly the correct characterization of the EU is, it is, without question, a public legal order of some kind, rather than an expression of some natural version of cosmopolitan right. The lack of a conceptual space for a natural version of cosmopolitan right raises a puzzle about what it would mean for there to be a partial and incomplete realization of it. I do not deny that ideas of public right can be incompletely realized; the idea of a public legal order is shaped by the idea of an original contract, that is, a multitude of human beings united "under a will," which is an idea of reason, to which no actual legal order or indeed anything encountered in experience could be adequate.[126] So there is no difficulty with the idea of an imperfectly realized right of the state, or an imperfectly realized right of nations. So, too, there is no difficulty with the idea of an imperfectly realized cosmopolitan order. Nonetheless, there is a question about the way in which a cosmopolitan order could be imperfectly realized. The imperfect version of the right of the state and the right of nations seem to relate the same parties as the ideal case. The existing and imperfect legal orders in which we find ourselves represent the united will of human beings inhabiting a region of the earth's surface, however imperfectly. So, too, with an imperfect right of nations: structural features of the UN entail that certain kinds of bad conduct can be done with impunity because of the veto of permanent members of the Security Council. Nonetheless, it is an imperfect version of the Right of Nations because it has the same scope—all the nations of the world—in the same formal mandate—the protection of peace. An imperfect version of cosmopolitan right would likewise seem to apply to all human beings in relation to all nations on earth, everyone that is enclosed "together within determinate limits (by the spherical shape of the place they live in, a *globus terraqueous*).[127]

A partly localized cosmopolitanism is imperfect in a different way. As Wolthuis and Corrias point out, "EU internal market law raises interesting issues with respect to Kant's theory because it is an extensive system of positive cosmopolitan law between only a limited group of states" (Wolthuis and Corrias p. 211). That is the question of cosmopolitanism in one group of countries. Yet the measure that they propose for assessing it does not appear to ask about the extent to which it is adequate to its own normative principle. Instead, they characterize it as transitional, and assess it in terms of its place in a transition to be made from provisional to peremptory cosmopolitan law; that at least is how I understand their claim to be taking "an instance of positive cosmopolitan law and ask whether it contributes to the global legal condition" (Wolthuis and Corrias p. 211.)

As they point out, one of the puzzling things is that the EU combines elements of different types of legal ordering. They suggest that the European economic union governing the movement of persons, goods, and capital across borders is a cosmopolitan

[126] 6:371.
[127] 6:352.

idea. It gives rise to justiciable claims on the part of individual citizens of member states, whose claims are dependent upon their citizenship in member states, but not mediated through the states of which they are members. Instead, every individual has rights directly against each member state, not as a members of that individual's home state, but as a citizen of the EU.

By contrast, the EU's monetary union is transnational without being cosmopolitan. It relates states to each other and but not to individual human beings who are not members of those states. It limits the actions of member states in their fiscal and monetary policy, in particular by making it impossible for them to deal with debt by devaluing their currency. I take it that the point of contrast here is not that the economic union is good in some way in which the monetary union is bad (that may be true but is an entirely contingent matter) but rather that the monetary union, whatever its strengths or weaknesses, cannot be understood as an instance of cosmopolitan law. Because it does not relate individuals to states of which they are not members, it does not give them distinctive claims. The monetary union counsels prudent policy on the part of states participating in it, but, as Wolthuis and Corrias describe it, member states are "above the law" in the sense that there is no judicial mechanism for constraining their conduct in relation to it. They illustrate with the example of the Greek financial crisis. The other members of the Euro group were able to impose demands on Greece, simply by their will; giving effect to those demands was a matter of the alignment of interests, and so corresponds to the idea of direct, Athenian-type democracy. Kant criticizes that idea in the first "definitive" article of perpetual peace as the worst kind of despotism, lacking as it does any distinction between executive and legislative action, and thus official action is not subject to law, but only to plenary power-conferring rules. The power in question is entirely horizontal, exercised by the member states of the monetary union in relation to each other, in parallel with the powers exercised by individual citizens of ancient Athens over each other. Perhaps they go too far in saying that this is not law; as they describe it, it seems to be despotism, that is, force and law without freedom.

Corrias and Wolthuis also suggest that such an arrangement might be acceptable if it was fully voluntary, akin to the idea of a voluntary coalition that "can be dissolved at any time."[128] Pointing to the example of Greece, and the responses of the other Economic and Monetary Union members (EMU) to it in the financial crisis, they suggest that, in that case, Greece had no viable possibility of exit. Although I would not want to claim that it would have been viable for Greece to exit the EMU, in terms of the juridical structure of the situation, it would, of course, have been legally possible for it to do so; there was no legal restriction, and so it was not an indissoluble union. There was simply an enormous financial and political cost to doing so. Now it might well be said that things that have enormous financial or political costs are as bad as things that are legally prohibited, but if the task is to understand the juridical structure of the situation, the distinction between things that are prohibited and things that are not viable options matters centrally. One way of seeing the importance of this distinction is to note the parallel between this case and cases in which a nation suffers under

[128] 6:351.

an enormous debt burden to foreign creditors. Such debts are sometimes odious, that is, undertaken by corrupt officials for their private enrichment. In such circumstances, there may be reason to deny their enforceability as private relations between debtor and creditor. But Corrias and Wolthuis do not contend that anything like this was the source of Greece's difficulties in relation to the EMU; it was not a case in which corrupt officials had racked up debt in the name of a state. Rather, there were a series of financial decisions, collectively and legitimately taken, that put Greece in a financially untenable position, from which it could not extricate itself by devaluing its currency. I leave it to others to reflect on the ways in which such an outcome might have been prevented, as well as to propose alternative organizational structures for forms of monetary union. My point is only that the Greek crisis might be better understood as one that took place within a fundamentally private arrangement between the member states. Perhaps that just underscores their point that the monetary union is not cosmopolitan in the sense that is of interest to Kant. Here is a different way of understanding that point: the monetary union goes only to how things stand between separate states. The reason it is not a matter of cosmopolitan right is not that its organization is despotic (although that is objectionable) but rather that it does not concern the relation between a state and an individual human being who is not a member of it. Private relations are always potentially despotic, insofar as each is entitled to stand on their rights, and to negotiate with others on the basis of those rights. The shared "united will" required for entering into a condition of public right requires that the state provide protections against private forms of despotism. Perhaps a properly public form of monetary union would have to do the same thing. It would nonetheless still not be cosmopolitan, simply because its subject matter would concern a different form of relation.

As they present it, the economic union looks more cosmopolitan. The creation of an "Area of Freedom, Security, and Justice" allows citizens of member states to travel freely and even settle within other members states. As they note, these rights are justiciable through the European Commission and European Court of Justice. These rights therefore give individual citizens of member states rights against other states, which look like they form a structure of cosmopolitan right. But, as Corrias and Wolthuis note, they are more robust than the right to visit, and include most significantly the right to settle and to access government services. As such, it is less a cosmopolitan union than, as they put it, "an extension of the civil condition for its member states and these member states' citizens." As they point out, it does not restrict the rights of outsiders; it only gives additional rights to citizens of member states.

Although I agree that this structure at least holds out the hope that "To bring relations between states and foreigners under law by itself brings the aim of eternal peace closer," I am not sure that it does so under the idea of cosmopolitan right. The EU developed out of the European Common Market, the express aim of which was to remove economic incentives to war between France and Germany; economic integration raises the costs of war and reduces the expected benefits even if (contrary to fact) a war could be swift and decisive. These advantages are important, and are instances of the forces that underlie Kant's hopes for trade to serve the expansion of peace. For all that, however, they are not cosmopolitan in the sense that Kant intends in *Perpetual Peace* or the *Doctrine of Right*. In the world in which we find ourselves more economic integration will promote peace, and the division of the world into trading blocs or nations prepared to go it alone is a threat to peace. These factors are important, and wise leaders and global institutions must remember them, because they will make the use of force less advantageous and so, it is to be hoped, less tempting. But that does not make such factors constitutive of peace in the way in which Kant conceives of cosmopolitan right as an element of the complete determination of the idea of public law so that all human interaction is brought under law.

This brings me back to my hesitation about the contrast between provisional and peremptory versions of cosmopolitan right. Reducing the likelihood of regional or continental wars is obviously in the service of peace, and the EU has made great strides in securing peace in a region with a long history of war. That does not establish, however, that a more complete version of cosmopolitan right would grant to all human beings, in all nations, the rights that EU citizens enjoy in member countries. I thus hesitate to endorse Wolthuis and Corrias's view, adapted from Niesen, that EU law should be understood as an articulation of a possible version of positive cosmopolitan law for the entire world.

In order to underscore this contrast, let me say something more about Wolthuis and Corrias's treatment of the EU law governing migration. Mobility rights within the EU have generated EU-wide issues about the regulation of external borders. This is a type of problem that is specific to cosmopolitanism in one region, because a global cosmopolis would not have external borders; if all nations were to eventually join the EU, there would be no external borders at all. The current situation, however, restricted as it is to member states, necessarily generates a distinction between members and non-members. That distinction in turn opens up the possibility of an additional distinction, which EU law has drawn, between forced and voluntary migration. Forced migrants have internal rights during their stay, as well as rights against refoulment. Both of these are basic protections of cosmopolitan right; they are part of what it is to have a right to visit. As they point out, outsourcing these obligations—deeming forced migrants to be in a safe third country independently of conditions there—is inconsistent with cosmopolitan right. I would have put the point slightly differently than they did. I do not think that a procedure that imposes the judgment of a majority of members of a federation on a minority is always problematic. Provided that a legal order acts in conformity with the requirements of legality, the content of particular decisions can be subject to political negotiation and decision-making.

The difficulty that their discussion points to instead is that any such form of outsourcing is that something that is inherently a question of right to which only an affirmative answer is possible—the question of whether a human being is to be brought under law—is treated instead as a political question. Cosmopolitan right begins with the thought that everyone is always already a member of some legal order, always already in some kind of legal condition. If circumstances or the deeds of others make someone lose that membership, the place where that person arrives must take them in. Arrangements between one legal order (national or supranational) with respect to refugee resettlement is only consistent with right if they ensure that every human being is always a member of some legal order.

3. Concluding Remark

I want to thank all of the contributors once again for their careful and sympathetic engagement with my work, and the editors for their work in bringing together such distinguished contributors. People in other areas of life or other academic disciplines are sometimes surprised by the amount of disagreement in a field such as legal philosophy. As a philosopher, I can think of no greater honor than to have such distinguished commentators disagree with me. I hope I have made some progress in addressing some of their objections.

Index

For the benefit of digital users, indexed terms that span two pages (e.g., 52–53) may, on occasion, appear on only one of those pages.